Advances in the Molecular Mechanisms of Abscisic Acid and Gibberellins Functions in Plants 2.0

Advances in the Molecular Mechanisms of Abscisic Acid and Gibberellins Functions in Plants 2.0

Editor

Víctor Quesada

MDPI • Basel • Beijing • Wuhan • Barcelona • Belgrade • Manchester • Tokyo • Cluj • Tianjin

Editor
Víctor Quesada
Instituto de Bioingeniería
Universidad Miguel Hernández de Elche
Elche
Spain

Editorial Office
MDPI
St. Alban-Anlage 66
4052 Basel, Switzerland

This is a reprint of articles from the Special Issue published online in the open access journal *International Journal of Molecular Sciences* (ISSN 1422-0067) (available at: www.mdpi.com/journal/ijms/special_issues/AAAPlant_2).

For citation purposes, cite each article independently as indicated on the article page online and as indicated below:

LastName, A.A.; LastName, B.B.; LastName, C.C. Article Title. *Journal Name* **Year**, *Volume Number*, Page Range.

ISBN 978-3-0365-5024-4 (Hbk)
ISBN 978-3-0365-5023-7 (PDF)

Contents

About the Editor

Víctor Quesada

Dr. Quesada is an Associate Professor at the Universidad Miguel Hernández de Elche (UMH), Spain. He completed his undergraduate studies in Biology at the Universidad de Alicante, Spain, in 1993, and received his Ph.D. at the UMH working in the isolation and characterization of salt tolerant mutants in the plant model system Arabidopsis thaliana (1999). He then moved to the John Innes Centre (JIC) in Norwich, UK, from 2000 to 2003, to work in flowering time regulation in Arabidopsis. After finishing his post-doc, he moved to the Bioengineering Institute of the UMH to work in leaf morphogenesis. He became an Assistant Professor (December 2003) and an Associate Professor in 2009 at the UMH. Since 2009, his independent research activity is focused on the functional characterization of Arabidopsis nuclear genes involved in organellar gene expression, to shed light on the understanding of their roles in plant growth, development and adaptation to abiotic stress.

Preface to "Advances in the Molecular Mechanisms of Abscisic Acid and Gibberellins Functions in Plants 2.0"

Gibberellins (GA) and abscisic acid (ABA) are two phytohormones that regulate, in an antagonistic way, plant growth as well as several developmental processes from seed maturation and germination to flowering time, through hypocotyl elongation and root growth. In general, ABA and GA inhibit and promote, respectively, cell elongation and growth. Consequently, this mutual antagonism between GA and ABA governs many developmental decisions in plants.

In addition to its role as a growth and development modulator, ABA is primarily known for being a major player in the response and adaptation of plants to diverse abiotic stress conditions, including cold, heat, drought, salinity or flooding. Remarkably, different works have also recently pointed to a function for GA in the control of some biological processes in response to stress.

The selection of research and review papers of this book, mostly focused on ABA, covers a wide range of topics related to the most recent advances in the molecular mechanisms of ABA and GA functions in plants.

Víctor Quesada

Editor

International Journal of
Molecular Sciences

Editorial

Advances in the Molecular Mechanisms of Abscisic Acid and Gibberellins Functions in Plants 2.0

Víctor Quesada 

Instituto de Bioingeniería, Universidad Miguel Hernández, Campus de Elche, 03202 Elche, Spain; vquesada@umh.es; Tel.: +34-96-665-8812

Citation: Quesada, V. Advances in the Molecular Mechanisms of Abscisic Acid and Gibberellins Functions in Plants 2.0. *Int. J. Mol. Sci.* **2022**, *23*, 8524. https://doi.org/10.3390/ijms23158524

Received: 23 July 2022
Accepted: 27 July 2022
Published: 31 July 2022

Publisher's Note: MDPI stays neutral with regard to jurisdictional claims in published maps and institutional affiliations.

Abscisic acid (ABA) and gibberellins (GA) are two important hormones that antagonistically regulate many aspects of plant growth and development. These antagonistic regulatory functions of ABA and GA involve different plant development stages, i.e., seed maturation and dormancy, hypocotyl elongation, root growth and flowering [1]. In a broad sense, GA and ABA respectively promote and inhibit cell elongation and growth. Noteworthily, several factors mediate ABA and GA antagonism; for instance, regulating the transcription patterns of ABA and GA biosynthesis genes and, hence, the balance between ABA and GA [2]. Apart from playing a key role in regulating several physiological and developmental processes, ABA has also been traditionally considered the plant stress hormone par excellence. Accordingly, ABA directly responds to different adverse environmental conditions (i.e., low temperatures, drought, salinity or flooding) through the ABA signal transduction pathway, which results in massive changes in gene expression [3], stomatal closure [4] and lower transpiration rates [5,6]. Interestingly, recent data suggest that GA are also involved in plant response to environmental stress [7,8].

This Special Issue is the continuation of the previous Special Issue "Advances in the Molecular Mechanisms of Abscisic Acid and Gibberellins Functions in Plants". It contains six original research articles and four reviews published by field experts. These works cover outstanding advances in the molecular mechanisms by which GA and ABA control different aspects of plant physiology, development and for ABA, also in the response to abiotic stresses. In line with this, the original research articles investigate the effects of ABA on epicuticular wax metabolism regulation in a citrus fruit cultivar [9], the regulation of WNK kinases and their comprehensive interaction with ABA components [10], CCD family members in poplar and their expression patterns in response to ABA and abiotic stress [11], the molecular function of NAC transcription factor PwNAC11 in drought stress [12], the interaction between AFPs and DELLA proteins [13], and the molecular mechanism by which AGB1 regulates the GA pathway in Arabidopsis [14]. The reviews published in this Special Issue focus on how ABA regulates plant development and stress tolerance through alternative splicing [15], the mechanisms of ABA-mediated drought stress responses in plants [16], the roles of ABA and GA in stem/root tuber development [17] and the molecular mechanisms underlying ABA and GA regulation of seed development from embryogenesis to maturation [18]. In this editorial, I sum up the most significant findings of these insightfu works.

The research article by Romero and Lafuente [9] exhaustively investigated the effects of ABA on epicuticular wax metabolism regulation in a mutant citrus cultivar named Pinalate (*Citrus sinensis* L. Osbeck), defective in ABA-biosynthesis and, hence, with low ABA levels. To this end, Pinalate sweet orange fruits were harvested, treated with ABA and then exposed to water stress by storing them at low relative humidity. Romero and Lafuente [9] reported that ABA treatment modifies the composition and metabolism of the epicuticular wax of Pinalate fruit after detachment, which depends on whether fruit is exposed, or not, to postharvest dehydration, and also on water stress duration. Remarkably, the epicuticular wax load in Pinalate fruit increases after detachment, whereas ABA exposure is able to attenuate this increase. Finally, the study of the expression profile of the key genes involved in wax metabolism revealed that they are all influenced by ABA to a certain extent. Romero and Lafuente [9] highlighted that the obtained results can improve current knowledge about ABA regulation of cuticular wax metabolism in fruit and might be useful for industrial wax synthesis.

WNK [With-No-Lysine (K)] kinases are serine-threonine kinases. They are so named because the K residue present in subdomain II of most kinases is not conserved in WNK kinases. Instead, the K residue is replaced with a cysteine residue [19]. WNK kinases have been found in several eukaryotes, including plants where the WNK gene family is larger and more diverse than in animals. Plant WNK kinases are involved in not only the regulation of various physiological and developmental processes, but also in the response to different abiotic and biotic stresses. In line with this, some WNK family members have been implicated in the ABA-signaling pathway. Accordingly, *Arabidopsis thaliana* (hereafter Arabidopsis) WNK8 kinase functions as a negative regulator of ABA signaling. Interestingly, the presence of a conserved peptide encoded by an upstream open reading frame (uORF) called CPuORF58 has been reported in the 5′-UTR of *WNK8* [20]. Li et al. [10] thoroughly studied *WNK8* regulation and interaction with ABA components. They indicated that WNK8 negatively regulates ABA response during seed germination and post-germination development. Additionally, the use of several genetic constructs bearing different mutations in the 5′-UTR of *WNK8* revealed that CPuORF58 is essential for the translational suppression of WNK8, likely because this uORF causes ribosome stalling. Another analysis showed that *WNK8* 5′-UTR actually contains two short open reading frames, CPuORF58 and uORF1, and both are required for *WNK8* expression. When mutant *wnk8-1* plants are transformed with the native *WNK8* promoter that drives *WNK8* expression, they exhibit a similar phenotype to that of the wild type, whereas the plants transformed with a similar construct, but mutated in CPuORF58 translation initiation codon ATG, display much less sensitivity to ABA. Furthermore, WNK8 and its downstream target RACK1 (receptor for activated C kinase1) synergistically coordinate ABA signaling. Li et al. [10] concluded that *WNK8* post-transcriptional regulation by the CPuORF58 conserved peptide located in its 5′-UTR is required for accomplishing plant ABA requirements.

Carotenoid cleavage dioxygenases (CCDs) catalyze the cleavage of several carotenoids in higher plants, which results in the biosynthesis of biologically smaller apocarotenoids, including phytohormones such as ABA and signaling molecules such as β-cyclocitral, which play important roles in regulating plant growth, development and stress response [21]. Despite the significance of plant CCDs, the CCD gene family has not been hitherto studied in poplar (*Populus trichocarpa*), a species used as a model system in plant research with a significant ecological value. Wei et al. [11] reported an extensive genome-wide analysis of the CCD family of genes in poplar (*PtCCDs*). These authors systematically studied: *PtCCD* gene structures and conserved motifs; the expansion and contraction of *PtCCD* genes; the presence of *cis*-acting elements; the three-dimensional structures of PtCCD proteins and interaction networks. Wei et al. [11] evaluated the transcript profiles of *PtCCDs* in several tissues and determined the expression profile of *PtCCDs* in response to ABA and abiotic stress. They noted that *PtCCD* genes exhibit diverse tissue-specific expression patterns and, in response to abiotic stress and ABA, different *PtCCD* members display divergent response profiles. These results suggest that *PtCCD* genes may be involved in numerous physiological processes and in tolerance to both ABA and abiotic stress.

The NAC (NAM, ATAF1/2, and CUC2) transcription factors (TFs) in different plant species play important roles in the response to adverse environmental conditions. Notwithstanding, knowledge about NAC TFs involvement in the response of coniferous forests to abiotic stress is still scarce. To advance in this field, Yu et al. [12] performed a comprehensive functional analysis of the *NAC11* gene of the coniferous species *Picea wilsonii* in relation to drought tolerance. They reported that *PwNAC11* is significantly up-regulated by drought at 3 h [treatment with polyethylene glycol (PEG)] or ABA exposure in *P. wilsonii* seedlings, which suggests that PwNAC11 probably participates in early responses to stress. Later, Yu et al. [12] obtained *PwNAC11* overexpression (OE) lines in Arabidopsis and carried out some experimental assays to investigate the PwNAC11 function in drought stress. Along these lines, under drought conditions the Arabidopsis *PwNAC11* OE lines show enhanced drought tolerance and increased photoprotection and reactive oxygen species (ROS) scavenging ability. Moreover, *PwNAC11* OE enhances the expression of stress- and ABA-responsive genes under drought conditions. In line with the up-regulation of *PwNAC11* with exogenous ABA treatment in *P. wilsonii*, Yu et al. [12] found that *PwNAC11* OE increases ABA sensitivity and promotes ABA-induced stomatal closure in Arabidopsis. These findings indicate that *PwNAC11* OE lines are more sensitive to exogenous ABA. Furthermore, the authors searched for PwNAC11 potential interactors by performing yeast two-hybrid and bimolecular fluorescence complementation assay to find that PwNAC11 interacts in vivo with ABA-induced protein ABRE Binding Factor3 (ABF3) and ABA-independent DRE-BINDING PROTEIN 2A (DREB2A). Furthermore, the results obtained by Yu et al. [12] showed that PwNAC11 cooperates with ABF3 and DREB2 to activate *ERD1* transcription through the binding to the ABRE and DRE motifs of the *ERD1* promoter, respectively. These authors propose that PwNAC11 might mediate drought stress response via ABA-dependent and ABA-independent pathways.

The antagonistic effects of ABA and GA, respectively repressing and promoting seed dormancy and germination, are mediated by a complex network of positive and negative regulators of transcription. Two sets of negative regulators are the DELLA proteins and the AFP family of ABI5 [ABA-insensitive (ABI)]-binding proteins. DELLAs and AFPs are repressors of the GA and ABA response, respectively. Extensive interactions between ABI5, on the one hand, and AFPs and DELLAs on the other hand, have been reported. In their work, Finkelstein and Lynch [13] explored the possibility of an interaction between DELLAs and AFP proteins. By using yeast two-hybrid and bimolecular fluorescence complementation assays, direct interactions at different levels of intensity between DELLA and AFP family members were detected. Finkelstein and Lynch [13] also investigated potential genetic interactions between AFPs and DELLA proteins. They found that the overexpression (OE) of AFP proteins in a *sleepy1* (*sly1*) mutant background, a hyperdormant mutant where the GA response is repressed due to the overaccumulation of DELLA proteins, suppresses *sly1* hyperdormancy and hypersensitivity to ABA. However, AFP OE does not modify the dwarf and poor fertility phenotype of the *sly* mutant, but brings about a reduction in the accumulation of the seed storage proteins associated with the *sly* mutation. Finkelstein and Lynch [13] propose that the reported interactions are suggestive of additive effects of AFPs and DELLAs, which is consistent with these proteins acting in convergent pathways.

Unlike the above-discussed research manuscripts, which deal with different aspects of ABA, or the ABA and GA function, in plants, the work of Qi et al. [14] focused on GA, and specifically on two components of the GA pathway: Arabidopsis G-protein β subunit (AGB1) and the DNA-binding protein MYB62, a GA pathway suppressor. Heterotrimeric G proteins are transmembrane proteins that transduce signals from a wide range of extracellular stimuli and regulate several cellular and physiological functions in eukaryotes. In plants, G proteins are involved in developmental processes, stress responses and innate immunity [22]. Qi et al. [14] identified mutants *agb1-2* and N692967 that are affected in AGB1, which are dwarfs and contain significantly reduced GA levels compared to the wild type when undergoing GA_3 treatment. By using yeast-two hybrid, pull-down and firefly luciferase complementary imaging, Qi et al. [14] found that AGB1 physically interacts with the DNA-binding region of GA pathway suppressor MYB62. The results of the genetic analysis carried out by Qi et al. [14] are consistent with MYB62 acting downstream of AGB1 in the GA pathway. The expression analysis performed by RT-qPCR and competitive DNA binding assays revealed that MYB62 can bind MYB elements in the promoter of downstream gene *GA2ox7* (encoding a GA degradation enzyme) to induce *GA2ox7* transcription. Interestingly, the interaction of AGB1 with the DNA-binding region of MYB62 inhibits the binding of MYB62 on the *GA2ox7* promoter. These results reveal that AGB1 functions as a negative regulator of MYB62 activity and, consequently, positively regulates the GA pathway in Arabidopsis.

In eukaryotic cells, alternative splicing (AS) generates different mature mRNAs from the same mRNA precursor by selecting distinct combinations of splicing sites. In this way, AS increases the diversity and complexity of eukaryotic transcriptomes and proteomes. In plants, AS is extensively involved in not only the hasty regulation of plant growth and development, but also in adaptation to adverse environmental conditions. However, the mechanisms by which ABA can regulate plant development and tolerance to abiotic stress by mediating AS are not well understood. Yang et al. [15] reported in their review that ABA mediates plant abiotic stress tolerance and development, particularly through AS events. First, the authors briefly revised the main roles of ABA in abiotic stress response, the regulatory network of ABA signaling and ABA functions in development by focusing on seed germination and flowering. Next, they showed that ABA-induced AS occurs primarily in regulatory genes, such as transcription factors, protein kinases and splicing factors, as well as in genes of the ABA signaling pathway in Arabidopsis. Yang et al. [15] highlighted that ABA affects the splicing of genes of its own signaling pathway and other genes, especially by regulating the expression of splicing factors, such as SR genes and U2AF. They reported that this results in different splicing isoforms by, thereby, increasing protein abundance and improving plant adaptation to harsh environmental conditions.

Drought severely impairs plant growth and agriculture production, which are especially relevant under current climate change conditions. ABA is a key hormone that controls plant responses to drought through complex molecular signaling mechanisms. In line with this, ABA is able to synchronize a wide range of functions in plants by helping to overcome drought stress. Muhammad Aslam et al. [16] exhaustively worked on summarizing current knowledge about ABA-mediated drought responses through physiological, biochemical and root system alterations. In their review, Muhammad Aslam et al. [16] also highlighted the molecular mechanism of ABA-mediated drought regulation (i.e., ABA-dependent translational and post-translational modifications) as well as ABA crosstalk with other hormones, including auxins, GA, cytokinins, ethylene, salicylic acid and jasmonic acid, required to help plants to cope with drought and other abiotic stresses.

The remaining two reviews included in this Special Issue have focused on the roles of ABA and GA in stem/root tuber development [17] and seed germination [18] by highlighting molecular aspects. In the first case, the paper by Chen et al. [17] provides a comprehensive overview of recent advances in understanding ABA and GA functions in the stem/root tuber development of different tuber crops. Root and tuber crops, such as potato (*Solanum tuberosum*), sweet potato (*Ipomoea batatas*) and cassava (*Manihot esculenta*), are important for human food given their underground storage organs that are rich in water, carbohydrates and a range of proteins and secondary metabolites. Chen et al. [17] revised the roles of ABA and GA metabolism and signaling pathways in stem/root tuber development by showing that these phytohormones perform antagonistic functions in stem/root tuber development. In this way, ABA stimulates tuber formation, whereas GA represses tuber swelling. Therefore, a higher GA/ABA ratio results in longer stems and delayed tuberization, while a higher ABA/GA ratio leads to tuberization.

Seed development can be divided into two main phases, embryogenesis and maturation, which end with seeds entering dormancy to leave them ready to germinate when environmental conditions are favorable and after dormancy breaks. Complex gene regulatory networks govern seed development and germination, which involve several phytohormones. Of them, ABA and GA are the main hormones that antagonistically regulate seed development and germination. Although much progress has been made in understanding the molecular mechanism of ABA and GA regulation of seed maturation, much less is known about the role of both hormones in embryogenesis. Kozaki and Aoyanagi [18] thoroughly summarized current knowledge about the intricate molecular gene networks that regulate gene seed development from embryogenesis to maturation (including the accumulation of seed storage products, desiccation tolerance and induction and maintenance of primary seed dormancy) by focusing especially on the function of GA and ABA in all these processes. Kozaki and Aoyanagi [18] emphasized the intricate crosstalk among signaling phytohormones, besides ABA and GA, during seed formation, which deserves to be further investigated to provide a more complete picture of the mechanism governing seed development.

I am convinced that the high quality of the original research articles and reviews published in this Special Issue will undoubtedly contribute to improve our understanding of the molecular mechanisms of ABA and GA functions in plants. I wish to thank all the authors for their contributions and the reviewers for their critical assessments of these articles. I also thank Assistant Editor Ms. Reyna Li for offering me the opportunity to serve "Advances in the Molecular Mechanisms of Abscisic Acid and Gibberellins Functions in Plants 2.0" as a Guest Editor.

Funding: This research received no external funding.

Conflicts of Interest: The author declares no conflict of interest.

References

1. Shu, K.; Zhou, W.; Chen, F.; Luo, X.; Yang, W. Abscisic Acid and Gibberellins Antagonistically Mediate Plant Development and Abiotic Stress Responses. *Front. Plant Sci.* **2018**, *9*, 416. [CrossRef]
2. Shu, K.; Chen, Q.; Wu, Y.R.; Liu, R.J.; Zhang, H.W.; Wang, P.F.; Li, Y.; Wang, S.; Tang, S.; Liu, C.; et al. ABI4 mediates antagonistic effects of abscisic acid and gibberellins at transcript and protein levels. *Plant J.* **2016**, *85*, 348–361. [CrossRef] [PubMed]
3. Sah, S.K.; Reddy, K.R.; Li, J. Abscisic Acid and Abiotic Stress Tolerance in Crop Plants. *Front. Plant Sci.* **2016**, *7*, 571. [CrossRef] [PubMed]
4. Munemasa, S.; Hauser, F.; Park, J.; Waadt, R.; Brandt, B.; Schroeder, J.I. Mechanisms of abscisic acid-mediated control of stomatal aperture. *Curr. Opin. Plant Biol.* **2015**, *28*, 154–162. [CrossRef] [PubMed]
5. Kuromori, T.; Sugimoto, E.; Shinozaki, K. Arabidopsis mutants of *AtABCG22*, an ABC transporter gene, increase water transpiration and drought susceptibility. *Plant J.* **2011**, *67*, 885–894. [CrossRef] [PubMed]
6. McAdam, S.A.; Brodribb, T.J.; Banks, J.A.; Hedrich, R.; Atallah, N.M.; Cai, C.; Geringer, M.A.; Lind, C.; Nichols, D.S.; Stachowski, K.; et al. Abscisic acid controlled sex before transpiration in vascular plants. *Proc. Natl. Acad. Sci. USA* **2016**, *113*, 12862–12867. [CrossRef]
7. Hamayun, M.; Hussain, A.; Khan, S.A.; Kim, H.Y.; Khan, A.L.; Waqas, M.; Irshad, M.; Iqbal, A.; Rehman, G.; Jan, S.; et al. Gibberellins producing endophytic fungus *Porostereum spadiceum* AGH786 rescues growth of salt affected soybean. *Front. Microbiol.* **2017**, *8*, 686. [CrossRef]
8. Urano, K.; Maruyama, K.; Jikumaru, Y.; Kamiya, Y.; Yamaguchi-Shinozaki, K.; Shinozaki, K. Analysis of plant hormone profiles in response to moderate dehydration stress. *Plant J.* **2017**, *90*, 17–36. [CrossRef]
9. Romero, P.; Lafuente, M.T. The Combination of Abscisic Acid (ABA) and Water Stress Regulates the Epicuticular Wax Metabolism and Cuticle Properties of Detached Citrus Fruit. *Int. J. Mol. Sci.* **2021**, *22*, 10242. [CrossRef]
10. Li, Z.; Fu, Y.; Shen, J.; Liang, J. Upstream Open Reading Frame Mediated Translation of WNK8 Is Required for ABA Response in Arabidopsis. *Int. J. Mol. Sci.* **2021**, *22*, 10683. [CrossRef]
11. Wei, H.; Movahedi, A.; Liu, G.; Li, Y.; Liu, S.; Yu, C.; Chen, Y.; Zhong, F.; Zhang, J. Comprehensive Analysis of Carotenoid Cleavage Dioxygenases Gene Family and Its Expression in Response to Abiotic Stress in Poplar. *Int. J. Mol. Sci.* **2022**, *23*, 1418. [CrossRef] [PubMed]
12. Yu, M.; Liu, J.; Du, B.; Zhang, M.; Wang, A.; Zhang, L. NAC Transcription Factor PwNAC11 Activates *ERD1* by Interaction with ABF3 and DREB2A to Enhance Drought Tolerance in Transgenic Arabidopsis. *Int. J. Mol. Sci.* **2021**, *22*, 6952. [CrossRef] [PubMed]
13. Finkelstein, R.R.; Lynch, T.J. Overexpression of ABI5 Binding Proteins Suppresses Inhibition of Germination Due to Overaccumulation of DELLA Proteins. *Int. J. Mol. Sci.* **2022**, *23*, 5537. [CrossRef] [PubMed]
14. Qi, X.; Tang, W.; Li, W.; He, Z.; Xu, W.; Fan, Z.; Zhou, Y.; Wang, C.; Xu, Z.; Chen, J.; et al. Arabidopsis G-Protein β Subunit AGB1 Negatively Regulates DNA Binding of MYB62, a Suppressor in the Gibberellin Pathway. *Int. J. Mol. Sci.* **2021**, *22*, 8270. [CrossRef]
15. Yang, X.; Jia, Z.; Pu, Q.; Tian, Y.; Zhu, F.; Liu, Y. ABA Mediates Plant Development and Abiotic Stress via Alternative Splicing. *Int. J. Mol. Sci.* **2022**, *23*, 3796. [CrossRef] [PubMed]
16. Muhammad Aslam, M.; Waseem, M.; Jakada, B.H.; Okal, E.J.; Lei, Z.; Saqib, H.S.A.; Yuan, W.; Xu, W.; Zhang, Q. Mechanisms of Abscisic Acid-Mediated Drought Stress Responses in Plants. *Int. J. Mol. Sci.* **2022**, *23*, 1084. [CrossRef]
17. Chen, P.; Yang, R.; Bartels, D.; Dong, T.; Duan, H. Roles of Abscisic Acid and Gibberellins in Stem/Root Tuber Development. *Int. J. Mol. Sci.* **2022**, *23*, 4955. [CrossRef] [PubMed]
18. Kozaki, A.; Aoyanagi, T. Molecular Aspects of Seed Development Controlled by Gibberellins and Abscisic Acids. *Int. J. Mol. Sci.* **2022**, *23*, 1876. [PubMed]

19. Xu, B.; English, J.M.; Wilsbacher, J.L.; Stippec, S.; Goldsmith, E.J.; Cobb, M.H. WNK1, a novel mammalian serine/threonine protein kinase lacking the catalytic lysine in subdomain II. *J. Biol. Chem.* **2000**, *275*, 16795–16801.
20. Hayden, C.A.; Jorgensen, R.A. Identification of novel conserved peptide uORF homology groups in Arabidopsis and rice reveals ancient eukaryotic origin of select groups and preferential association with transcription factor-encoding genes. *BMC Biol.* **2007**, *5*, 32. [CrossRef]
21. Felemban, A.; Braguy, J.; Zurbriggen, M.D.; Al-Babili, S. Apocarotenoids involved in plant development and stress response. *Front. Plant Sci.* **2019**, *10*, 1168. [CrossRef] [PubMed]
22. Maruta, N.; Trusov, Y.; Jones, A.M.; Botella, J.R. Heterotrimeric G Proteins in Plants: Canonical and Atypical Gα Subunits. *Int. J. Mol. Sci.* **2021**, *22*, 1184. [CrossRef] [PubMed]

International Journal of
Molecular Sciences

Article

The Combination of Abscisic Acid (ABA) and Water Stress Regulates the Epicuticular Wax Metabolism and Cuticle Properties of Detached Citrus Fruit

Paco Romero * 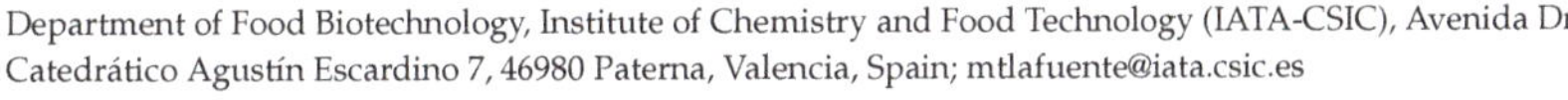 and María Teresa Lafuente

Department of Food Biotechnology, Institute of Chemistry and Food Technology (IATA-CSIC), Avenida Dr. Catedrático Agustín Escardino 7, 46980 Paterna, Valencia, Spain; mtlafuente@iata.csic.es
* Correspondence: promero@iata.csic.es

Abstract: The phytohormone abscisic acid (ABA) is a major regulator of fruit response to water stress, and may influence cuticle properties and wax layer composition during fruit ripening. This study investigates the effects of ABA on epicuticular wax metabolism regulation in a citrus fruit cultivar with low ABA levels, called Pinalate (*Citrus sinensis* L. Osbeck), and how this relationship is influenced by water stress after detachment. Harvested ABA-treated fruit were exposed to water stress by storing them at low (30–35%) relative humidity. The total epicuticular wax load rose after fruit detachment, which ABA application decreased earlier and more markedly during fruit-dehydrating storage. ABA treatment changed the abundance of the separated wax fractions and the contents of most individual components, which reveals dependence on the exposure to postharvest water stress and different trends depending on storage duration. A correlation analysis supported these responses, which mostly fitted the expression patterns of the key genes involved in wax biosynthesis and transport. A cluster analysis indicated that storage duration is an important factor for the exogenous ABA influence and the postharvest environment on epicuticular wax composition, cuticle properties and fruit physiology. Dynamic ABA-mediated reconfiguration of wax metabolism is influenced by fruit exposure to water stress conditions.

Keywords: ABA deficiency; fruit dehydration; gene expression; hormone application; Pinalate; postharvest

Citation: Romero, P.; Lafuente, M.T. The Combination of Abscisic Acid (ABA) and Water Stress Regulates the Epicuticular Wax Metabolism and Cuticle Properties of Detached Citrus Fruit. *Int. J. Mol. Sci.* **2021**, *22*, 10242. https://doi.org/10.3390/ijms221910242

Academic Editor: Víctor Quesada

Received: 23 July 2021
Accepted: 21 September 2021
Published: 23 September 2021

Publisher's Note: MDPI stays neutral with regard to jurisdictional claims in published maps and institutional affiliations.

1. Introduction

Abscisic acid (ABA) was first known as abscisin II [1,2] because it was identified as a substance that regulated cotton fruit and leaf abscission. Later studies revealed its role in both biotic and abiotic stresses, and highlighted the involvement of this hormone in the regulation of the molecular mechanisms underlying the response to dehydrating conditions in plants [3]. In fruit, the participation of ABA in fruit water stress regulation has also been demonstrated in both climacteric and non-climacteric fruit [4]. More recently, a relationship between ABA and cuticle, the first barrier to limit water loss, has been suggested in model plants [5,6] and during fruit ripening [7–10]. ABA treatment increased wax load in Arabidopsis plants, mainly due to changes in the alkane fraction [5]. Also, ABA deficiency increased cuticle permeability and resulted in thinner cuticles during tomato leaf development [6,9]. In contrast, deficient ABA levels in tomato fruit do not have a marked effect on wax content or composition, while ABA spraying increased wax load and cuticle thickness in cherry fruit [10]. In addition, a relationship between ABA and cuticle biology has been proposed in citrus fruit [7,8], although the putative ABA-mediated regulation of the cuticle composition and properties in fruit exposed to abiotic postharvest stresses remains elusive and needs to be further investigated.

The cuticle layer covers aerial plant parts and acts as the first barrier for the interaction with the environment. Given its lipophilic nature, it is a major water retention determinant in different plant organs, and is also involved in the regulation of temperature fluctuations and gas diffusion in addition to protection from pathogen invasion, among others [11–13]. Cuticular waxes are mixtures of cyclic compounds (e.g., triterpenoids) and aliphatic compounds (e.g., alkanes, fatty acids (FA), alcohols and aldehydes) derived from very-long-chain fatty acids (VLCFA) [14]. They can be divided into intracuticular and epicuticular waxes depending on whether their deposition is embedded in or on the cutin scaffold, respectively. Previous reports point out that fruit cuticles, and more specifically the epicuticular wax layer composition, is associated with fruit quality maintenance after harvest, in regard to susceptibility to water loss, pathogen infection, loss of firmness and the development of peel disorders [15–21].

Fruit cuticle composition is dependent on the species, and even on the cultivar, and is also determined by internal factors such as fruit development and ripening [7–9,22–34]. The effect of plant regulators, such as gibberellic acid, 2,4-dichlorophenoxy acetic acid, ethylene and 1-MCP, on epicuticular wax composition has been investigated [35–39], and a role for ABA in modulating the epicuticular wax metabolism and cuticle properties of fruit has been proposed [7,8,10]. Fruit cuticle composition and properties are also sensitive to external environmental factors such as humidity, temperature and light radiation [16,40,41]. Of these, water stress during the cultivation of or after harvesting fruit is considered the major factor that limits plant productivity and fruit quality after detachment, respectively. Cutin and intracuticular waxes have frequently been considered the main cuticle components responsible for reducing water permeability. Lately, however, the number of research works that correlate epicuticular wax load and composition with fruit water loss has grown [7,8,18,41–43]. Conversely, reports on how environmental humidity affects fruit cuticular wax metabolism are limited to tomatoes, grapes and pears [43–45], and whether ABA influences fruit cuticle properties as well as epicuticular wax content and composition during postharvest water stress remains elusive.

Citrus fruit is an important crop grown all over the world. As in most non-climacteric fruits, ABA plays a key role in citrus fruit development and ripening [46–52], but also in response to water stress [53–58]. Indeed, several studies have highlighted the relevance of endogenous ABA levels and an operational ABA signaling network in the molecular response of citrus fruit to dehydrating conditions determining their external fruit quality [53,55,59]. The availability of an impaired ABA biosynthesis mutant named Pinalate (*Citrus sinensis* L. Osbeck) has been crucial for such scientific progress to be made. A Pinalate orange is not a knockout mutant, but a spontaneous bud mutation that presents severe fruit-specific blockage of the carotenoid biosynthetic pathway, which results in yellow coloration, high susceptibility to postharvest dehydration and drastically low ABA levels in the flavedo (the outer colored peel part) [53,60–62]. It has also been reported that the epicuticular wax composition of this fruit differs from that in other orange cultivars with higher ABA levels during fruit ripening, despite the total wax load not being affected by low ABA content in the fully mature stage [8]. These findings converge with those in other citrus cultivars, pointing to a role for ABA in the regulation of cuticle biology during citrus fruit maturation [7]. However, no ABA feeding experiments have been performed and the relationship between ABA and citrus cuticles remains elusive. In addition, the effects of ABA or water stress on citrus cuticles after harvesting the fruit have never been investigated. By making the most of the markedly reduced ABA levels in Pinalate fruit, this study investigated whether ABA treatment drives changes in fruit cuticle properties as well as epicuticular wax load and composition after detachment, and if these modifications are influenced by exposure to water stress conditions during postharvest storage. We addressed these questions by comparing the wax composition as well as cuticle permeability and thickness, together with the fruit weight loss and firmness measurements, of the Pinalate fruit, either treated or not with ABA and left under high relative humidity (RH) or postharvest water stress conditions that favor fruit dehydration. Correlation analyses allowed us to identify the relations among ABA content, wax constituents, cuticle properties and fruit quality maintenance during storage. To dig more further into the molecular mechanisms underlying these relations, an expression analysis of the key genes involved in the biosynthesis and transport of wax components was performed.

2. Results

2.1. Effects of ABA and Water Stress on Wax Content and Wax Fraction Distribution

Pinalate oranges were treated, or not, with 1 mM ABA solution before being stored under the control (90–95% RH) or water stress (30–35% RH) environmental conditions. As shown in Figure 1, the total wax content in Pinalate increased by two-fold in the first week of storage and slightly decreased thereafter regardless of the environmental conditions. ABA treatment reduced the total wax load increment observed by 1 week, which was more marked under the dehydrating conditions. By 3 weeks, this trend was observed only in the fruit stored under the postharvest water stress conditions (Figure 1).

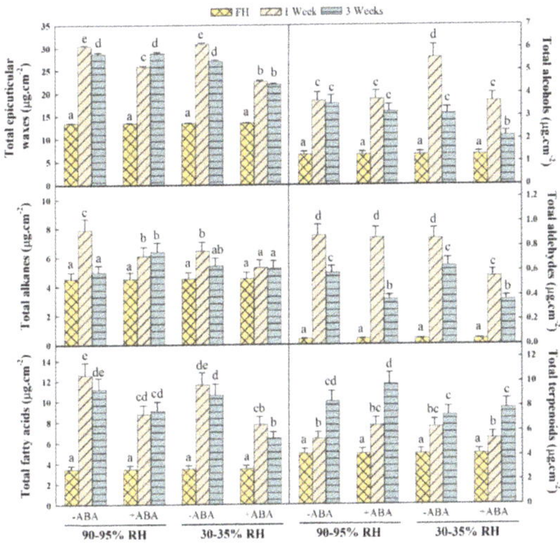

Figure 1. Effect of ABA and water stress on the total epicuticular wax and individual fractions content in Pinalate fruit. The effect of ABA (1 mM) treatment was evaluated together with the influence of high (90–95%) or low (30–35%) RH conditions on the total content (µg.cm^{-2}) of epicuticular wax and the individual fractions up to 3 weeks of leaving Pinalate fruit at 20 °C. Bars are means ±SD of four replicates per condition. FH: Freshly harvested fruit. For each panel, the different letters above the bars indicate significant ($p < 0.05$) differences among conditions according to an ANOVA analysis followed by a Tukey test ($p < 0.05$).

Noticeable responses to the total content of five major wax fractions were detected in response to ABA treatment and the postharvest water stress conditions. These fractions were alkanes (C_{22}-C_{34}), FA (C_{12}-C_{28}), alcohols (C_{22}-C_{32}), aldehydes (C_{24}, C_{26}) and terpenoids (C_{15}, C_{29} and C_{30}) (Figure 1). The total alkane content increased by 1 week during the control or dehydrating fruit storage. Independently of the water stress environment, ABA treatment reduced this increase. Similarly, the increase observed in the total FA content after harvest was reduced by applying ABA to Pinalate fruit, regardless of storage conditions. In contrast, an effect of ABA on alcohol content was only observed in water-stressed fruit. Thus, the alcohol accumulation pattern in the Pinalate fruit under water stress remained after the ABA treatment, but the total achieved content lowered by about 30%. Aldehyde content increased by about nine-fold 1 week after fruit detachment and decreased thereafter during storage under both the control and water stress conditions. A major effect of ABA on aldehyde content was observed in the Pinalate fruit exposed to postharvest water stress, as this fraction accumulation lowered by about 25% and 30% by week one and week three, respectively, compared to untreated fruit. The total terpenoid content also increased after harvesting the fruit. No differences in terpenoid accumulation were observed between the ABA-treated and untreated fruit stored at either high or low RH, but the total content achieved in the ABA-treated fruit was higher at high RH than during water stress storage.

The effects of ABA treatment on wax composition were dependent on exposure to postharvest water stress and storage duration (Figure 2). At harvest, the most abundant components were alkanes (33.7%), followed by terpenoids (30.6%), FA (25.7%), alcohols (9.7%) and aldehydes (0.3%). In response to harvest, alkane abundance lowered, and a transitory decrease in terpenoids was observed by 1 week. The percentage of FA and aldehydes increased 1 week after storage and remained almost steady thereafter under either environmental condition (Figure 2). Nevertheless, by 1 week of postharvest water stress, the abundance of alkanes and FA was lower, and that of alcohols and terpenoids was higher than in the fruit left at high RH. By 3 weeks, however, these trends were inversed and postharvest water stress brought about a decrease in terpenoid abundance and an increase in the alkane proportion. Hormone application enhanced the initial decrease in alkane abundance in the Pinalate fruit stored at high RH, while the opposite was found under the fruit-dehydrating condition. ABA treatment slightly modified the alcohol proportion by 1 week under either the control or postharvest water stress storage. The increase in the FA proportion after fruit detachment was attenuated by ABA application, and was more marked by 3 weeks under the fruit-dehydrating condition. Terpenoid abundance was increased by ABA independently of storage and period. However, the effects of ABA on the aldehyde proportion varied depending on postharvest water stress exposure and duration (Figure 2).

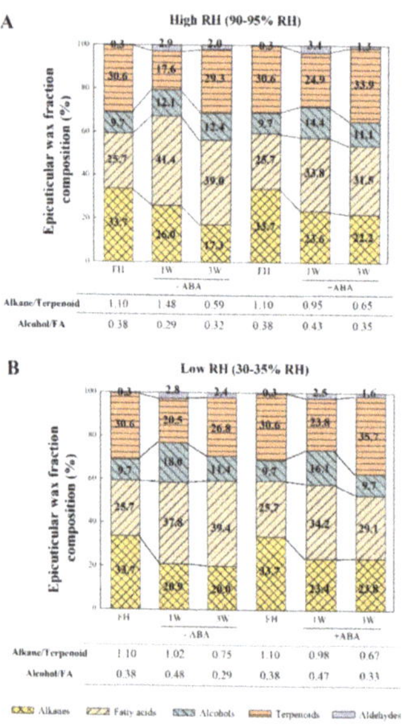

Figure 2. Effect of ABA and water stress on the proportion of epicuticular wax fractions in Pinalate fruit. The effect of ABA (1 mM) treatment on the percentage (%) of epicuticular wax fractions by 1 and 3 weeks (W) of Pinalate fruit storage at 20 °C was evaluated together with the influence of the (**A**) high (90–95%) or (**B**) low (30–35%) RH conditions. FH: Freshly harvested fruit. The values on the bars indicate the percentage of each fraction. The alkane/terpenoid and alcohol/FA ratios are indicated per condition. The same legend is used for panels (**A**,**B**).

As a result of these changes, the alkane/terpenoid and alcohol/FA ratios, which are inversely related to fruit water retention, were modified in response to ABA application and/or fruit exposure to postharvest water stress. The alkane/terpenoid ratio transitorily increased 1 week after fruit detachment when stored at high RH, but continuously decreased in response to ABA treatment under this condition. In contrast, this ratio continuously lowered in the fruit exposed to postharvest water stress, and no influence of ABA was observed. The alcohol/FA ratio slightly varied during fruit storage. The effects of ABA were also minor. If any, an increase in the alcohol/FA ratio was observed in the ABA-treated fruit stored for 1 week at high RH compared to the untreated fruit (Figure 2).

2.2. Epicuticular Wax Composition Analysis

Heptacosane (C_{27}), nonacosane (C_{29}) and hentriacontane (C_{31}) were the most abundant long-chain alkanes in Pinalate fruit, irrespective of ABA treatment and storage conditions (Figure 3A,E). The accumulation of tricosane (C_{23}) and pentacosane (C_{25}) was also noticeable among the odd-chain alkanes. Even-chain-length compounds (C_{22}-C_{34}) were a minority among alkane components and were not consistently detected under the assayed experimental conditions. The accumulation profiles of the odd long chain alkanes were differently affected by ABA treatment depending on exposure or not to postharvest water stress. Some differences in their accumulation patterns were found when comparing high RH and fruit-dehydrating conditions. Thus, ABA treatment lowered the increments detected for the odd long-chain-length compounds by 1 week at high and low RH. In contrast, hormone application increased these compounds by 3 storage weeks at high RH, but did not modify their contents in the water-stressed fruit during this period. Their accumulation was higher at high rather than low RH by 1 week, but no differences caused by environmental conditions were found after 3 storage weeks in the untreated fruit (Figure 3A,E).

At harvest, the most abundant FA were lignoceric (C_{24}) and cerotic (C_{26}) acids, followed by palmitic (C_{16}) and stearic (C_{18}) acids. All these compounds, as well as montanic acid (C_{28}), increased after harvesting fruit, but specific accumulation patterns were found in response to ABA application in addition to the storage period and conditions. Thus, the content of both lignoceric and cerotic acids increased during storage, and ABA treatment attenuated their accumulation independently of the environment (Figure 3B,F). In contrast, palmitic and stearic acids increased in response to ABA application by 1 week, but only at high RH. This effect was lost by 3 weeks. No effects of ABA on these FA were detected under the fruit-dehydrating condition by 1 week, but hormone treatment reduced their accumulation by 3 weeks. Similarly, exogenous ABA did not modify the montanic acid content under the water stress conditions, but its accumulation lowered at high RH by 1 and 3 weeks. In addition, hormone treatment reduced montanic acid accumulation under the fruit-dehydrating conditions by 3 weeks (Figure 3F).

The most abundant primary alcohols were docosanol (C_{22}), tetracosanol (C_{24}) and pentacosanol (C_{25}). Triacontanol (C_{30}) and dotriacontanol (C_{32}) were not detected at harvest, but accumulated during storage (Figure 3C,G). All these compounds increased after harvest independently of the environmental conditions, but the effect of ABA treatment on the content of each compound depended on RH and storage duration. Thus, docosanol content, the most abundant alcohol in Pinalate fruit, was not affected by ABA treatment at high RH, but lowered under water stress by 1 week. In contrast, hormone application decreased the docosanol content by 3 weeks regardless of the storage conditions. Our analyses identified two aldehydes, tetracosanal (C_{24}) and hexacosanal (C_{26}), which increased after detachment, but did not show any remarkable differences between storage conditions (Figure 3C,G). ABA decreased their accumulation by 3 weeks. This trend was evident in tetracosanal content by 1 week, but only under postharvest water stress.

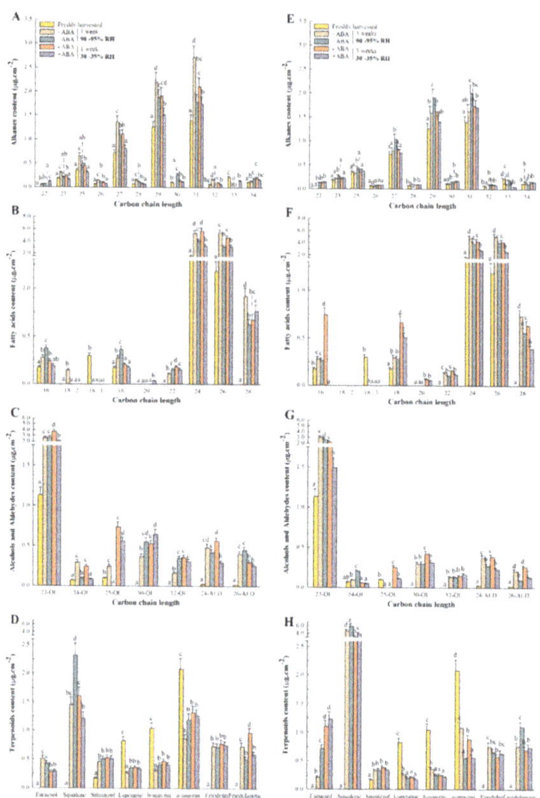

Figure 3. Effect of ABA and water stress on epicuticular wax constituents during Pinalate fruit storage. The effects of ABA (1 mM) treatment on the content of the epicuticular wax components were evaluated together with the influence of high (90–95%) or low (30–35%) RH by 1 (**A–D**) and 3 (**E–H**) weeks at 20 °C. (**A,E**) Alkanes; (**B,F**) fatty acids; (**C,G**) alcohols and aldehydes; and (**D,H**) terpenoids. Bars are the means ±SD of four replicates per condition. Different letters above the bars indicate significant ($p < 0.05$) differences among conditions according to an ANOVA analysis followed by a Tukey test ($p < 0.05$) for each component and storage time separately.

Among terpenoids, only α- and β-amyrins, lupenone and sitosterol were detected at harvest, although farnesol, squalene, friedelin and friedelanone accumulated thereafter (Figure 3D,H). Lupenone, α- and β-amyrins decreased in Pinalate fruit in response to fruit detachment and independently of storage conditions. The other terpenoids, however, increased after harvest. The response to ABA application was diverse among these compounds and depended on RH and storage duration. Sitosterol, friedelin and farnesol contents did not vary after ABA treatment, except for farnesol content, which increased in the ABA-treated fruit exclusively when stored for 3 weeks at high RH. Squalene content, the most abundant terpenoid in Pinalate fruit, was inversely regulated by ABA application between the control and fruit-dehydrating conditions by 1 week. By 3 weeks, its content had considerably increased and ABA induced its accumulation regardless of the environmental conditions.

2.3. Variations in Cuticle Properties and Fruit ABA Content, Firmness and Weight Loss

The ABA content in Pinalate fruit slightly increased after detachment irrespective of storage conditions. Likewise, exogenous ABA application increased the hormone content in both the fruit stored at high RH and under postharvest water stress by about five-fold (Figure 4A). Cuticle thickness continuously decreased after detachment. By 1 week, this decrease was slight under postharvest water stress, but more drastic at high RH (Figure 4B). When the experiment ended (3 weeks), cuticles were thicker in the fruit left under the postharvest dehydration conditions than those left at high RH. ABA treatment did not significantly affect these patterns independently of storage. Nevertheless, the thinnest cuticles were those from the ABA-treated fruit stored for 3 weeks at high RH. Cuticle permeability (estimated as the cuticle transpiration rate) barely changed in response to the dehydrating environment and/or ABA treatment. In fact, an increase in cuticle permeability was detected only in the ABA-treated fruit stored for 3 weeks at high RH (Figure 4C). Cumulative weight loss per surface area continuously increased in Pinalate fruit during storage, and it was about four-fold higher when fruit were kept under the postharvest dehydration conditions (Figure 4D). A slight, but statistically significant, reduction in cumulative weight loss was observed by 1 and 3 weeks in the ABA-treated fruit versus the untreated fruit left under water stress. Fruit firmness remained unchanged by 1 week after detachment and decreased thereafter in the fruit left at high RH, and no effect of ABA was observed. Postharvest water stress, however, brought about a marked decrease in fruit firmness regardless of ABA treatment (Figure 4E).

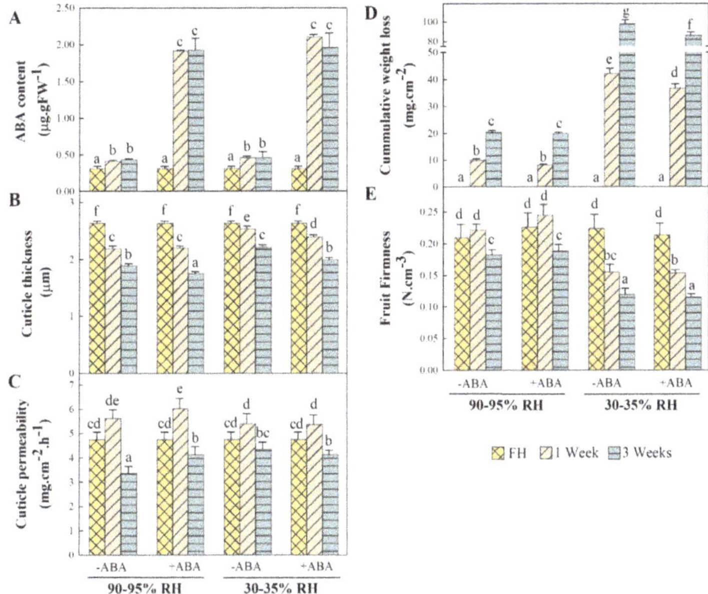

Figure 4. Effect of ABA and water stress on cuticle properties and fruit physiological parameters in Pinalate fruit. (**A**) ABA content is expressed as g per g of fresh weight (FW) of the flavedo. Bars are the means ±SD of three replicates of five fruit each. (**B**) Cuticle thickness. Bars represent the means ±SD of about 50 measurements for all three biological replicates analyzed per condition. (**C**) Cuticle permeability. Bars are the means ±SD of three replicates per condition. (**D**) Cumulative weight loss of Pinalate fruit calculated as fruit weight loss per surface area. Bars are the means ±SD of three replicates of 10 fruit each. (**E**) Fruit firmness was determined according to the intact fruit compression resistance load (N) and normalized by fruit size (cm^3). Bars indicate the means ±SD of three replicates of 10 fruit each. For each studied parameter, different letters indicate the statistical ($p < 0.05$) differences in all the conditions together according to an ANOVA analysis followed by a Tukey test ($p < 0.05$).

2.4. Relationships between ABA Content, Cuticle Composition and Properties and Fruit Physiology

In order to elucidate the relationships among the experimental conditions as regards ABA application and exposure to postharvest water stress, the total epicuticular wax content, the content of each wax fraction, ABA levels, cuticle thickness and permeability, and fruit weight loss and firmness were used to perform a cluster analysis (Figure 5A). According to the HCA, the freshly harvested (FH) and stored fruit were the most clearly discriminated groups. Later, samples were clustered in a storage-period-dependent manner. During each storage period, the samples that were grouped depending on exposure to postharvest water stress followed a distinction between the ABA-treated and untreated fruit. The dendrogram corresponding to cuticle composition and properties, ABA content and fruit physiological parameters can be divided into three major groups. The first includes two subclades composed of cuticle thickness, permeability and fruit firmness in addition to total alkane content. Second, ABA content, total terpenoids and fruit weight loss were clustered together. The last group contained the total wax load and the total content of all the cuticle fractions, except alkanes and terpenoids (Figure 5A).

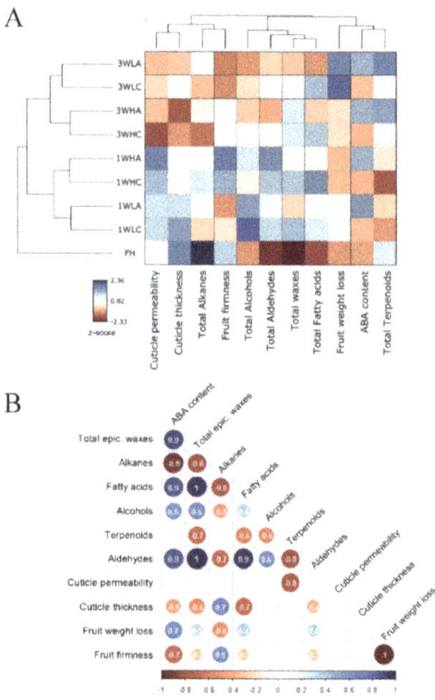

Figure 5. Clustering and correlation analyses of epicuticular wax composition, ABA content, cuticle properties and fruit physiological parameters in Pinalate fruit. (**A**) Hierarchical clustering analysis of the Pinalate fruit left at high (H, 90–95%) or low (L, 30–35%) RH for 1 week or 3 weeks (W), and treated (A, ABA) or not (C, control) with 1mM ABA, based on the chemical composition of their epicuticular wax layer, ABA content, cuticular properties and fruit physiology. The colors in the heatmap indicate the z-score value for each parameter and condition according to the scale in the legend. (**B**) Correlation matrix among the abundance of individual epicuticular wax fractions, ABA content, cuticle properties and fruit physiological parameters in the Pinalate fruit left under the above-described conditions. Numbers indicate the regression coefficient value, and only the statistically significant ones are colored according to the scale in the legend.

The effects of ABA application and exposure to water stress after fruit detachment on cuticle composition and properties, as well as on fruit weight loss and firmness, were further studied by a statistical correlation analysis (Figure 5B). ABA content correlated positively with total epicuticular wax (r = 0.9), FA (r = 0.9), alcohol (r = 0.6) and aldehyde (r = 0.9) proportions and fruit weight loss (r = 0.7). Inversely, ABA content negatively correlated with alkane proportion (r = 0.9), cuticle thickness (r = −0.5) and fruit firmness (r = −0.7). The alkane proportion showed a negative correlation with fruit weight loss (r = −0.6), but correlated positively with cuticle thickness (r = 0.7) and fruit firmness (r = 0.6). Inversely, FA and aldehyde abundance related positively to fruit weight loss (r = 0.4), but negatively to fruit firmness (r = −0.3) and cuticle thickness (r = −0.7 and r = −0.4, respectively). The percentage of terpenoids was the only wax fraction to show a correlation with cuticle permeability (r = −0.8) (Figure 5B). These relationships agreed with the correlations found among the different epicuticular wax fractions as alkanes negatively correlated with FA and aldehydes (r = −0.8 and r = −0.7, respectively) and FA related positively to the aldehyde proportion (r = 0.9). In turn, fruit weight loss negatively related to fruit firmness (r = −0.97) (Figure 5B).

2.5. Effects of ABA and Water Stress on the Transcriptional Regulation of Epicuticular Wax-Related Genes

In order to further investigate a putative epicuticular wax metabolism reconfiguration mediated by ABA and/or exposure to postharvest dehydration, a transcriptional analysis on key genes involved in epicuticular wax biosynthesis, transport and regulation was performed (Figure 6). The expression of *CsCER3*, involved in the synthesis of alkanes, remained almost steady during fruit storage at high RH, but significantly decreased with exposure to fruit dehydration. The effect of ABA on *CsCER3* accumulation depended on the storage environment. Indeed, ABA treatment did not statistically affect *CsCER3* expression at high RH, but counteracted the decreased transcript accumulation pattern observed under water stress in the untreated fruit. The expression levels of *CsCER4/FAR3*, which encode an FA reductase involved in the synthesis of primary alcohols, transitorily peaked by 1 week at high RH, but continuously lowered when fruit were exposed to water stress after detachment. ABA treatment generally reduced the accumulation of transcripts regardless of storage conditions, although this effect was more marked at high RH. The gene expression levels of *CsSQS*, involved in squalene synthesis, increased after harvest, independently of storage, but this increment was transitory when fruit were left in a dehydrating environment. Independently of being exposed or not to postharvest water stress, ABA application reduced *CsSQS* transcript accumulation throughout the storage period. *CsCER6/KCS6*, which encode a β-ketoacyl-CoA synthase involved in the synthesis of VLCFA precursors, remained steady in the fruit left at high RH, but dropped by three-fold when fruit were exposed to dehydration for 3 weeks. The effects of ABA on the accumulation of *CsCER6/KCS6* transcripts differed depending on storage, because hormone treatment lowered the gene expression levels at high RH, but slightly increased the transcript levels by 3 weeks under postharvest water stress. Epicuticular wax transporters *CsABCG11/WBC11* and *CsABCG12/WBC12* were similarly regulated in regard to ABA and the water stress response. Both transporters were continuously induced during storage at high RH, and remained almost steady after an initial increase by 1 week when fruit were exposed to water stress. Transcript levels were also lower after water stress storage.

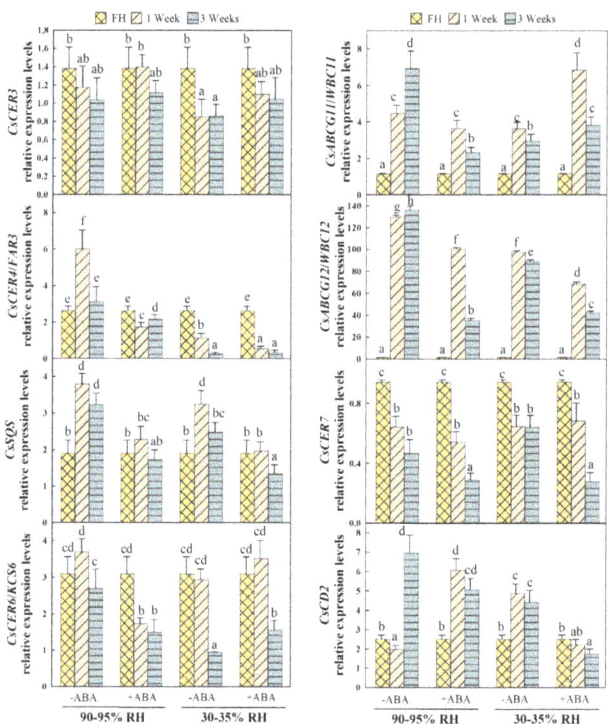

Figure 6. Effect of ABA and water stress on the transcriptional regulation of epicuticular wax metabolism. Relative expression levels of the genes related to the transcriptional (*CsCD2*) and post-transcriptional (*CsCER7*) regulation of the biosynthesis (*CsCER6/KCS6*, *CsCER3*, *CsCER4* and *CsSQS*) and transport (*CsABCG11/WBC11*, *CsABCG12/WBC12*) of the wax components in the Pinalate fruit treated (+ABA) or not (−ABA) with ABA (1 mM), and left at high (90–95%) or low (30–35%) RH for up to 3 weeks at 20 °C. Gene expression values are expressed as fold change levels of all conditions as compared to the freshly harvested (FH) fruit. Values are the means of three biological replicates per condition. Different letters indicate statistical ($p < 0.05$) differences among all the conditions together according to an ANOVA analysis followed by a Tukey test ($p < 0.05$) for each gene individually.

After ABA treatment, the *CsABCG11/WBC11* and *CsABCG12/WBC12* expression levels peaked by 1 week regardless of the water stress exposure. It is worth noting that ABA enhanced *CsABCG11/WBC11* accumulation, but lowered that of *CsABCG12/WBC12* under water stress compared to the untreated fruit. The expression levels of the *CsCER7* post-transcriptional regulator bottomed down in response to harvest by 1 week and remained steady thereafter independently of the environment. ABA also repressed transcript accumulation by 3 weeks at either high or low RH. Last, the *CsCD2* transcription factor expression showed a significant influence for both ABA and postharvest water stress exposure. Thus, *CsCD2* gene expression transiently decreased by 1 week to sharply increase thereafter at high RH, but increased in response to fruit dehydration in the first week of storage. In addition, ABA treatment counteracted the initial decrease in gene expression at high RH, while inhibiting the increment in the *CsCD2* transcript levels when fruit were exposed to water stress (Figure 6).

3. Discussion

The cuticle is extremely sensitive to surrounding fluctuations and, specifically, the epicuticular wax composition adjusts in response to environmental signals [63–66]. The study of the ABA-mediated regulation of cuticle properties and composition has been a hot research topic for years in both Arabidopsis model plants and several horticultural crops. However, research into this relationship in fruit has been limited to tomatoes, cherries and citruses [7–10]. In addition, whether ABA regulates epicuticular wax composition after fruit detachment or if such effects depend on fruit being exposed to dehydrating conditions remains elusive, despite ABA being the main hormone to regulate the fruit water stress response [4]. Moreover, cuticle properties influence fruit water loss and, hence, fruit quality during postharvest [11,15,16,40,66]. In citrus fruit, the association between epicuticular waxes and fruit water retention has been established [7,67]. There are also reports that moderated water stress (70–75% RH) during postharvest causes fruit dehydration and triggers ABA-mediated signals in order to reduce fruit water loss and, hence, alleviate external quality loss [53]. An important advance contributed by the present research is that the effect of increasing the fruit ABA content by hormone feeding on fruit epicuticular wax metabolism was studied in combination with exposure to water stress, which increased fruit weight loss after detachment. To this end, Pinalate, an ABA-biosynthesis-impaired citrus mutant cultivar, that displays sharply reduced hormone levels in the flavedo, was treated with ABA and exposed to postharvest water stress.

The major finding in the present work is that ABA treatment dynamically modifies the epicuticular wax composition and metabolism of Pinalate fruit after detachment depending on whether the fruit is exposed, or not, to postharvest dehydration, and also on the duration of stress. This is deduced from the results, which showed that the increased total epicuticular wax load observed after fruit detachment was diminished by ABA treatment only after 1 week at high RH, but more markedly and during complete storage (up to 3 weeks) under fruit water stress (Figure 1). These ABA application effects were also evident on the total contents of individual wax fractions, such as alkanes, alcohols and aldehydes, and best-fitted the accumulation profile of FA, but not terpenoids (Figure 1). The compositional data analysis of each wax fraction further supported our statement as different responses to ABA treatment were observed by 1 and 3 weeks, which depended on the storage conditions for most individual wax components (Figure 3). In this context, the effects of ABA treatment on modifying wax composition were more evident when focusing on the proportion of each separate fraction. The most important changes can be summarized as attenuated FA accumulation and the enhanced increase in the terpenoid proportion in the ABA-treated fruit compared to those untreated, regardless of the postharvest storage conditions (Figure 2). In fact, even though ABA content correlated with the total epicuticular wax load and to all the wax fraction abundances, except terpenoids (Figure 5B), the contents of several terpenoid compounds, such as squalene, lupenone, friedelanone and α- and β-amyrins, were regulated by ABA application (Figure 3D,H). However, this ABA-mediated regulation differed between high RH and water stress, and also between 1 and 3 storage weeks, which probably modified their abundance in this fraction, and might influence both cuticle permeability and fruit weight loss [36,68–71]. This agrees with the fact that ABA treatment increased cuticle permeability by 3 weeks at high RH, but reduced Pinalate fruit weight loss during postharvest dehydration (Figure 4C,D).

Currently, research into the effects of postharvest water stress on wax composition is limited to pear and tomato; Korla pear wax content and composition adjusts to different RH storage conditions, while the wax chemical profile of tomato fruit, including a long-shelf-life cultivar, does not modify in response to water stress after harvesting the fruit [43,45]. In the present study, orange fruit displaying sharply low ABA levels showed increased total wax load after fruit detachment, without influence from the water stress of the postharvest environment. On the contrary, the total contents of the alkanes and alcohols in Pinalate were influenced by storage conditions by 1 week (Figure 1). Furthermore, under the postharvest dehydration conditions, the abundance of alkanes and FA was lower, and that of alcohols and terpenoids was higher after this period. Moreover, by 3 weeks, these trends were reversed as postharvest water stress brought about a decrease in terpenoid abundance and an increase in alkane proportion. Consequently, the alkane/terpenoid and alcohol/FA ratios varied with both RH and storage duration (Figure 2). Several research works converge with the idea that the proportion of hydrocarbons (i.e., alkanes), unlike terpenoids and cyclic compounds, is the most effective fraction for water retention. Thus, a drop in the alkane/terpenoid ratio would increase the proportion of amorphous structures and, consequently, cuticle permeability and/or fruit water loss [36,68–72]. Results from our chemical analysis partially agree with these findings, as the alkane/terpenoid ratio dropped in response to water stress by 1 week when a four-fold increase in cumulative fruit weight loss was noted (Figures 2 and 4). Notwithstanding, ABA application lowered the alkane/terpenoid ratio after 1 week of storage at high RH, while no differences in fruit weight loss or cuticle permeability were found between the ABA-treated and untreated fruit left under this condition. Conversely, fruit weight loss statistically decreased as a result of ABA treatment under postharvest dehydration, but the alkane/terpenoid ratio in these fruit barely changed compared to the untreated fruit (Figures 2 and 4). Besides, a higher alcohol/FA ratio has been associated with greater fruit weight loss [45], which agrees with our results showing that the dehydrating condition increased the alcohol/FA ratio compared to remaining at a high RH for 1 week. Minor ABA effects on this ratio were observed. If any, ABA treatment increased the alcohol/FA ratio by about 50% by 1 week at high RH, which was due mainly to a reduction in FA abundance (Figures 2 and 4). Altogether, the results presented herein partially diverge from the idea of a straight correlation between these ratios and the fruit cuticle permeability and susceptibility to weight loss. In addition, the attenuated total epicuticular wax accumulation as a result of ABA treatment after Pinalate fruit detachment contrasts with not only reported increments in wax content in cherry fruit after on-vine spraying with ABA [10], but also with the fact that ABA-deficient tomato fruit mutants do not display differences in total wax content compared to their parental lines [9]. The epicuticular wax load of the fully mature Pinalate fruit did not differ from that in other citrus cultivars with higher endogenous ABA contents [8]. In this context, we should bear in mind that Pinalate fruit presents low ABA levels in the peel, but it is not a knockout mutant, and ABA treatment applied to orange fruit with higher ABA levels has not led to changes in either susceptibility to fruit weight loss or the molecular mechanisms underlying fruit dehydration, probably because endogenous levels of the hormone might suffice to trigger cellular processes to cope with stress [53,60,62]. These facts indicate that the ABA-mediated regulation of cuticle metabolism is species-/cultivar-dependent, and other intrinsic factors, such as endogenous hormone levels, climacteric or non-climacteric fruit ripening and/or fruit sensitivity to hormone application when on-vine or detached, might influence this relationship.

The expression profiles of the key genes involved in the synthesis, transport and transcriptional regulation of epicuticular waxes generally support the notion that the effect of ABA on this metabolism depends on both RH and the duration of the imposed water stress after harvesting the fruit. Indeed, the expression patterns of all the studied genes were somewhat influenced by ABA. Despite the lack of statistical differences in the *CsCER3* levels between the ABA-treated and untreated fruit stored in either RH environment, it was noteworthy that ABA application alleviated the diminished transcript accumulation of this gene, which is involved in alkane synthesis during postharvest dehydration (Figure 6). However, the effect of ABA decreasing the expression of *CsSQS* and *CsWBC12*, both involved in the synthesis of squalene and terpenoid precursors and in transporting cuticular components to the extracellular matrix, respectively, was observed regardless of storage duration and conditions. This was also true for *CsCER7*, a post-transcriptional regulator, accumulation despite the effect of ABA decreasing gene expression depending on storage duration, since it was only observed by 3 weeks (Figure 6). In contrast, the effects of ABA on *CsCER4*, *CsCER6*, *CsWBC11* and *CsCD2* regulation were dependent on both exposure to postharvest dehydration and stress duration. Moreover, the expression of the *CsWBC11* transporter and the *CsCD2* transcription factor was affected by ABA via a trend that did not vary between 1 and 3 weeks when fruit were stored under the dehydrating condition, but it differed between the storage periods at high RH. Inversely, *CsCER6* and

CsCER4, respectively involved in the synthesis of VLCA and primary alcohols, showed an ABA effect that did not depend on storage period at high RH, but varied between 1 and 3 weeks when fruit were exposed to water stress. Altogether these data reflect that the ABA-mediated transcriptional regulation of epicuticular wax metabolism is complex and depends on the ABA treatment and postharvest water stress combination, and this relation may change as stress exposure progresses.

Our results indicate that detachment from tree causes the accumulation of epicuticular wax load in citrus fruit with low ABA levels, and that applying ABA after harvesting the fruit attenuates this increase. This ABA-related effect substantially increases in fruit exposed to water stress during postharvest. In turn, this research reveals that ABA-driven changes in the epicuticular wax chemical profile and metabolism depend firstly on storage duration, and in a minor extent on the exposure to water stress conditions during postharvest. Therefore, this data provides clues for industrial wax synthesis purposes and improves the knowledge on the ABA-dependent regulation of cuticular wax metabolism in fruits

4. Materials and Methods

4.1. Fruit Materials and Experimental Design

Pinalate (*Citrus sinensis* L. Osbeck) sweet orange fruit were harvested from five adult trees grown in experimental orchards in the fully mature ripening stage according to the normal cultural practices applied by the Citrus Germplasm Bank at the IVIA (Moncada, Valencia) in Spain. Fruit were immediately delivered to the laboratory and those without peel damages or visual defects and with an average diameter of 6 cm were assigned to two groups. Fruit were treated with either ABA by dipping them for 1 min in an aqueous 1 mM ABA solution containing 0.7% ethanol to dissolve the hormone (treatment group, +ABA) or water containing 0.7% ethanol (control group, −ABA) following the same procedure. ABA treatment was repeated 2 weeks after harvest to ensure high ABA levels throughout the experiment. This dosage was selected according to previous experiments performed by our group which showed the effect of exogenous ABA on the response of citrus fruit to stressful conditions [53,55,59]. Both sets of fruit were divided into two subgroups and stored in the dark at 20 °C for up to 3 weeks in incubation chambers under control (90–95% RH, subgroup 1) or dehydrating (30–35% RH, subgroup 2) conditions. Samples were collected at harvest (freshly harvested fruit, FH), and the ABA-treated and untreated fruit were left for 1 and 3 weeks in the control or dehydrating environments (9 different conditions). For all the conditions, four biological replicates, each consisting of five fruit, were used for the wax content and composition analyses. Three additional biological replicates of 10 fruit each per condition were used to take fruit firmness and weight loss measurements. For all the conditions, three biological replicates of five fruit each were employed to determine cuticle thickness and permeability, and to collect flavedo samples, which were frozen and homogenized in liquid nitrogen, and kept at −80 °C for the ABA and gene expression analyses. Therefore, all nine samples composing the experimental design consisted of 65 fruit each.

4.2. Cuticular Wax Analysis

Epicuticular waxes from intact fruit of known surface areas were extracted by dipping fruit for 1 min in two successive chloroform baths, the first of which contained 100 µg of tetracosane as the internal standard. Waxes were derivatized using BSTFA and resuspended in 100 µL of chloroform after the evaporation of excess BSTFA, as previously described [43]. Wax extracts were injected into a gas chromatograph (GC) 7890B system (Agilent) equipped with an HP-5MS UI (30 m × 250 µm × 0.25 µm) column (Agilent Technologies, Santa Clara, CA, USA) and a 5977A simple quadrupole detector (Agilent Technologies) at the SCSIE-UV Gas Chromatography Facility (Valencia, Spain) following the conditions in [8]. Briefly, the oven temperature was held at 70 °C for 2 min before being raised by 10 °C min^{-1} to 200 °C, by 3 °C min^{-1} to 300 °C and finally held for 20 min. The injector temperature was

250 °C. The comparison of the relative retention times with those of commercial standards was used to identify most of the wax components. Computer matching against commercial (Nist, wiley7n) libraries and by MS literature data was also used for identification.

4.3. Cuticle Permeability and Thickness

Peel disks of 3 cm diameter were excised from the equatorial zone of intact fruit. Cuticles were enzymatically isolated and their permeability was measured in gravimetric chambers as described in detail in [8]. Briefly, isolated cuticles were placed face-up in 3D-printed chambers which exposed a constant cuticle surface area, which acted as the only separation barrier between the known amount of water inside the chamber and the dehydrating environment of a desiccation container (25 °C, 0% RH). Folded chambers were stored for up to 1 week and weight loss was measured daily. Cuticle permeability was estimated as the weight loss per hour and per unit of cuticle surface area. At the end of the assay, the integrity of cuticles was verified by 0.01% aqueous solution (w/v) of Toluidine Blue O (Merck, Darmstadt, Germany) staining. One disk of all five fruit, composing a biological replicate per sample, was used to obtain the isolated cuticles, and three biological replicates per sample were analyzed. Cuticle thickness was determined by light microscopy. Pericarp cubes were excised from fruit, and tissue fixation and embedding were performed as in [9]. A solution of Oil Red O (Alfa Aesar, Kandel, Germany) in isopropyl alcohol was applied to 10 μm sections. The stained slides were visualized under an Eclipse 90i Nikon microscope (Nikon corporation, Tokyo, Japan) with a 40X objective and which used the Nis Elements BR 3.2 software (Nikon corporation, Japan). The distance between the outer cuticle part and the top of the most external epidermal cell was used to measure cuticle thickness by the Fiji software (ImageJ 1.49q Software, National Institutes of Health, Bethesda, MD, USA), as previously described by [8]. One pericarp cube from the equatorial zone of all five fruit, composing a biological replicate per sample, was excised, and three biological replicates per sample were analyzed. About five measurements were taken in each section (roughly 75 values per sample).

4.4. ABA Analysis

As previously described [51], ABA was extracted from 1 g of fresh weight frozen flavedo with 80% acetone, containing 0.5 g L^{-1} of citric acid and 100 mg L^{-1} of butylated hydroxytoluene. After centrifuging, the supernatant was three-serial diluted in ice-cold TBS (6.05 g Tris, 8.8 g L^{-1} of NaCl and 0.2 mg L^{-1} of Mg Cl$_2$, pH 7.8) and three samples for each dilution were analyzed by indirect ELISA.

4.5. Fruit Weight Loss and Firmness Determinations

During storage, Pinalate fruit were weighed daily to determine the amount of water loss. Fruit cumulative weight loss was calculated as the amount of water loss per surface area. Fruit surface area was determined by measuring three diameters per fruit for every biological replicate. Fruit firmness was analyzed as in [73] with minor modifications. Fruit compression resistance, based on a 5 mm deformation at two points of the fruit equator, was measured by a 4502 Instron Testing Machine (Instron).

4.6. Clustering Analysis

A hierarchical cluster analysis (HCA) was performed to group the different conditions in the experimental design according to their similarities in terms of both cuticle composition and properties in addition to the measured fruit physiological parameters. Average linkage clustering and Euclidian distance methods were followed to plot the dendrogram and heatmap (www.heatmapper.ca (accessed on 5 April 2021)).

4.7. RNA Extraction, cDNA Synthesis and RT-qPCR

RNA extraction, cDNA synthesis and RT-qPCR were performed following previously described well-established protocols [50]. Total RNA was isolated from flavedo samples, and 2 µg were used for the first-strand cDNA synthesis with the "Maxima H Minus First Strand cDNA Synthesis kit with dsDNase" (Thermo Scientific). The specific primer pairs for the genes of interest (*CsCER3*, *CsCER4/FAR3*, *CsCER6/KCS6*, *CsSQS*, *CsABCG11/WBC11*, *CsABCG12/WBC12*, *CsCD2* and *CsCER7*) and those employed for data normalization (*CsACT* and *CsTUB*) (Supplementary Materials Table S1) were mixed with SYBR Green to monitor cDNA amplification in a LyghCycler480 System (Roche Diagnostic). Amplicon specificity was determined by a melting curve analysis. Fold change relative gene expression values of the target genes were obtained by the Relative Expression Software Tool (REST, rest.gene-quantification.info), as previously described in [50]. Three independent biological replicates and two technical replicates were performed per sample.

4.8. Statistical Analyses

Statistical analyses were performed using the INFOSTAT software. Data of the parametric variables were subjected to an analysis of variance (ANOVA), and the significance of differences was determined by Tukey's test ($p < 0.05$) on the mean values. Correlation analyses, carried out with the R software, established the relations among wax fractions and components, cuticle properties and fruit quality parameters. The statistical significance of the positive and negative correlations was considered at $p < 0.05$.

Supplementary Materials: The following are available online at https://www.mdpi.com/article/10.3390/ijms221910242/s1, Table S1: Primers used for qPCR analysis.

Author Contributions: Conceptualization, P.R. and M.T.L.; formal Analysis, P.R.; funding acquisition, P.R. and M.T.L.; investigation, P.R.; methodology, P.R.; project administration, P.R.; software, P.R.; supervision, P.R. and M.T.L.; visualization, P.R.; writing—original draft, P.R.; writing—review and editing, P.R. and M.T.L. Both authors have read and agreed to the published version of the manuscript.

Funding: This research was funded by the 3F: FutureFreshFruit Project as part of the Marie Skłodowska-Curie Actions and the European Horizon 2020 programme, grant number H2020-MSCA-IF-656127.

Acknowledgments: The technical assistance of R. Sampedro is gratefully acknowledged. We also thank G. Ancillo (IVIA) for allowing us to use the Spanish Citrus Germplasm Bank.

Conflicts of Interest: The authors declare no conflict of interest.

References

1. Ohkuma, K.; Lyon, J.L.; Addicott, F.T.; Smith, O.E. Abscisin II, an abscission-accelerating substance from young cotton fruit. *Science* **1963**, *142*, 1592–1593. [CrossRef]
2. Addicott, F.T.; Lyon, J.L.; Ohkuma, K.; Thiessen, W.E.; Carns, H.R.; Smith, O.E.; Cornforth, J.W.; Milborrow, B.V.; Ryback, G.; Wareing, P.F. Abscisic acid: A new name for Abscisin II (Dormin). *Science* **1968**, *159*, 1493. [CrossRef] [PubMed]
3. Nakashima, K.; Yamaguchi-Shinozaki, K.; Shinozaki, K. The transcriptional regulatory network in the drought response and its crosstalk in abiotic stress responses including drought, cold, and heat. *Front. Plant Sci.* **2014**, *5*, 170. [CrossRef] [PubMed]
4. Leng, P.; Yuan, B.; Guo, Y. The role of abscisic acid in fruit ripening and responses to abiotic stress. *J. Exp. Bot.* **2014**, *65*, 4577–4588. [CrossRef]
5. Kosma, D.K.; Bourdenx, B.; Bernard, A.; Parsons, E.P.; Lü, S.; Joubès, J.; Jenks, M.A. The impact of water deficiency on leaf cuticle lipids of *Arabidopsis*. *Plant Physiol.* **2009**, *151*, 1918–1929. [CrossRef] [PubMed]
6. Curvers, K.; Seifi, H.; Mouille, G.; de Rycke, R.; Asselbergh, B.; Van Hecke, A.; Vanderschaeghe, D., Hofte, H.; Callewaert, N.; Van Breusegem, F.; et al. Abscisic acid deficiency causes changes in cuticle permeability and pectin composition that influence tomato resistance to *Botrytis cinerea*. *Plant Physiol.* **2010**, *154*, 847–860. [CrossRef]
7. Wang, J.; Sun, L.; Xie, L.; He, Y.; Luo, T.; Sheng, L.; Luo, Y.; Zeng, Y.; Xu, J.; Deng, X.; et al. Regulation of cuticle formation during fruit development and ripening in 'Newhall' navel orange (*Citrus sinensis* Osbeck) revealed by transcriptomic and metabolomic profiling. *Plant Sci.* **2016**, *243*, 131–144. [CrossRef]
8. Romero, P.; Lafuente, M.T. Abscisic acid deficiency alters epicuticular wax metabolism and morphology that leads to increased cuticle permeability during sweet orange (*Citrus sinensis*) fruit ripening. *Front. Plant Sci.* **2020**, *11*, 1914. [CrossRef]
9. Martin, L.B.B.; Romero, P.; Fich, E.A.; Domozych, D.S.; Rose, J.K.C. Cuticle biosynthesis in tomato leaves is developmentally regulated by abscisic acid. *Plant Physiol.* **2017**, *174*, 1384–1398. [CrossRef]
10. Correia, S.; Santos, M.; Glińska, S.; Gapińska, M.; Matos, M.; Carnide, V.; Schouten, R.; Silva, A.P.; Gonçalves, B. Effects of exogenous compound sprays on cherry cracking: Skin properties and gene expression. *J. Sci. Food Agric.* **2020**, *100*, 2911–2921. [CrossRef] [PubMed]

11. Martin, L.B.B.; Rose, J.K.C. There's more than one way to skin a fruit: Formation and functions of fruit cuticles. *J. Exp. Bot.* **2014**, *65*, 4639–4651. [CrossRef]
12. Yeats, T.H.; Rose, J.K.C. The formation and function of plant cuticles. *Plant Physiol.* **2013**, *163*, 5–20. [CrossRef]
13. Bhanot, V.; Fadanavis, S.V.; Panwar, J. Revisiting the architecture, biosynthesis and functional aspects of the plant cuticle: There is more scope. *Environ. Exp. Bot.* **2021**, *183*, 104364. [CrossRef]
14. Joubès, J.; Domergue, F. Biosynthesis of the Plant Cuticle. In *Hydrocarbons, Oils and Lipids: Diversity, Origin, Chemistry and Fate*; Springer International Publishing: Berlin, Germany, 2018; pp. 1–19.
15. Lara, I.; Heredia, A.; Domínguez, E. Shelf life potential and the fruit cuticle: The unexpected player. *Front. Plant Sci.* **2019**, *10*, 770. [CrossRef]
16. Tafolla-Arellano, J.C.; Báez-Sañudo, R.; Tiznado-Hernández, M.E. The cuticle as a key factor in the quality of horticultural crops. *Sci. Hortic.* **2018**, *232*, 145–152. [CrossRef]
17. Parsons, E.P.; Popopvsky, S.; Lohrey, G.T.; Lü, S.; Alkalai-Tuvia, S.; Perzelan, Y.; Paran, I.; Fallik, E.; Jenks, M.A. Fruit cuticle lipid composition and fruit post-harvest water loss in an advanced backcross generation of pepper (*Capsicum* sp.). *Physiol. Plant.* **2012**, *146*, 15–25. [CrossRef] [PubMed]
18. Chu, W.; Gao, H.; Chen, H.; Fang, X.; Zheng, Y. Effects of cuticular wax on the postharvest quality of blueberry fruit. *Food Chem.* **2018**, *239*, 68–74. [CrossRef]
19. Chai, Y.; Li, A.; Chit Wai, S.; Song, C.; Zhao, Y.; Duan, Y.; Zhang, B.; Lin, Q. Cuticular wax composition changes of 10 apple cultivars during postharvest storage. *Food Chem.* **2020**, *324*, 126903. [CrossRef]
20. Saladié, M.; Matas, A.J.; Isaacson, T.; Jenks, M.A.; Goodwin, S.M.; Niklas, K.J.; Xiaolin, R.; Labavitch, J.M.; Shackel, K.A.; Fernie, A.R.; et al. A reevaluation of the key factors that influence tomato fruit softening and integrity. *Plant Physiol.* **2007**, *144*, 1012–1028. [CrossRef]
21. Lownds, N.; Banaras, M.; Bosland, P.W. Relationships between postharvest water loss and physical properties of pepper fruit (*Capsicum annuum* L.). *HortScience* **1993**, *28*, 1182–1184. [CrossRef]
22. Tafolla-Arellano, J.C.; Zheng, Y.; Sun, H.; Jiao, C.; Ruiz-May, E.; Hernández-Oñate, M.A.; González-León, A.; Báez-Sañudo, R.; Fei, Z.; Domozych, D.; et al. Transcriptome analysis of mango (*Mangifera indica* L.) fruit epidermal peel to identify putative cuticle-associated genes. *Sci. Rep.* **2017**, *7*, 46163. [CrossRef] [PubMed]
23. Huang, H.; Burghardt, M.; Schuster, A.-C.; Leide, J.; Lara, I.; Riederer, M. Chemical composition and water permeability of fruit and leaf cuticles of *Olea europaea* L. *J. Agric. Food Chem.* **2017**, *65*, 8790–8797. [CrossRef]
24. Diarte, C.; Xavier de Souza, A.; Staiger, S.; Deininger, A.-C.; Bueno, A.; Burghardt, M.; Graell, J.; Riederer, M.; Lara, I.; Leide, J. Compositional, structural and functional cuticle analysis of *Prunus laurocerasus* L. sheds light on cuticular barrier plasticity. *Plant Physiol. Biochem.* **2021**, *158*, 434–445. [CrossRef]
25. Vichi, S.; Cortés-Francisco, N.; Caixach, J.; Barrios, G.; Mateu, J.; Ninot, A.; Romero, A. Epicuticular wax in developing olives (*Olea europaea*) is highly dependent upon cultivar and fruit ripeness. *J. Agric. Food Chem.* **2016**, *64*, 5985–5994. [CrossRef] [PubMed]
26. Belge, B.; Llovera, M.; Comabella, E.; Graell, J.; Lara, I. Fruit cuticle composition of a melting and a nonmelting peach cultivar. *J. Agric. Food Chem.* **2014**, *62*, 3488–3495. [CrossRef]
27. Parsons, E.P.; Popopvsky, S.; Lohrey, G.T.; Alkalai-Tuvia, S.; Perzelan, Y.; Bosland, P.; Bebeli, P.J.; Paran, I.; Fallik, E.; Jenks, M.A. Fruit cuticle lipid composition and water loss in a diverse collection of pepper (*Capsicum*). *Physiol. Plant.* **2013**, *149*, 160–174. [CrossRef] [PubMed]
28. Sala, J.M. Content, chemical composition and morphology of epicuticular wax of Fortune mandarin fruits in relation to peel pitting. *J. Sci. Food Agric.* **2000**, *80*, 1887–1894. [CrossRef]
29. Domínguez, E.; Fernández, M.D.; Hernández, J.C.L.; Parra, J.P.; España, L.; Heredia, A.; Cuartero, J. Tomato fruit continues growing while ripening, affecting cuticle properties and cracking. *Physiol. Plant.* **2012**, *146*, 473–486. [CrossRef]
30. Peschel, S.; Franke, R.; Schreiber, L.; Knoche, M. Composition of the cuticle of developing sweet cherry fruit. *Phytochemistry* **2007**, *68*, 1017–1025. [CrossRef] [PubMed]
31. Knoche, M.; Beyer, M.; Peschel, S.; Oparlakov, B.; Bukovac, M.J. Changes in strain and deposition of cuticle in developing sweet cherry fruit. *Physiol. Plant.* **2004**, *120*, 667–677. [CrossRef] [PubMed]
32. D'Angeli, S.; Altamura, M. Unsaturated lipids change in olive tree drupe and seed during fruit development and in response to cold-stress and acclimation. *Int. J. Mol. Sci.* **2016**, *17*, 1889. [CrossRef] [PubMed]
33. Baker, E.; Bukovac, M.; Hunt, G. Composition of tomato fruit cuticle as related to fruit growth and development. In *The Plant Cuticle*; Cutler, D., Alvin, K., Price, C., Eds.; London Academic Press: London, UK, 1982; pp. 33–44.
34. Oliveira Lino, L.; Quilot-Turion, B.; Dufour, C.; Corre, M.-N.; Lessire, R.; Génard, M.; Poëssel, J.-L. Cuticular waxes of nectarines during fruit development in relation to surface conductance and susceptibility to *Monilinia laxa*. *J. Exp. Bot.* **2020**, *71*, 5521–5537. [CrossRef] [PubMed]
35. Cajuste, J.F.; González-Candelas, L.; Veyrat, A.; García-Breijo, F.J.; Reig-Armiñana, J.; Lafuente, M.T. Epicuticular wax content and morphology as related to ethylene and storage performance of 'Navelate' orange fruit. *Postharvest Biol. Technol.* **2010**, *55*, 29–35. [CrossRef]
36. El-Otmani, M.; Coggins, C.W.J. Fruit age and growth regulator effects on the quantity and structure of the epicuticular wax of Washington Navel orange fruit. *J. Am. Soc. Hortic. Sci.* **1985**, *110*, 371–378.

37. Li, F.; Min, D.; Song, B.; Shao, S.; Zhang, X. Ethylene effects on apple fruit cuticular wax composition and content during cold storage. *Postharvest Biol. Technol.* **2017**, *134*, 98–105. [CrossRef]

38. Curry, E. Effects of 1-MCP applied postharvest on epicuticular wax of apples (*Malus domestica* Borkh.) during storage. *J. Sci. Food Agric.* **2008**, *88*, 996–1006. [CrossRef]

39. Klein, B.; Ribeiro, Q.M.; Thewes, F.R.; de OliveiraAnese, R.; de Candido deOliveira, F.; Santos, I.D.; Ribeiro, S.R.; Donadel, J.Z.; Brackmann, A.; Barin, J.S.; et al. The isolated or combined effects of dynamic controlled atmosphere (DCA) and 1-MCP on the chemical composition of cuticular wax and metabolism of 'Maxi Gala' apples after long-term storage. *Food Res. Int.* **2020**, *140*, 109900. [CrossRef]

40. Lara, I. The fruit cuticle: Actively tuning postharvest quality. In *Preharvest Modulation of Postharvest Fruit and Vegetable Quality*; Siddiqui, M.W., Ed.; Elsevier: Amsterdam, The Netherlands, 2018; pp. 93–120. ISBN 978-0-12-809807-3.

41. Trivedi, P.; Nguyen, N.; Hykkerud, A.L.; Häggman, H.; Martinussen, I.; Jaakola, L.; Karppinen, K. Developmental and environmental regulation of cuticular wax biosynthesis in fleshy fruits. *Front. Plant Sci.* **2019**, *10*, 431. [CrossRef] [PubMed]

42. Jetter, R.; Riederer, M. Localization of the transpiration barrier in the epi- and intracuticular waxes of eight plant species: Water transport resistances are associated with fatty acyl rather than alicyclic components. *Plant Physiol.* **2016**, *170*, 921–934. [CrossRef]

43. Romero, P.; Rose, J.K.C. A relationship between tomato fruit softening, cuticle properties and water availability. *Food Chem.* **2019**, *295*, 300–310. [CrossRef] [PubMed]

44. Brizzolara, S.; Minnocci, A.; Yembaturova, E.; Tonutti, P. Ultrastructural analysis of berry skin from four grapes varieties at harvest and in relation to postharvest dehydration. *OENO One* **2020**, *54*, 1121–1131. [CrossRef]

45. Wang, Y.; Mao, H.; Lv, Y.; Chen, G.; Jiang, Y. Comparative analysis of total wax content, chemical composition and crystal morphology of cuticular wax in Korla pear under different relative humidity of storage. *Food Chem.* **2021**, *339*, 128097. [CrossRef]

46. Tadeo, F.R.; Cercós, M.; Colmenero-Flores, J.M.; Iglesias, D.J.; Naranjo, M.A.; Ríos, G.; Carrera, E.; Ruiz-Rivero, O.; Lliso, I.; Morillon, R.; et al. Molecular physiology of development and quality of citrus. In *Advances in Botanical Research*; Academic Press: Cambridge, MA, USA, 2008; Volume 47, pp. 147–223. ISBN 9780123743275.

47. Zhang, Y.-J.; Wang, X.-J.; Wu, J.-X.; Chen, S.-Y.; Chen, H.; Chai, L.-J.; Yi, H.-L. Comparative transcriptome analyses between a spontaneous late-ripening sweet orange mutant and its wild type suggest the functions of ABA, sucrose and JA during citrus fruit ripening. *PLoS ONE* **2014**, *9*, e116056. [CrossRef] [PubMed]

48. Lado, J.; Gambetta, G.; Zacarias, L. Key determinants of citrus fruit quality: Metabolites and main changes during maturation. *Sci. Hortic.* **2018**, *233*, 238–248. [CrossRef]

49. Terol, J.; Nueda, M.J.; Ventimilla, D.; Tadeo, F.; Talon, M. Transcriptomic analysis of *Citrus clementina* mandarin fruits maturation reveals a MADS-box transcription factor that might be involved in the regulation of earliness. *BMC Plant Biol.* **2019**, *19*, 47. [CrossRef] [PubMed]

50. Romero, P.; Lafuente, M.T.; Rodrigo, M.J. A sweet orange mutant impaired in carotenoid biosynthesis and reduced ABA levels results in altered molecular responses along peel ripening. *Sci. Rep.* **2019**, *9*, 9813. [CrossRef]

51. Lafuente, M.T.; Martínez-Téllez, M.A.; Zacarías, L. Abscisic acid in the response of 'Fortune' mandarins to chilling. Effect of maturity and high-temperature conditioning. *J. Sci. Food Agric.* **1997**, *73*, 494–502. [CrossRef]

52. Romero, P.; Lafuente, M.; Rodrigo, M.J. The *Citrus* ABA signalosome: Identification and transcriptional regulation during sweet orange fruit ripening and leaf dehydration. *J. Exp. Bot.* **2012**, *63*, 4931–4945. [CrossRef]

53. Romero, P.; Rodrigo, M.J.; Alférez, F.; Ballester, A.-R.; González-Candelas, L.; Zacarías, L.; Lafuente, M.T. Unravelling molecular responses to moderate dehydration in harvested fruit of sweet orange (*Citrus sinensis* L. Osbeck) using a fruit-specific ABA-deficient mutant. *J. Exp. Bot.* **2012**, *63*, 2753–2767. [CrossRef] [PubMed]

54. Romero, P.; Lafuente, M.T.; Alférez, F. A transcriptional approach to unravel the connection between phospholipases A$_2$ and D and ABA signal in citrus under water stress. *Plant Physiol. Biochem.* **2014**, *80*, 23–32. [CrossRef]

55. Romero, P.; Gandía, M.; Alférez, F. Interplay between ABA and phospholipases A$_2$ and D in the response of citrus fruit to postharvest dehydration. *Plant Physiol. Biochem.* **2013**, *70*, 287–294. [CrossRef] [PubMed]

56. Rodrigo, M.J.; Alquezar, B.; Zacarías, L. Cloning and characterization of two 9-cis-epoxycarotenoid dioxygenase genes, differentially regulated during fruit maturation and under stress conditions, from orange (*Citrus sinensis* L. Osbeck). *J. Exp. Bot.* **2006**, *57*, 633–643. [CrossRef] [PubMed]

57. Agustí, J.; Zapater, M.; Iglesias, D.J.; Cercós, M.; Tadeo, F.R.; Talón, M. Differential expression of putative 9-cis-epoxycarotenoid dioxygenases and abscisic acid accumulation in water stressed vegetative and reproductive tissues of citrus. *Plant Sci.* **2007**, *172*, 85–94. [CrossRef]

58. Xian, L.; Sun, P.; Hu, S.; Wu, J.; Liu, J.-H. Molecular cloning and characterization of *CrNCED1*, a gene encoding 9-cis-epoxycarotenoid dioxygenase in *Citrus reshni*, with functions in tolerance to multiple abiotic stresses. *Planta* **2014**, *239*, 61–77. [CrossRef]

59. Romero, P.; Rodrigo, M.J.; Lafuente, M.T. Differential expression of the *Citrus sinensis* ABA perception system genes during postharvest fruit dehydration. *Postharvest Biol. Technol.* **2013**, *76*, 65–73. [CrossRef]

60. Rodrigo, M.J.; Marcos, J.F.; Alférez, F.; Mallent, M.D.; Zacarías, L. Characterization of Pinalate, a novel *Citrus sinensis* mutant with a fruit-specific alteration that results in yellow pigmentation and decreased ABA content. *J. Exp. Bot.* **2003**, *54*, 727–738. [CrossRef]

61. Alférez, F.; Sala, J.M.; Sanchez-Ballesta, M.T.; Mulas, M.; Lafuente, M.T.; Zacarias, L. A comparative study of the postharvest performance of an ABA-deficient mutant of oranges: I. Physiological and quality aspects. *Postharvest Biol. Technol.* **2005**, *37*, 222–231. [CrossRef]

62. Rodrigo, M.J.; Lado, J.; Alós, E.; Alquézar, B.; Dery, O.; Hirschberg, J.; Zacarías, L. A mutant allele of ζ-carotene isomerase (Z-ISO) is associated with the yellow pigmentation of the "Pinalate" sweet orange mutant and reveals new insights into its role in fruit carotenogenesis. *BMC Plant Biol.* **2019**, *19*, 465. [CrossRef] [PubMed]

63. Koch, K.; Hartmann, K.D.; Schreiber, L.; Barthlott, W.; Neinhuis, C. Influences of air humidity during the cultivation of plants on wax chemical composition, morphology and leaf surface wettability. *Environ. Exp. Bot.* **2006**, *56*, 1–9. [CrossRef]

64. Lewandowska, M.; Keyl, A.; Feussner, I. Wax biosynthesis in response to danger: Its regulation upon abiotic and biotic stress. *New Phytol.* **2020**, *227*, 698–713. [CrossRef] [PubMed]

65. Shaheenuzzamn, M.; Shi, S.; Sohail, K.; Wu, H.; Liu, T.; An, P.; Wang, Z.; Hasanuzzaman, M. Regulation of cuticular wax biosynthesis in plants under abiotic stress. *Plant Biotechnol. Rep.* **2021**, *15*, 1–12. [CrossRef]

66. Zarrouk, O.; Pinheiro, C.; Misra, C.S.; Fernández, V.; Chaves, M.M. Fleshy fruit epidermis is a protective barrier under water stress. In *Water Scarcity and Sustainable Agriculture in Semiarid Environment*; Academic Press: Cambridge, MA, USA, 2018; pp. 507–533. ISBN 9780128131640.

67. Wang, J.; Hao, H.; Liu, R.; Ma, Q.; Xu, J.; Chen, F.; Cheng, Y.; Deng, X. Comparative analysis of surface wax in mature fruits between Satsuma mandarin (*Citrus unshiu*) and 'Newhall' navel orange (*Citrus sinensis*) from the perspective of crystal morphology, chemical composition and key gene expression. *Food Chem.* **2014**, *153*, 177–185. [CrossRef] [PubMed]

68. Isaacson, T.; Kosma, D.K.; Matas, A.J.; Buda, G.J.; He, Y.; Yu, B.; Pravitasari, A.; Batteas, J.D.; Stark, R.E.; Jenks, M.A.; et al. Cutin deficiency in the tomato fruit cuticle consistently affects resistance to microbial infection and biomechanical properties, but not transpirational water loss. *Plant J.* **2009**, *60*, 363–377. [CrossRef] [PubMed]

69. Leide, J.; Hildebrandt, U.; Reussing, K.; Riederer, M.; Vogg, G. The developmental pattern of tomato fruit wax accumulation and its impact on cuticular transpiration barrier properties: Effects of a deficiency in a beta-ketoacyl-coenzyme A synthase (*LeCER6*). *Plant Physiol.* **2007**, *144*, 1667–1679. [CrossRef]

70. Riederer, M.; Schreiber, L. Protecting against water loss: Analysis of the barrier properties of plant cuticles. *J. Exp. Bot.* **2001**, *52*, 2023–2032. [CrossRef] [PubMed]

71. Vogg, G.; Fischer, S.; Leide, J.; Emmanuel, E.; Jetter, R.; Levy, A.A.; Riederer, M. Tomato fruit cuticular waxes and their effects on transpiration barrier properties: Functional characterization of a mutant deficient in a very-long-chain fatty acid β-ketoacyl-CoA synthase. *J. Exp. Bot.* **2004**, *55*, 1401–1410. [CrossRef] [PubMed]

72. Kosma, D.K.; Jenks, M.A. Eco-physiological and molecular-genetic determinants of plant cuticle function in drought and salt stress tolerance. In *Advances in Molecular Breeding toward Drought and Salt Tolerant Crops*; Jenks, M.A., Hasegawa, P.M., Jain, S.M., Eds.; Springer: Dordrecht, The Netherlands, 2007; pp. 91–120.

73. Lafuente, M.T.; Alférez, F.; Romero, P. Postharvest ethylene conditioning as a tool to reduce quality loss of stored mature sweet oranges. *Postharvest Biol. Technol.* **2014**, *94*, 104–111. [CrossRef]

International Journal of **Molecular Sciences**

MDPI

Article

Upstream Open Reading Frame Mediated Translation of WNK8 Is Required for ABA Response in *Arabidopsis*

Zhiyong Li [1,2,3,*], Yajuan Fu [1], Jinyu Shen [1] and Jiansheng Liang [1,2,*]

1 Department of Biology, Southern University of Science and Technology, Shenzhen 518055, China; 11930723@mail.sustech.edu.cn (Y.F.); shenjy@mail.sustech.edu.cn (J.S.)
2 Key Laboratory of Molecular Design for Plant Cell Factory of Guangdong Higher Education Institutes, Department of Biology, Southern University of Science and Technology, Shenzhen 518055, China
3 Academy for Advanced Interdisciplinary Studies, Southern University of Science and Technology, Shenzhen 518055, China
* Correspondence: lizy6@sustech.edu.cn (Z.L.); liangjs@sustech.edu.cn (J.L.)

Abstract: With no lysine (K) (WNK) kinases comprise a family of serine/threonine kinases belonging to an evolutionary branch of the eukaryotic kinome. These special kinases contain a unique active site and are found in a wide range of eukaryotes. The model plant *Arabidopsis* has been reported to have 11 WNK members, of which WNK8 functions as a negative regulator of abscisic acid (ABA) signaling. Here, we found that the expression of WNK8 is post-transcriptionally regulated through an upstream open reading frame (uORF) found in its 5′ untranslated region (5′-UTR). This uORF has been predicted to encode a conserved peptide named CPuORF58 in both monocotyledons and dicotyledons. The analysis of the published ribosome footprinting studies and the study of the frameshift CPuORF58 peptide with altered repression capability suggested that this uORF causes ribosome stalling. Plants transformed with the native *WNK8* promoter driving WNK8 expression were comparable with wild-type plants, whereas the plants transformed with a similar construct with mutated CPuORF58 start codon were less sensitive to ABA. In addition, WNK8 and its downstream target RACK1 were found to synergistically coordinate ABA signaling rather than antagonistically modulating glucose response and flowering in plants. Collectively, these results suggest that the WNK8 expression must be tightly regulated to fulfill the demands of ABA response in plants.

Keywords: upstream open reading frame; translation; abscisic acid; protein kinase WNK8

Citation: Li, Z.; Fu, Y.; Shen, J.; Liang, J. Upstream Open Reading Frame Mediated Translation of WNK8 Is Required for ABA Response in *Arabidopsis*. *Int. J. Mol. Sci.* **2021**, 22, 10683. https://doi.org/10.3390/ijms221910683

Academic Editor: Víctor Quesada

Received: 16 August 2021
Accepted: 27 September 2021
Published: 1 October 2021

Publisher's Note: MDPI stays neutral with regard to jurisdictional claims in published maps and institutional affiliations.

1. Introduction

Protein kinases comprise a superfamily of enzymes that regulate a wide range of cellular processes through induced phosphorylation of downstream protein substrates in all living cells. WITH NO LYSINE (WNK) kinases belong to a superfamily of serine/threonine protein kinases uniquely characterized by the lack of a catalytic lysine residue in the kinase subdomain II, which is crucial for coordinating ATP and catalyzing phosphoryl transfers within other protein kinase superfamily and downstream targets [1].

WNK kinases have been identified in various eukaryotes. They were first discovered in mammals while screening for new members of the mitogen-activated protein, serine/threonine kinase (MAPK) family. Humans have four different WNK kinases, namely WNK1, WNK2, WNK3, and WNK4 [2]. Studies have reported four WNK members in mice, two in *Xenopus*, and a single WNK homolog member in *Drosophila melanogaster*, *Caenorhabditis elegans*, and *Giardia lambia* [3–5]. In animals, WNK kinases play critical roles in ion homeostasis by regulating numerous ion channels, including sodium chloride cotransporter (NCC) and sodium potassium chloride cotransporter (NKCC1), as well as in angiogenesis, tumor cell growth, and neuropathic disorders [6–12]. In plants, the WNK gene family is larger and more diverse than in animals. A total of 11 and 9 WNK members have been identified in the model plant *Arabidopsis* and *Oryza sativa*, respectively [13,14].

An even larger number of WNK members have been found in peach (18) and soybean (26) [15,16]. Plant WNK family members have been implicated to play important roles in many plant physiological and developmental processes, including the regulation of flowering time [13,17], fruit development and ripening [16], internal circadian rhythms [18–20], root system architecture [15], cellular pH homeostasis, [21] and various abiotic and biotic stress [22–27].

The plant hormone abscisic acid (ABA) is known to play important roles in development and stress responses. Some WNK family members have been found to be involved in ABA-signaling pathway. WNK8 and WNK9 from *Arabidopsis* have been shown to regulate salt and osmotic stress responses through an ABA-dependent pathway [23,24]. Studies have shown that *AtWNK8* expression is enhanced under salt and sorbitol treatments and that the *wnk8* mutant lines have higher proline content and exhibit significant catalase (CAT) and peroxidase (POD) activities, which lead to in high salinity and osmotic stress tolerance [23]. Furthermore, *AtWNK8* has been proposed to negatively modulate ABA signaling by interacting with ABA-signaling core components, including the ABA receptor, PYR1, and type 2C protein phosphatase (PP2CA) [28]. In contrast, *AtWNK9* positively regulates ABA signaling and enhances drought tolerance in transgenic plants [24]. *OsWNK9*, a member of rice WNK gene family, has been functionally well characterized, and its overexpression in *Arabidopsis* has been shown to confer high tolerance to drought and salt stress in an ABA-dependent manner [25]. Moreover, a previous study showed that the soybean *GmWNK1* gene reduces sensitivity to both ABA and mannitol treatment with increased endogenous ABA in *Arabidopsis* [15]. Furthermore, this study showed that GmWNK1 directly interacted with a crucial ABA catabolism enzyme, GmCYP707A1. Collectively, these findings revealed the diverse roles of plant WNKs in ABA response and the relevance of their interaction with ABA in both signaling pathways and ABA metabolic processes. However, many questions regarding the regulation of WNK kinases and their comprehensive interaction with ABA components remain to be explored.

In the present study, we investigated *Arabidopsis* WNK8 found that two individual T-DNA insertion lines of *WNK8* showed similar ABA response phenotypes. However, no significant transcript regulation of *WNK8* was observed under ABA treatment compared to the control. Further studies showed that a conserved peptide encoded by an upstream open reading frame (uORF), called uORF58, has been identified in the 5′-UTR of *WNK8*. We showed that the translation of WNK8 was repressed by CPuORF58 in vivo and that this element may act by a ribosome stalling mechanism, independently of the main open reading frame (mORF) downstream of the uORF. Moreover, we showed that such an ingenious regulation is necessary for plants to fulfill the demands of ABA response caused by the uncontrolled expression of WNK8. Finally, we found that WNK8 coordinates ABA signaling with RACK1, a downstream target of WNK8.

2. Results

2.1. WNK8 Negatively Regulates ABA Response in Seed Germination and Post-germination Development

A previous study showed that different T-DNA insertion mutants of *WNK8* led to opposite ABA responses during seed germination [28]. To learn more about the roles of WNK8 in response to ABA, two T-DNA insertion mutants, SALK_206987C (*wnk8-1*) and SALK_103318C (*wnk8-2*), were used in this study (Figure 1A). The T-DNA insertions were confirmed to be present in the fourth exon of the *WNK8* gene by genotyping and sequencing (Figure 1B). An RT-PCR analysis using region-specific primers detected no *WNK8* transcript in either allele (Figure 1B).

We further carried out a seed germination assay for each genotype following ABA treatment. The germination rates of the different genotypes were similar in the absence of ABA. In the presence of 1.5 μM ABA, both *wnk8-1* and *wnk8-2* seeds showed higher ABA sensitivity with much lower germination rates than the wild-type seeds (Figure 1C). In addition, the *wnk8-1* and *wnk8-2* mutants showed significantly lower cotyledon greening rates after germination in the medium with ABA for 7 days (Figure 1D). Taken together, these results suggest that WNK8 negatively regulates the ABA response during seed germination and post-germination development.

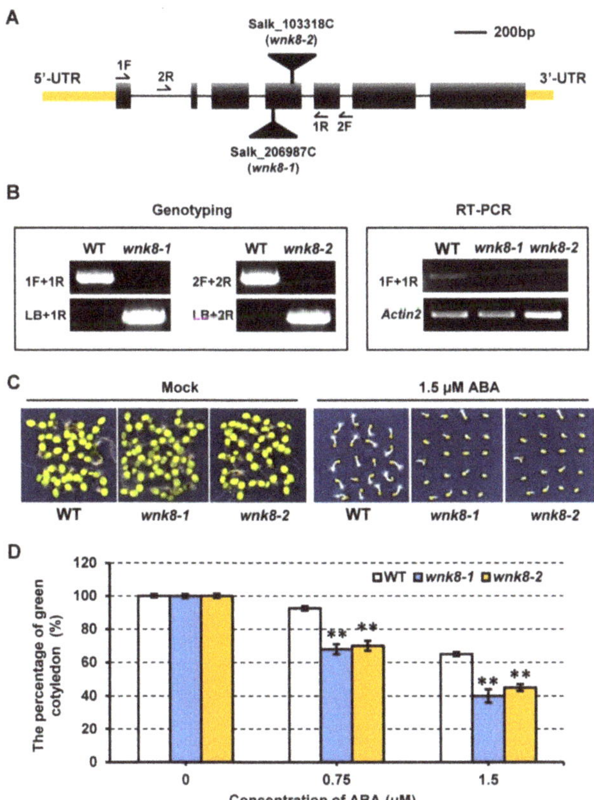

Figure 1. WNK8 negatively regulates ABA response during seed germination and post-germination development. (**A**) Schematic presentation of the *WNK8* gene. 5′ and 3′-UTR are highlighted with yellow boxes. Introns are depicted by solid lines and exons are indicated by black boxes. Two black triangles represented the T-DNA insertion sites of respective *wnk8* alleles. Positions of oligonucleotides used for genotyping and RT-PCRs in (**B**) are indicated by arrows. (**B**) Genotyping and RT-PCRs analysis in Col-0 wild-type (WT) and *wnk8* alleles. *Actin2* was used as the internal control. (**C**) Representative images for seed germination of Col-0 and *wnk8* alleles in the absence or presence of ABA. (**D**) Cotyledon greening rates of indicated genotypes when grown on media supplemented with different concentration of ABA. Data indicate repeat experiments (*n* = 4). Asterisks indicate significant differences to WT.

2.2. WNK8 Has a Conserved Open Reading Frame in Its 5′-UTR

To study the expression pattern of *WNK8* in the presence of ABA, we quantified *WNK8* mRNA accumulation under high ABA concentration (50 µM) and found only slightly increase of *WNK8* mRNA levels (Figure S1). This indicated that the regulatory mechanism of *WNK8* at high ABA concentrations is unlikely to be involved in transcription.

To learn more about the expression of WNK8, transcription and translation were further investigated using the ATHENA database (http://athena.proteomics.wzw.tum.de accessed on 16 August 2021), a collection of many protein and transcript expression profiles from *Arabidopsis thaliana* (Col-0) plants [29]. The ATHENA search analysis showed that similar transcription levels of *WNK8* in different organs or tissues, except for higher transcriptional expression in pollen (Figure S2A). Consistent with the data from ATHENA, the qRT-PCR results confirmed higher expression levels of *WNK8* in open flowers and mature pollens (Figure S2B). In contrast, the translation levels of WNK8 varied greatly among the organs or tissues, in which it was almost undetectable, such as in cotyledons (CT), hypocotyl (HY), rosette leaf (LFs), and carpel (CP). This suggests a post-transcriptional regulation mechanism for *WNK8*.

The presence of an encoded peptide in the 5′-UTR of *WNK8* upstream of its mORF has been previously reported [30]. We further analyzed the published ribosome profiling data for the distribution of ribosomes along the *WNK8* sequence [31–33]. The data clearly indicated that the uORF had a high frequency of ribosome occupancy, with an average higher ribosome density compared to that of the mORF (Figure 2A). We further performed an in silico analysis of WNK8 homologs within representative plant species to determine the functionality of such a peptide. The basic local alignment search tool BLASTP retrieved 12 nucleotide sequences encoding WNK8 homologs from both monocots and dicots, using the WNK8 protein sequence as the query. These homologous sequences were further analyzed for the presence of ORFs, and this analysis led to the identification of 12 uORFs (Figure 2B). Further alignment of these sequences showed a certain degree of conservation in both monocots and dicots (Figure 2C). More peptide conservation was observed in monocots than in dicots (Figures 2D,E and S3).

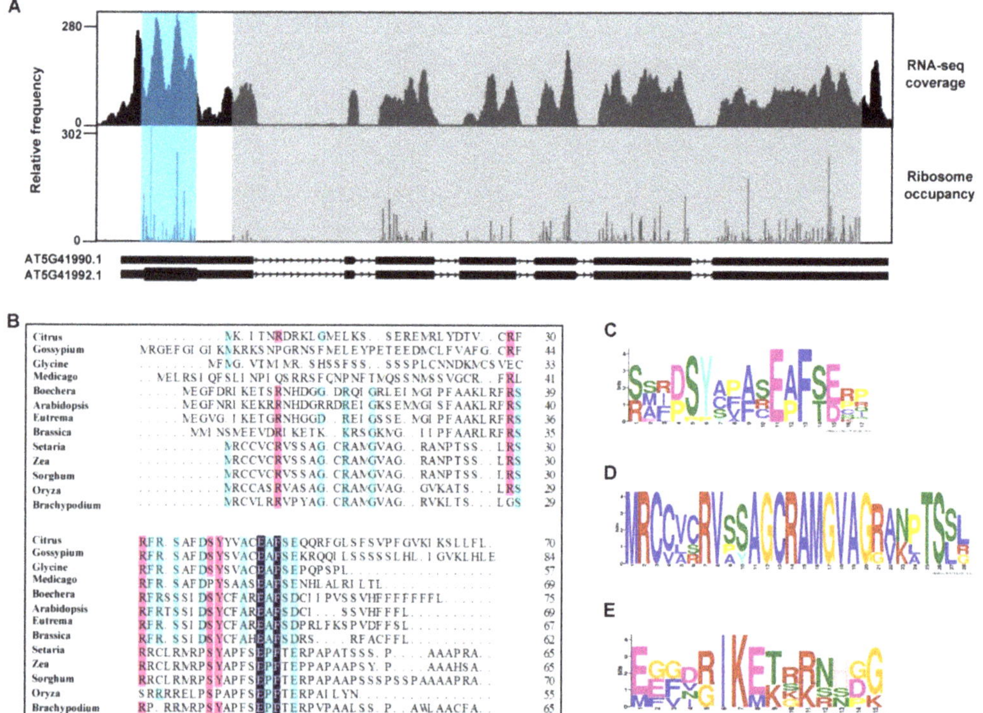

Figure 2. Conserved open reading frame is found in the 5′-UTR of WNK8. (**A**) Comparative ribosome footprinting profile of *WNK8*. The coverage of RNA-seq reads is shown in upper panel and the ribosome occupancy is shown in lower panel. Light blue: CPuORF58; Gray: main ORF; White: UTR. (**B**) Sequence alignment of the predicted amino acid sequences of CPuORF58 from different plant species. The sequence logos are resulted from the sequence alignment of the uORFs from all selected plant species (**C**), monocts (**D**), and dicots (**E**). Letter height indicates the frequency in the alignment.

2.3. CPuORF58 Is Essential for the Translational Suppression of the WNK8

5′-UTRs play a significant role in translational regulation [33]. We presumed that the CPuORF58 peptide located in the 5′-UTR could play a translational regulatory role in controlling the expression of WNK8. To address this hypothesis, we generated mutations to replace the start codon (ATG) of CPuORF58 with AAG. Two different genetic constructs were generated with the expression of the *GUS* reporter gene controlled by the *WNK8* promoter including its 5′-UTR with the native (*pWNK8:GUS*) or mutated CPuORF58 (*pWNK8m:GUS*). These constructs were transformed into Col-0 *Arabidopsis* plants. Several independent lines were generated for both constructs, and homozygous lines were established. Subsequently, the lines showing similar *GUS* transcript levels were selected (Figure S4) and assayed histochemically for GUS expression in the same tissues (cotyledons, rosette leaves, inflorescence, and anthers). The results showed that weak GUS expression was detected in the plants transformed with *pWNK8:GUS*, whereas a strong GUS color was detected in all selected tissues and organs of the plants transformed with *pWNK8m:GUS* (in which CPuORF58 was mutated) (Figure 3A).

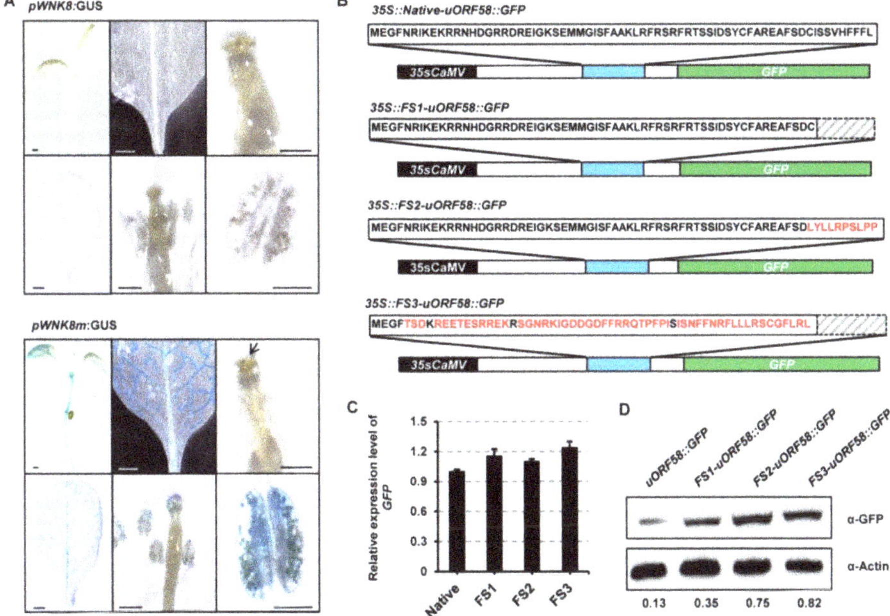

Figure 3. Mutated CPuORF58 enhance the translation of different downstream mORFs. (**A**) GUS expression analyzed by histochemical detection in different stages of seedlings or plants transformed with native promoter *pWNK8:GUS* or site-mutated promoter *pWNK8m:GUS*. (**B**) Schematic representation of the constructs for *Arabidopsis* transformation. The amino acid sequences of the native and mutated uORF with different ones in red were shown at the bottom. FS is short for frame shift; the black box represented 35S CaMV promoter; the white boxes are for *WNK8* 5′-UTR; the light blue boxes are shown as CPuORF58; the green boxes are for GFP ORFs. The expression levels of *GFP* were analyzed using qRT-PCR (**C**) and immunoblotting (**D**) of plants transformed with indicated genetic constructs. Actin2 was used as the internal control.

The ribosome profiling data presented some peak distributions in the CPuORF58 peptide of WNK8, indicating ribosome stalling (Figure 2A). A previous study showed that the nascent peptides encoded by uORFs cause ribosomal arrest during mRNA translation [34]. Therefore, we decided to examine the importance of the CPuORF58 sequence followed by the generation of several genetic constructs with the truncated or frameshift uORF. A schematic representation of these constructs presented in Figure 3B. In *35S:FS1-uORF58:GFP* and *35S:FS2-uORF58:GFP*, the C-terminus of uORF58 was truncated and shifted, respectively. In *35S:FS3-uORF58:GFP*, the nucleotide sequence of *uORF58* exhibited little change, whereas the deduced amino acid sequence was completely changed. These constructs were transformed into Col-0 *Arabidopsis* plants and homozygous lines were established. Stable transgenic lines with these constructs showing similar *GFP* transcript levels were selected (Figure 3C). The GFP expression was analyzed by 'immunoblotting. The results showed that both truncated and frameshift mutations in uORF58 led to increased GFP levels, suggesting the involvement of uORF58 in translational suppression of WNK8 (Figure 3D).

2.4. Two Independent Regions in the 5′-UTR of WNK8 Are Required for Gene Expression

A further analysis showed that the 5′-UTR of *WNK8* contains two short open reading frames (here named uORF1 and CPuORF58) (Figure 4A). Regarding the presence of the two uORFs, we constructed a list of site-mutated *WNK8* 5′-UTR to evaluate responsible role of these regions in gene expression regulation (Figure 4B). The GUS expression driven by 35S CaMV promoter was applied as the control. The wild-type (UTR^{WT}) and site-mutated of *WNK8* 5′-UTR were inserted between the 35S promoter and the GUS reporter gene. All constructs were introduced into GV3101, and a transient expression assay of GUS expression was performed within tobacco leaves, and followed by GUS histochemistry. The result showed that GUS expression under 35S promoter was strong in control. In contrast, the wild-type of *WNK8* 5′-UTR (UTR^{WT}) significantly repressed GUS expression (Figure 4C). The individual site-mutated of *WNK8* 5′-UTR was also able to repress GUS expression driven but less effect in compared with the wild-type 5′-UTR, similar to the inhibition seen from the CPuORF58 in transgenic *Arabidopsis* plants. Unexpectedly, the extent of repression in GUS expression was not enhanced in the double site-mutated UTR^{m1+m2} of *WNK8* 5′-UTR.

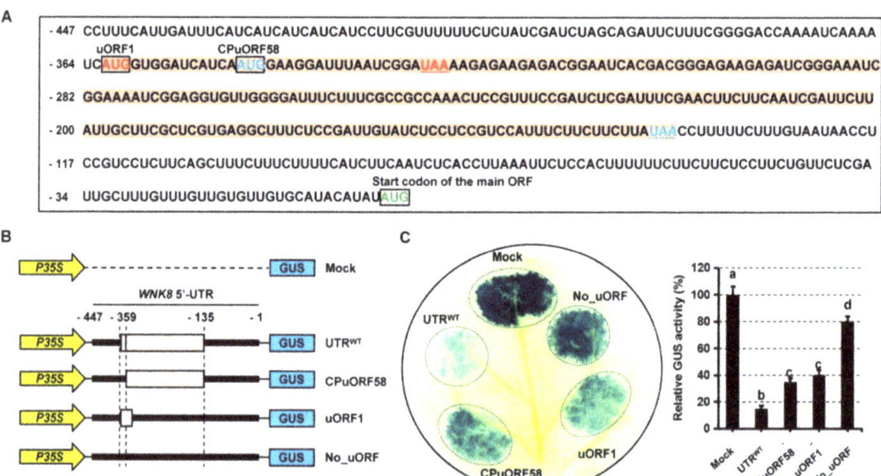

Figure 4. Effects of uORF disruptions in *WNK8* 5′-UTR on downstream gene expression. (**A**) Nucleotide sequence of the *WNK8* 5′-UTR. Sequences in yellow represent uORFs. The start codons for two uORFs and the main ORF are colored and in frame with boxes. Stop codons of uORFs are underlined and highlighted in color same to that of corresponding start codons. (**B**) Schematic representations of the constructs for transient expression. Thick black lines represent the *WNK8* 5′-UTR and pink boxes represents the uORFs, respectively. The constructs at the right represent point mutations (AUG to AAG) in the start codons of *WNK8* 5′-UTR. (**C**) Transient expression analysis in tobacco leaves with the DNA constructs. Illustrative photograph of tobacco leaves transformed with the indicated constructs and visualized with GUS histochemistry. In addition, relative reporter activities are shown at right and different groups marked with letters represent the 0.05 significance level.

2.5. The Absence of a Strictly Regulated Expression of WNK8 Altered ABA Sensitivity

Two different genetic constructs were generated to study the role of CPuORF58 in vivo. In the first construct, WNK8 expression was controlled by a 2.3-kb region upstream of the start codon, including its 5′-UTR (*pWNK8:WNK8*), and in the second construct, a point mutation (G→C) was introduced in the start codon ATG of CPuORF58 (*pWNK8m:WNK8*). In both constructs, the Myc tag was directly attached to the mORF of *WNK8*. These constructs were used to transform *wnk8-1* mutant plants. As stated above, independent transgenic lines were generated for these constructs, after which homozygous lines were established. The WNK8 protein was rarely detected in plants transformed with *pWNK8:WNK8*, which was used as the complementation material hereafter (Figure 5A). This was similar to the weak expression of GUS driven, by the *WNK8* promoter. In contrast, strong WNK8 expression was detected in plants transformed with *pWNK8m:WNK8* (Figure 5B).

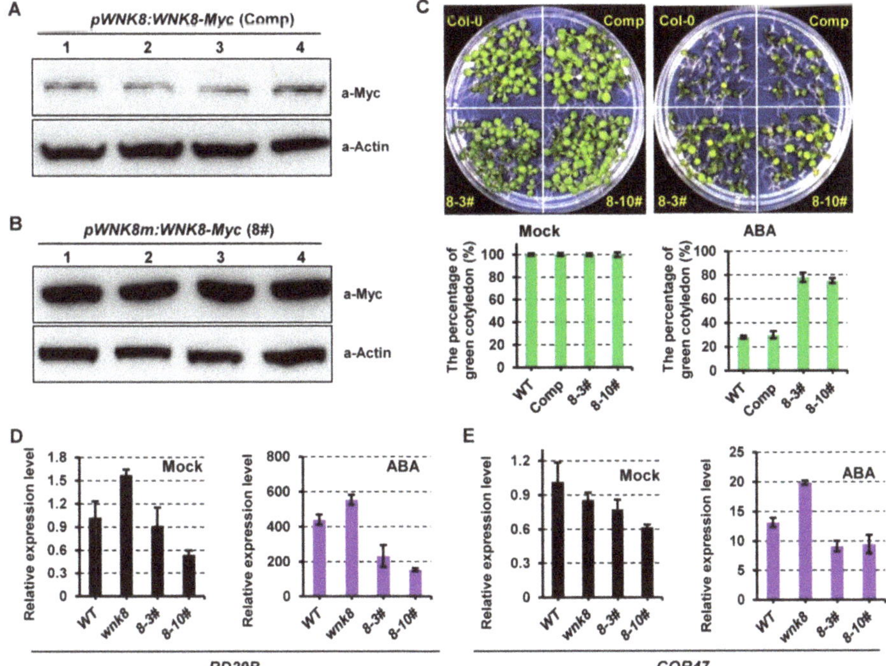

Figure 5. ABA sensitivity is altered under abnormal expression of WNK8. (**A,B**) The protein accumulation analysis of WNK8 in transgenic lines. Total proteins were extracted from the individual stable transgenic seedlings transformed with *pWNK8:WNK8* (**A**) or *pWNK8m:WNK8* (**B**), and further probed with antibodies α–Myc and α–Actin. (**C**) Photographs of seedlings grown on different media in the absence or presence of ABA at day 10 after stratification. Two independent transgenic lines transformed with *pWNK8m:WNK8* (8-3# and 8-10#) and the complementation of WNK8 in *wnk8-1* line (Comp) were analyzed. Cotyledon greening percentages were recorded after germination for 10 days. Three independent experiments were conducted, and over 50 seeds of each genotype were applied in each replicate. The standard deviation of three replicates were indicated with error bars. (**D,E**) The expression analysis of ABA response genes in different genotypes. The expression profiles of ABA-signaling-related genes *RD29B* (**D**) and *COR47* (**E**) were examined in one-week-old seedlings in the absence or presence of 50 μM ABA for 3 h. The transcriptional levels of ABA response genes were normalized to that of *Actin2*.

These lines were further selected to evaluate the role of WNK8 in the ABA response. As expected, the complementation of WNK8 in *wnk8-1* (Comp) showed a phenotype similar to that of the wild-type phenotype. However, plants transformed with *pWNK8m:WNK8* exhibited much lower sensitivity to ABA (Figure 5C). We further investigated the expression patterns of the ABA response genes *RD29B* and *COR47* in these genetic backgrounds. Under normal conditions, the expression levels of the two ABA response genes were similar in all selected plants, but slightly lower in plants transformed with *pWNK8m:WNK8* (Figure 5D,E). After ABA treatment, the transcription levels of both *RD29B* and *COR47* were significantly increased, with the highest in the *wnk8* mutant. The expression of the two ABA response genes was also induced in plants transformed with *pWNK8m:WNK8* but was lower than their expression in Col-0 and *wnk8* mutant plants. This suggested that WNK8 negatively affected the expression of ABA response genes, which is consistent with a previous study [28].

2.6. WNK8 and RACK1 Coordinate ABA Signaling

Receptor for activated C kinase1 (RACK1) has been suggested to act downstream of WNK8 to regulate flowering and glucose responsiveness as well as to play a negative role in ABA signaling [17,35]. Hence, we examined whether WNK8 and RACK1 function together in coordinating ABA signaling. Seed germination analysis showed that both *wnk8-1* and *rack1a-2* were more sensitive to ABA than Col-0 plants (Figure 6A). This was different from the opposing flowering and glucose phenotypes of *rack1a-2* and *wnk8-1* single mutants [17]. Interestingly, the double mutant *rack1a-2 wnk8-1* was even more sensitive to ABA in seed germination than the single mutant. This suggests that WNK8 and RACK1 function in the same pathway to regulate the ABA response.

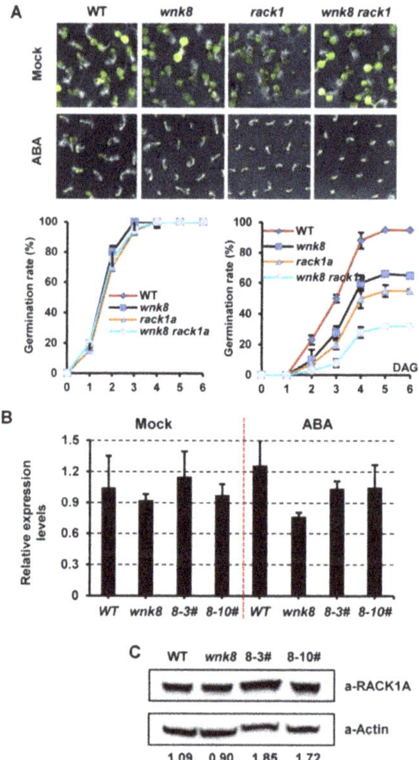

Figure 6. WNK8 and RACK1A coordinate ABA signaling. (**A**) WNK8 genetically interacts with RACK1A in response to ABA. Photographs of cotyledon expansion for indicated genotypes were shown in the absence or presence of 1.0 μM ABA. Time course of seed germination were recorded at the lower panel. The expression of RACK1A in WNK8 related genetic background was examined at the transcriptional (**B**) and translational levels (**C**).

WNK8-mediated phosphorylation negatively affects the stability of the RACK1 protein [17]. Thus, we examined the expression of *RACK1* in different genetic backgrounds related to WNK8. The results showed that transcription of *RACK1A* was constant in Col-0, *wnk8-1* mutant, and plants transformed with *pWNK8m:WNK8* (Figure 6B). The protein levels of RACK1A were also similar in *Col-0* and *wnk8-1* mutant plants (Figure 6C). However, the accumulation of RACK1A was higher in plants with high WNK8 expression. This indicated the occurrence of other kinds of posttranslational modifications that may contribute to the stability of the RACK1A protein and counteract the effect of WNK8-mediated phosphorylation.

3. Discussion

Plants have developed diverse environmental adaptation strategies for their development and survival. In general, they adapt to alterations in environmental conditions by controlling their gene expression profiles through both transcriptional regulation and post-transcriptional regulation. It is widely known that different regulatory elements are present in the 5′-UTR of mRNAs. uORFs appear in the 5′-UTR as translational control elements, which generally attenuate the translation of the downstream mORF in most cases [36–39]. uORFs have been found in approximately 20% to 50% of eukaryotic transcripts. Approximately 50% of protein-encoding genes possess one or more uORFs in humans [39]. In plants, 24–30% of the 5′-UTR of mRNAs contains uORFs; however, only a few uORFs have been identified and characterized [40,41]. Therefore, it is necessary to further explore uORFs in plants and investigate their translational control mechanisms, which could provide a deeper understanding of plant development and adaptations to changing environments.

uORFs have been identified in many transcripts responsible for plant development, including transcription factors (TFs) and RNA processing factors [42]. In particular, genes encoding TFs are over-represented. Several studies have demonstrated that uORFs play important roles in the translational regulation of many pathways in plants including metabolic, plant morphogenesis, disease resistance, and nutrient absorption pathways [43–47]. For instance, several enzymes, such as S-adenosylmethionine decarboxylase (AdoMetDC) and flavin-containing polyamine oxidases (PAO), are controlled by uORFs in a small metabolite-dependent manner, which are important for the biosynthesis of polyamine (PA) and phosphocholine (PCho), respectively [48–52]. Some proteins are tightly regulated by uORFs, though other signals including light and plant hormones. Both *Arabidopsis* TFs AtHB1 (HOMEOBOX 1) and PHYTOCHROME-INTERACTING FACTOR 3 (AtPIF3) are light-dependent and regulated by uORFs [53–55]. One uORF in the 5′-UTR of the Brassinosteroid receptor protein AtBRI1 (Brassinosteroid insensitive 1) has been to be essential for the stability of BR (Brassinosteroid) levels in vivo [56]. Here, we demonstrated that CPuORF58, a highly conserved genetic element in the 5′-UTR of WNK8, is important for ABA-signaling response in plants. Our results showed that the protein level of WNK8 is closely related to the ABA response in *Arabidopsis*. The two T-DNA insertion lines *wnk8-1* and *wnk8-2*, which lacked *WNK8* transcripts, were more ABA-sensitive (Figure 1). In contrast, high expressed WNK8 expression conferred significantly less ABA sensitivity when CPuORF58 was mutated in plants (Figure 5). Thus, WNK8 expression must be tightly regulated to fulfill the demands of ABA signaling in plants.

Although uORFs are distributed in a considerable number of mRNAs, only a small number of these regulatory elements are conserved between species. Here, we showed that in all retrieved DNA sequences encoding *WNK8* homologs from both monocot and dicot species that contain known 5′-UTRs, and CPuORF58 was predicted (Figure 2B). Although the conservation of the total amino acids is not very high due to the unusual longer size (ranging from 55 aa–84 aa), it is important to note that higher conservation in some specific amino acids is species–specific (Figure 3C–E). Clear coverage peaks upstream of the uORF stop codon for the 5′-UTR of WNK8 were observed with ribosome footprinting analyses, indicating ribosome stalling in this region (Figure 2A). Here, we demonstrated the role of CPuORF58 in translational repression in vivo, which is most likely due to ribosome stalling. With frameshift mutations, the repressive activity of CPuORF58 was lost in plants, although little change was observed in the RNA sequence (Figure 3B–D). In addition, another uORF (uORF1) was identified in the 5′-UTR of *WNK8* overlapping with CPuORF58 (Figure 4A). The transient expression assay showed that either of the two uORFs was effective in triggering the translation repression of the downstream mORF, while the extent of the reduction was stronger with CPuORF58. As uORF1 partially overlapped with CPuORF58, the translated uORF1 prevented the translation of CPuORF58. Thus, the extent of suppression of reporter expression was not enhanced when both uORFs were mutated, but when they acted as mimics of the mutated uORF1 (Figure 4B–D). This indicates that at least two post-transcriptional regulatory mechanisms contribute to tight control of the WNK8 accumulation in vivo.

RACK1A has been suggested to act downstream of WNK8 and play a negative role in ABA responses [17,35]. Here, we found that the double mutant *rack1a-2 wnk8-1* was more sensitive to ABA than either of the single mutants (Figure 6A). This indicated that WNK8 and RACK1A coordinate the ABA-signaling pathway differently from the glucose response pathway and flowering. The enhanced ABA activity in the *wnk8* mutant may lead to the repression of the RACK1 expression of, as ABA treatment resulted in rapid down-regulation of *RACK1* [35]. However, no significant change in the transcription of *RACK1A* was detected in either the control or *wnk8* mutant plants (Figure 6B). A similar expression of *RACK1A* was also found in plants with high WNK8 expression. It has been suggested that WNK8 affects the protein stability of RACK1 by phosphorylation modification [17]. However, no difference in RACK1A protein abundance was observed in either *wnk8* mutant or the wild-type plants, which is consistent with the results of a previous study. Instead, more RACK1A proteins were detected under high WNK8 expression. It is possible that the dominant effect of certain modifications contributes to stable RACK1A expression, even under high WNK8 expression. ABA represses *RACK1* at the transcriptional level, but also enhances the sumoylation of RACK1, which increases the stability of RACK1 in vivo [57]. As the ABA core receptor, PYR1 could be phosphorylated by WNK8 at several residues close to the ABA-binding site and may affect the ABA-affinity, leading to more ABA accumulation in cells [28]. It has also been suggested that WNK proteins are helpful in maintaining ABA homeostasis. GmWNK1 interacts with the putative ABA 8′-hydroxylase protein GmCYP707A1, and enhanced levels of endogenous ABA were observed in transgenic lines [22]. We suspected an accumulation of ABA when WNK8 proteins were highly expressed, therefore enhancing the protein stability of RACK1A. It remains to be investigated whether RACK1 sumoylation increases with increase in WNK8 expression. Moreover, it would be interesting to examine the ABA levels in these genetic lines in future studies.

4. Materials and Methods

4.1. Plant Materials and Growth Conditions

All plants used for transformation in this study are with the *Arabidopsis* ecotype Columbia-0 (Col-0). The mutants *wnk8-1* (SALK_206987C) and *wnk8-2* (SALK_103318C) in the Col-0 ecotype background, were supported by the Arabidopsis Biological Resource Center (http://www.arabidopsis.org accessed on 16 August 2021). The mutant *rack1a-2* (SALK_073786C) was the gift from Jiafu Jiang in Nanjing Agricultural University. Seeds were surfaced sterilized and grown in Murashige and Skoog (MS) medium with 1% sucrose and 0.8% phyto agar. Plates were kept for 3 days at 4 °C in the dark before moved to a culture room with 22 °C under a long-day (16/8 h light/dark) photoperiod. Seven-day-old seedlings were grown in the culture room with same conditions as previously stated. The *wnk8-1* was crossed with *rack1a-2* for generation of the double mutant *wnk8-1 rack1a-2*.

For the seed germination assay, seeds were surfaced sterilized and placed on the half strength MS medium in the absence or presence of various concentrations of ABA, and transferred to the culture room. The seed germination was calculated after stratification, and cotyledon expansion and greening were scored at the indicated time intervals.

4.2. RNA/DNA Extraction and Analysis

For genotyping of the T-DNA insertion mutants, leaves from two-week-old seedlings were harvested for DNA extraction. Edwards buffer (20 mM Tris-HCl (pH 7.5), 250 mM NaCl, 25 mM EDTA, and 0.5% SDS) was applied for quick one-step DNA extraction following the protocol. The genotyping PCR using gene-specific primers and a T-DNA primer LBa1 for mutant lines were performed using PCR mix (Vazyme, Nanjing, China). The insertion site was confirmed by DNA sequencing.

For quantitative reverse transcription-PCR (qRT-PCR), the total RNA was extracted from the *Arabidopsis* seedlings or different organs and tissues using ReliaPrep™ RNA Miniprep System (Promega, Madison, WI, USA). For cDNA synthesis, the first-strand cDNA was synthesized using 2 µg total RNA with NovoScript®Plus All-in-one 1st Strand cDNA Synthesis SuperMix (Novoprotein, Shanghai, China) following the instructions. The qRT-PCR was carried out with SYBR Green Master Mix (YEASEN, Shanghai, China) on Step-one Plus™ (Applied Biosystems, Carlsbad, CA, USA). The $2^{-\Delta\Delta CT}$ method was applied to quantify the gene expression levels [58]. *Actin2* was used as an internal control. All primers used are listed in Table S1.

4.3. Plasmids Construction and Plant Transformation

4.3.1. *pWNK8:GUS* and *pWNK8m:GUS*

The native version of the *WNK8* promoter (2.3 kb upstream of WNK8 translation start site) was amplified and the mutated version was assembled by overlapping PCR amplification using specific oligonucleotides (Supplemental Table S1). The PCR products were cloned into the *Hind*III and *Nco*I sites of *pCAMBIA1301*, replacing the 35S CaMV. The correct insertion was verified by sequencing.

4.3.2. *pWNK8:WNK8* and *pWNK8m:WNK8*

To generate the plasmid for the native expression of WNK8, the genomic region including 2.3 kb upstream of ATG plus introns and exons were cloned with nuclear sequence (GAGCAGAAACTCATCTCTGAAGAGGATCTG) for Myc tag (EQKLISEEDL) included in the reverse primer. The mutated version of the *WNK8* was assembled overlapping PCR. Both PCR products were cloned into pDONOR221 using Gateway™ BP Clonase™ Enzyme Mix (Thermo Scientific, Waltham, MA, USA), and further cloned into the pGWB4 plant expression vector using Gateway™ LR Clonase™ II Enzyme mix (Thermo Scientific, Waltham, MA, USA).

4.3.3. *35S:Native-uORF:GFP* and *35S:FS-uORF:GFP*

The WNK8 5′-UTR was amplified, and the indicated mutations were introduced by overlapping using specific oligonucleotides. All the PCR products were introduced into pDONOR221 using Gateway™ BP Clonase™ Enzyme Mix, and further cloned into the pGWB405 to produce C-terminal GFP-tagged fusion proteins under the control of the 35S promoter using Gateway™ LR Clonase™ II Enzyme mix. All the constructs were confirmed by sequencing.

For plasmids construction, all genes were amplified using Tks Gflex™ DNA polymerase (Takara, Shiga, Japan) with corresponding primers listed in Table S1. The plant expression constructs were introduced into the *Agrobacterium tumefaciens* GV3101 strain. Plant transformation was applied with the floral dipping method [59]. Further selection of transformed plants was performed in petri dishes with MS medium supplemented with the appropriate selector chemical (50 mg L^{-1} kanamycin or 25 mg L^{-1} hygromycin). Three or four homozygous T3 and T4 independent lines for each construct were further reproduced and used to analyze the gene expression levels and phenotypes.

4.4. Histochemical GUS Staining and GUS Activity Measurement

GUS staining was performed with commercial kit by following the protocol (Coolaber, Beijing, China). In brief, young seedlings, mature leaves, and flowers were immersed into fresh prepared GUS staining buffer, further with vacuum for 5 min, and then plants were kept at 37°C for 24 h. Chlorophyll was removed from the plant tissues with 70% ethanol.

To examine the GUS activity, a plant-GUS ELISA kit was applied following the protocol (Abmart, Shanghai, China). The fresh plant tissue samples were grinded with 0.75% NaCl solution. After centrifuge for 10 min with 3000 rpm, the supernatant was collected and further incubated with reagents provided in kit. The OD values from the standard and test samples were measured with a wavelength of 450 nm. The GUS activity in each sample can be normalized per unit tissue weight.

4.5. Immunoblotting

Seedlings of stable transgenic lines expressed the corresponding GFP or Myc fusions were freeze grounded into powder and homogenized in total protein buffer (20 mmol/L Tris-HCl (pH 7.5), 150 mmol/L NaCl, 2 mmol/L EDTA) with protease inhibitor cocktail in DMSO (YEASEN, Shanghai, China). Lysates were incubated on ice for 20 min and clarified by centrifugation at 15,000 g for 15 min at 4°C. For immunoblotting, samples were separated on 12% SDS polyacrylamide gel and transferred to PVDF membranes. The membranes were then blocked with 5% (g/v) defatted milk in TBST buffer (10 mM Tris-HCl (pH 7.4), 150 mM NaCl, 0.05% Tween 20) and probed with using appropriate antibodies including 1:6000 dilution α-GFP conjugated with HRP (MBL, Nagano, Japan), 1:5000 dilution of α-RACK1 (PhytoAB, SAN JOSE, CA, USA), α-Myc (MBL, Nagano, Japan) and α-Actin2 (Sangon, Shanghai, China) overnight at 4°C. Then the samples were washed with TBST buffer for three times and visualized using the ECL (Amersham™, Boston, MA, USA). Actin protein was used as the internal control. Image J2 was applied for quantification the intensity of signals [60].

4.6. Database Screening Analysis

The BLASTP program was applied for a search for *Arabidopsis* WNK proteins on the whole genome sequences from the Phytozome database (https://phytozome.jgi.doe.gov accessed on 16 August 2021), with AtWNK8 as the query sequence. Corresponding genomic DNA (gDNA) and coding DNA (cDNA) sequences of the WNK8s from the selected plant genome were retrieved. The uORF nucleotides and predicted amino acid sequences were aligned using the ClustalW software (version 2.0.12) [61]. The sequence logos were generated with MEME (https://meme-suite.org/meme/tools/meme accessed on 16 August 2021) with default setting. The GWIPS-viz (https://gwips.ucc.ie/ accessed on 16 August 2021) was applied to visualize the RNA-seq and ribosome profiling data of WNK8. The data exploration tool ATHENA (http://athena.proteomics.wzw.tum.de: 5002/master_arabidopsisshiny/ accessed on 16 August 2021) was used for comparing the proteome and transcriptome data of WNK8.

5. Conclusions

In this study, we found that t*Arabidopsis* WNK8 is post-transcriptionally regulated by the conserved peptide (CPuORF58) located in its 5′ untranslated region (5′-UTR). Together with another uORF, CPuORF58 represses WNK8 expression to fulfill the demands of ABA response in plants. Moreover, WNK8 and its downstream target RACK1 synergistically coordinate ABA signaling rather than antagonistically modulating plant glucose response and flowering. This study provides a better understanding of the role of uORFs in plant development and adaptation to changing environments.

Supplementary Materials: The following are available online at https://www.mdpi.com/article/10 .3390/ijms221910683/s1.

Author Contributions: Conceptualization, Z.L. and J.L.; Data curation, Z.L., Y.F. and J.S.; Formal analysis, Z.L.; Funding acquisition, Z.L. and J.L.; Investigation, Z.L. and Y.F.; Methodology, Z.L., Y.F. and J.S.; Project administration, Z.L. and J.L.; Resources, Z.L.; Software, Z.L., Y.F. and J.S.; Supervision, Z.L. and J.L.; Validation, Z.L. and Y.F.; Visualization, Z.L. and Y.F.; Writing—original draft, Z.L. and

Y.F.; Writing—review and editing, Z.L. and J.L. All authors have read and agreed to the published version of the manuscript.

Funding: The project was supported by Shenzhen government for fundamental research (JCYJ2017081 7104523456), Shenzhen Science and Technology Program (Grant No. KQTD20190929173906742), Key Laboratory of Molecular Design for Plant Cell Factory of Guangdong Higher Education Institutes (2019KSYS006) and the National Natural Science Foundation of China (31801277).

Institutional Review Board Statement: Not applicable.

Informed Consent Statement: Not applicable.

Data Availability Statement: The whole genome sequences for *WNK8* and the homologs can be found in the Phytozome database (https://phytozome.jgi.doe.gov accessed on 16 August 2021). The RNA-seq and ribosome profiling data for genes from *Arabidopsis* can be visualized in GWIPS-viz (https://gwips.ucc.ie/ accessed on 16 August 2021). And proteome and transcriptome for genes from *Arabidopsis* can be obtained from ATHENA (http://athena.proteomics.wzw.tum.de: 5002/master_arabidopsisshiny/ accessed on 16 August 2021).

Acknowledgments: We thank Jiafu Jiang from Nanjing Agricultural University for providing us the gift of *rack1a-2* mutant (SALK_073786C).

Conflicts of Interest: The authors declare no conflict of interest.

References

1. Xu, B.E.; English, J.M.; Wilsbacher, J.L.; Stippec, S.; Goldsmith, E.J.; Cobb, M.H. WNK1, a novel mammalian serine/threonine protein kinase lacking the catalytic lysine in subdomain II. *J. Biol. Chem.* **2000**, *275*, 16795–16801. [CrossRef] [PubMed]
2. Verissimo, F.; Jordan, P. WNK kinases, a novel protein kinase subfamily in multi-cellular organisms. *Oncogene* **2001**, *20*, 5562–5569. [CrossRef] [PubMed]
3. Gamba, G. Role of WNK kinases in regulating tubular salt and potassium transport and in the development of hypertension. *Am. J. Physiol. Renal Physiol.* **2005**, *288*, F245–F252. [CrossRef] [PubMed]
4. Wilson, F.H. Human hypertension caused by mutations in WNK kinases. *Science* **2001**, *293*, 1107–1112. [CrossRef]
5. McCormick, J.A.; Ellison, D.H. The WNKs: Atypical protein kinases with pleiotropic actions. *Physiol. Rev.* **2011**, *9*, 177–219. [CrossRef]
6. Huang, C.L.; Cheng, C.J. A unifying mechanism for WNK kinase regulation of sodium-chloride cotransporter. *Pfluger. Arch. Pflug. Arch. Eur. J. Phy.* **2015**, *467*, 2235–2241. [CrossRef]
7. Richardson, C.; Rafiqi, F.H.; Karlsson, H.K.; Moleleki, N.; Vandewalle, A.; Campbell, D.G.; Morrice, N.A.; Alessi, D.R. Activation of the thiazide-sensitive Na$^+$-Cl$^-$ cotransporter by the WNK-regulated kinases SPAK and OSR1. *J. Cell. Sci.* **2008**, *121*, 675–684. [CrossRef]
8. Yang, C.L.; Angell, J.; Mitchell, R.; Ellison, D.H. WNK kinases regulate thiazide-sensitive Na-Cl cotransport. *J. Clin. Investig.* **2003**, *111*, 1039–1045. [CrossRef]
9. Rinehart, J.; Kahle, K.T.; de Los Heros, P.; Vazquez, N.; Meade, P.; Wilson, F.H.; Hebert, S.C.; Gimenez, I.; Gamba, G.; Lifton, R.P. WNK3 kinase is a positive regulator of NKCC2 and NCC, renal cation-Cl- cotransporters required for normal blood pressure homeostasis. *Proc. Natl. Acad. Sci. USA* **2005**, *102*, 16777–16782. [CrossRef]
10. Sie, Z.L.; Li, R.Y.; Sampurna, B.P.; Hsu, P.J.; Liu, S.C.; Wang, H.D.; Huang, C.L.; Yuh, C.H. WNK1 kinase stimulates angiogenesis to promote tumor growth and metastasis. *Cancers* **2020**, *12*, 575. [CrossRef]
11. Lai, J.G.; Tsai, S.M.; Tu, H.C.; Chen, W.C.; Kou, F.J.; Lu, J.W.; Wang, H.D.; Huang, C.L.; Yuh, C.H. Zebrafish WNK lysine deficient protein kinase 1 (wnk1) affects angiogenesis associated with VEGF signaling. *PLoS ONE* **2014**, *9*, e106129. [CrossRef]
12. Tang, B.L. (WNK) ing at death: With-no-lysine (Wnk) kinases in neuropathies and neuronal survival. *Brain. Res. Bull.* **2016**, *125*, 92–98. [CrossRef]
13. Wang, Y.; Liu, K.; Liao, H.; Zhuang, C.; Ma, H.; Yan, X. The plant WNK gene family and regulation of flowering time in *Arabidopsis*. *Plant Biol.* **2008**, *10*, 548–562. [CrossRef]
14. Manuka, R.; Saddhe, A.A.; Kumar, K. Genome-wide identification and expression analysis of WNK kinase gene family in rice. *Comp. Biol. Chem.* **2015**, *59*, 56–66. [CrossRef]
15. Wang, Y.; Suo, H.; Zheng, Y.; Liu, K.; Zhuang, C.; Kahle, K.T.; Ma, H.; Yan, X. The soybean root-specific protein kinase GmWNK1 regulates stress-responsive ABA signaling on the root system architecture. *Plant J.* **2010**, *64*, 230–242. [CrossRef]
16. Cao, S.; Hao, P.; Shu, W.; Wang, G.; Xie, Z.; Gu, C.; Zhang, S. Phylogenetic and expression analyses of With-No-Lysine kinase genes reveal novel gene family diversity in fruit trees. *Hortic. Plant J.* **2019**, *5*, 7–58. [CrossRef]
17. Urano, D.; Czarnecki, O.; Wang, X.; Jones, A.M.; Chen, J.G. *Arabidopsis* receptor of activated C kinase1 phosphorylation by WITH NO LYSINE8 KINASE. *Plant Physiol.* **2015**, *167*, 507–516. [CrossRef]

18. Nakamichi, N.; Murakami-Kojima, M.; Sato, E.; Kishi, Y.; Yamashino, T.; Mizuno, T. Compilation and characterization of a novel WNK family of protein kinases in *Arabidopsis thaliana* with reference to circadian rhythms. *Biosci. Biotechnol. Biochem.* **2002**, *66*, 2429–2436. [CrossRef]

19. Murakami-Kojima, M.; Nakamichi, N.; Yamashino, T.; Mizuno, T. The APRR3 component of the clock-associated APRR1/TOC1 quintet is phosphorylated by a novel protein kinase belonging to the WNK family, the gene for which is also transcribed rhythmically in *Arabidopsis thaliana*. *Plant Cell Physiol.* **2002**, *43*, 675–683. [CrossRef]

20. Kumar, K.; Rao, K.P.; Biswas, D.K.; Sinha, A.K. Rice WNK1 is regulated by abiotic stress and involved in internal circadian rhythm. *Plant Signal. Behav.* **2011**, *6*, 316–320. [CrossRef]

21. Hong-Hermesdorf, A.; Brüx, A.; Grüber, A.; Grüber, G.; Schumacher, K. A WNK kinase binds and phosphorylates V-ATPase subunit C. *Febs. Lett.* **2006**, *580*, 932–939. [CrossRef]

22. Wang, Y.; Suo, H.; Zhuang, C.; Ma, H.; Yan, X. Overexpression of the soybean GmWNK1 altered the sensitivity to salt and osmotic stress in *Arabidopsis*. *J. Plant. Physiol.* **2011**, *168*, 2260–2267. [CrossRef]

23. Zhang, B.; Liu, K.; Zheng, Y.; Wang, Y.; Wang, J.; Liao, H. Disruption of AtWNK8 enhances tolerance of *Arabidopsis* to salt and osmotic stresses via modulating proline content and activities of catalase and peroxidase. *Int. J. Mol. Sci.* **2013**, *14*, 7032–7047. [CrossRef]

24. Xie, M.; Wu, D.; Duan, G.; Wang, L.; He, R.; Li, X.; Tang, D.; Zhao, X.; Liu, X. AtWNK9 is regulated by ABA and dehydration and is involved in drought tolerance in *Arabidopsis*. *Plant Physiol. Biochem.* **2014**, *77*, 73–83. [CrossRef]

25. Manuka, R.; Karle, S.B.; Kumar, K. OsWNK9 mitigates salt and drought stress effects through induced antioxidant systems in *Arabidopsis*. *Plant Physiol. Rep.* **2019**, *24*, 168–181. [CrossRef]

26. Manuka, R.; Saddhe, A.A.; Srivastava, A.K.; Kumar, K.; Penna, S. Overexpression of rice OsWNK9 promotes arsenite tolerance in transgenic *Arabidopsis* plants. *J. Biotechnol.* **2021**, *332*, 114–125. [CrossRef]

27. Dunker, F.; Trutzenberg, A.; Rothenpieler, J.S.; Kuhn, S.; Pröls, R.; Schreiber, T.; Tissier, A.; Kemen, A.; Kemen, E.; Hückelhoven, R.; et al. Oomycete small RNAs bind to the plant RNA-induced silencing complex for virulence. *Elife* **2020**, *9*, 1–23. [CrossRef]

28. Waadt, R.; Jawurek, E.; Hashimoto, K.; Li, Y.; Scholz, M.; Krebs, M.; Czap, G.; Hong-Hermesdorf, A.; Hippler, M.; Grill, E.; et al. Modulation of ABA responses by the protein kinase WNK8. *Febs Lett.* **2019**, *593*, 339–351. [CrossRef]

29. Mergner, J.; Frejno, M.; List, M.; Papacek, M.; Chen, X.; Chaudhary, A.; Samaras, P.; Richter, S.; Shikata, H.; Messerer, M.; et al. Mass-spectrometry-based draft of the *Arabidopsis* proteome. *Nature* **2020**, *579*, 409–414. [CrossRef]

30. Hayden, C.A.; Jorgensen, R.A. Identification of novel conserved peptide uORF homology groups in *Arabidopsis* and rice reveals ancient eukaryotic origin of select groups and preferential association with transcription factor-encoding genes. *BMC Biol.* **2007**, *5*, 32. [CrossRef]

31. Ingolia, N.T.; Ghaemmaghami, S.; Newman, J.R.; Weissman, J.S. Genome-wide analysis in vivo of translation with nucleotide resolution using ribosome profiling. *Science* **2009**, *10*, 218–223. [CrossRef] [PubMed]

32. Michel, A.M.; Fox, G.M.; Kiran, A.; De Bo, C.; O'Connor, P.B.; Heaphy, S.M.; Mullan, J.P.; Donohue, C.A.; Higgins, D.G.; Baranov, P.V. GWIPS-viz: Development of a ribo-seq genome browser. *Nucleic Acids Res.* **2014**, *42*, D859–D864. [CrossRef] [PubMed]

33. Hinnebusch, A.G.; Ivanov, I.P.; Sonenberg, N. Translational control by 5′-untranslated regions of eukaryotic mRNAs. *Science* **2016**, *352*, 1413–1416. [CrossRef] [PubMed]

34. Hayashi, N.; Sasaki, S.; Takahashi, H.; Yamashita, Y.; Naito, S.; Onouchi, H. Identification of *Arabidopsis thaliana* upstream open reading frames encoding peptide sequences that cause ribosomal arrest. *Nucleic Acids Res.* **2017**, *45*, 8844–8858. [CrossRef]

35. Guo, J.; Wang, J.; Xi, L.; Huang, W.D.; Liang, J.; Chen, J.G. RACK1 is a negative regulator of ABA responses in *Arabidopsis*. *J. Exp. Bot.* **2009**, *60*, 3819–3833. [CrossRef]

36. Morris, D.R.; Geballe, A.P. Upstream open reading frames as regulators of mRNA translation. *Mol. Cell. Biol.* **2020**, *20*, 8635–8642. [CrossRef]

37. Calvo, S.E.; Pagliarini, D.J.; Mootha, V.K. Upstream open reading frames cause widespread reduction of protein expression and are polymorphic among humans. *Proc. Natl. Acad. Sci. USA* **2009**, *106*, 7507–7512. [CrossRef]

38. Jeon, S.; Kim, J. Upstream open reading frames regulate the cell cycle-dependent expression of the RNA helicase Rok1 in *Saccharomyces cerevisiae*. *FEBS Lett.* **2010**, *584*, 4593–4598. [CrossRef]

39. Kwon, H.S.; Lee, D.K.; Lee, J.J.; Edenberg, H.J.; Ahn, Y.H.; Hur, M.W. Posttranscriptional regulation of human ADH5/FDH and Myf6 gene expression by upstream AUG codons. *Arch. Biochem. Biophys.* **2001**, *386*, 163–171. [CrossRef]

40. Shashikanth, M.; Krishna, A.R.; Ramya, G.; Devi, G.; Ulaganathan, K. Genome-wide comparative analysis of *Oryza sativa (japonica)* and *Arabidopsis thaliana* 5′-UTR sequences for translational regulatory signals. *Plant Biotech.* **2008**, *25*, 553–563. [CrossRef]

41. Niu, R.; Zhou, Y.; Zhang, Y.; Mou, R.; Tang, Z.; Wang, Z.; Zhou, G.; Guo, S.; Yuan, M.; Xu, G. uORFlight: A vehicle toward uORF-mediated translational regulation mechanisms in eukaryotes. *Database* **2020**, *2020*, baaa007. [CrossRef]

42. Kalyna, M.; Simpson, C.G.; Syed, N.H.; Lewandowska, D.; Marquez, Y.; Kusenda, B.; Marshall, J.; Fuller, J.; Cardle, L.; McNicol, J.; et al. Alternative splicing and nonsense-mediated decay modulate expression of important regulatory genes in *Arabidopsis*. *Nucleic Acids Res.* **2012**, *40*, 2454–2469. [CrossRef]

43. Rosado, A.; Li, R.; van de Ven, W.; Hsu, E.; Raikhel, N.V. *Arabidopsis* ribosomal proteins control developmental programs through translational regulation of auxin response factors. *Proc. Natl. Acad. Sci. USA* **2012**, *109*, 19537–19544. [CrossRef]

44. Yang, S.Y.; Lu, W.C.; Ko, S.S.; Sun, C.M.; Hung, J.C.; Chiou, T.J. Upstream open reading frame and phosphate-regulated expression of rice OsNLA1 controls phosphate transport and reproduction. *Plant Physiol.* **2020**, *182*, 393–407. [CrossRef]

45. Xu, G.; Yuan, M.; Ai, C.; Liu, L.; Zhuang, E.; Karapetyan, S.; Wang, S.; Dong, X. uORF-mediated translation allows engineered plant disease resistance without fitness costs. *Nature* **2017**, *545*, 491–494. [CrossRef]
46. Tanaka, M.; Takano, J.; Chiba, Y.; Lombardo, F.; Ogasawara, Y.; Onouchi, H.; Naito, S.; Fujiwara, T. Boron-dependent degradation of NIP5;1 mRNA for acclimation to excess boron conditions in *Arabidopsis*. *Plant Cell* **2011**, *23*, 3547–3559. [CrossRef]
47. Pajerowska-Mukhtar, K.M.; Wang, W.; Tada, Y.; Oka, N.; Tucker, C.L.; Fonseca, J.P.; Dong, X. The HSF-like transcription factor TBF1 is a major molecular switch for plant growth-to-defense transition. *Curr. Biol.* **2012**, *22*, 103–112. [CrossRef]
48. Hanfrey, C.; Elliott, K.A.; Franceschetti, M.; Mayer, M.J.; Illingworth, C.; Michael, A.J. A dual upstream open reading frame-based autoregulatory circuit controlling polyamine-responsive translation. *J. Biol. Chem.* **2005**, *280*, 39229–39237. [CrossRef]
49. Kamada-Nobusada, T.; Hayashi, M.; Fukazawa, M.; Sakakibara, H.; Nishimura, M. A putative peroxisomal polyamine oxidase, AtPAO4, is involved in polyamine catabolism in *Arabidopsis thaliana*. *Plant Cell Physiol.* **2008**, *49*, 1272–1282. [CrossRef]
50. Guerrero-González, M.L.; Rodríguez-Kessler, M.; Jiménez-Bremont, J.F. uORF, a regulatory mechanism of the *Arabidopsis* polyamine oxidase 2. *Mol. Biol. Rep.* **2014**, *41*, 2427–2443. [CrossRef]
51. Cruz-Ramírez, A.; López-Bucio, J.; Ramírez-Pimentel, G.; Zurita-Silva, A.; Sánchez-Calderon, L.; Ramírez-Chávez, E.; González-Ortega, E.; Herrera-Estrella, L. The xipotl mutant of *Arabidopsis* reveals a critical role for phospholipid metabolism in root system development and epidermal cell integrity. *Plant Cell* **2004**, *16*, 2020–2034. [CrossRef]
52. Alatorre-Cobos, F.; Cruz-Ramirez, A.; Hayden, C.A.; Pérez-Torres, C.A.; Chauvin, A.L.; Ibarra-Laclette, E.; Alva-Cortés, E.; Jorgensen, R.A.; Herrera-Estrella, L. Translational regulation of *Arabidopsis* XIPOTL1 is modulated by phosphocholine levels via the phylogenetically conserved upstream open reading frame 30. *J. Exp. Bot.* **2012**, *63*, 5203–5221. [CrossRef]
53. Ribone, P.A.; Capella, M.; Arce, A.L.; Chan, R.L. A uORF represses the transcription factor AtHB1 in aerial tissues to avoid a deleterious phenotype. *Plant Physiol.* **2017**, *175*, 1238–1253. [CrossRef]
54. Capella, M.; Ribone, P.A.; Arce, A.L.; Chan, R.L. *Arabidopsis thaliana* HomeoBox 1 (AtHB1), a Homedomain-Leucine Zipper I (HD-Zip I) transcription factor, is regulated by PHYTOCHROME-INTERACTING FACTOR 1 to promote hypocotyl elongation. *New Phytol.* **2015**, *207*, 669–682. [CrossRef]
55. Dong, J.; Chen, H.; Deng, X.W.; Irish, V.F.; Wei, N. Phytochrome B induces intron retention and translational inhibition of PHYTOCHROME-INTERACTING FACTOR3. *Plant Physiol.* **2020**, *182*, 159–166. [CrossRef]
56. Zhang, H.; Si, X.; Ji, X.; Fan, R.; Liu, J.; Chen, K.; Wang, D.; Gao, C. Genome editing of upstream open reading frames enables translational control in plants. *Nat. Biotechnol.* **2018**, *36*, 894–898. [CrossRef]
57. Guo, R.; Sun, W. Sumoylation stabilizes RACK1B and enhance its interaction with RAP2.6 in the abscisic acid response. *Sci. Rep.* **2017**, *7*, 44090. [CrossRef]
58. Pfaffl, M.W. A new mathematical model for relative quantification in real-time RT-PCR. *Nucleic Acids Res.* **2001**, *29*, e45. [CrossRef]
59. Clough, S.J.; Bent, A.F. Floral dip: A simplified method for Agrobacterium-mediated transformation of *Arabidopsis thaliana*. *Plant J.* **1998**, *16*, 735–743. [CrossRef]
60. Rueden, C.T.; Schindelin, J.; Hiner, M.C.; DeZonia, B.E.; Walter, A.E.; Arena, E.T.; Eliceiri, K.W. ImageJ2: ImageJ for the next generation of scientific image data. *BMC Bioinfor.* **2017**, *18*, 529. [CrossRef]
61. Larkin, M.A.; Blackshields, G.; Brown, N.P.; Chenna, R.; McGettigan, P.A.; McWilliam, H.; Valentin, F.; Wallace, I.M.; Wilm, A.; Lopez, R.; et al. Clustal W and Clustal X version 2.0. *Bioinformatics* **2007**, *23*, 2947–2948. [CrossRef] [PubMed]

International Journal of
Molecular Sciences

Article

Comprehensive Analysis of Carotenoid Cleavage Dioxygenases Gene Family and Its Expression in Response to Abiotic Stress in Poplar

Hui Wei [1,2,†], Ali Movahedi [3,4,†], Guoyuan Liu [1,2], Yixin Li [1,2], Shiwei Liu [1,2], Chunmei Yu [1,2], Yanhong Chen [1,2], Fei Zhong [1,2] and Jian Zhang [1,2,*]

[1]   Key Laboratory of Landscape Plant Genetics and Breeding, School of Life Sciences, Nantong University, Nantong 226000, China; 15850682752@163.com (H.W.); cjqm1989@126.com (G.L.); liyixinlyx0@163.com (Y.L.); 2009110160@stmail.ntu.edu.cn (S.L.); ychmei@ntu.edu.cn (C.Y.); chenyh@ntu.edu.cn (Y.C.); fzhong@ntu.edu.cn (F.Z.)

[2]   Key Lab of Landscape Plant Genetics and Breeding, Nantong 226000, China

[3]   Co-Innovation Center for Sustainable Forestry in Southern China, Key Laboratory of Forest Genetics & Biotechnology, Ministry of Education, College of Biology and the Environment, Nanjing Forestry University, Nanjing 210037, China; ali_movahedi@njfu.edu.cn

[4]   College of Arts and Sciences, Arlington International University, Wilmington, DE 19804, USA

*   Correspondence: yjnkyy@ntu.edu.cn

†   Authors contributed equally to this work.

Abstract: Carotenoid cleavage dioxygenases (CCDs) catalyzes the cleavage of various carotenoids into smaller apocarotenoids which are essential for plant growth and development and response to abiotic stresses. CCD family is divided into two subfamilies: 9-cis epoxycarotenoid dioxygenases (NCED) family and CCD family. A better knowledge of carotenoid biosynthesis and degradation could be useful for regulating carotenoid contents. Here, 23 *CCD* genes were identified from the *Populus trichocarpa* genome, and their characterizations and expression profiling were validated. The PtCCD members were divided into PtCCD and PtNCED subfamilies. The PtCCD family contained the PtCCD1, 4, 7, and 8 classes. The PtCCDs clustered in the same clade shared similar intron/exon structures and motif compositions and distributions. In addition, the tandem and segmental duplications resulted in the *PtCCD* gene expansion based on the collinearity analysis. An additional integrated collinearity analysis among poplar, Arabidopsis, rice, and willow revealed the gene pairs between poplar and willow more than that between poplar and rice. Identifying tissue-special expression patterns indicated that *PtCCD* genes display different expression patterns in leaves, stems, and roots. Abscisic acid (ABA) treatment and abiotic stress suggested that many *PtCCD* genes are responsive to osmotic stress regarding the comprehensive regulation networks. The genome-wide identification of *PtCCD* genes may provide the foundation for further exploring the putative regulation mechanism on osmotic stress and benefit poplar molecular breeding.

Keywords: carotenoid; CCD; NCED; ABA; poplar

Citation: Wei, H.; Movahedi, A.; Liu, G.; Li, Y.; Liu, S.; Yu, C.; Chen, Y.; Zhong, F.; Zhang, J. Comprehensive Analysis of Carotenoid Cleavage Dioxygenases Gene Family and Its Expression in Response to Abiotic Stress in Poplar. *Int. J. Mol. Sci.* **2022**, *23*, 1418. https://doi.org/10.3390/ijms23031418

Academic Editor: Víctor Quesada

Received: 3 November 2021
Accepted: 24 November 2021
Published: 26 January 2022

1. Introduction

Terpenoids consist of various primary and secondary metabolites that function in all living organisms. Carotenoids and their apocarotenoids, considered C_{40} isoprenoids, participate in multiple essential biological functions in plants and animals [1,2]. For example, xanthophylls and violaxanthin are the critical components of plant light-harvesting protein complexes [3], and carotenoids in chloroplasts are involved in light absorption, electron transfer, and removal of triplet oxygen and superoxide anion in plant photosynthesis [4,5]. In addition, lycopene and β-carotene are important pigments that affect the color of plant flowers and fruits, and their contents are associated with fruit colors and qualities [6,7]. In addition, zeaxanthin aldehyde, a carotenoid derivative, can be catalyzed into abscisic acid (ABA). ABA, also known as a stress hormone, plays a vital role in regulating various physiological and developmental processes. It can not only act as a growth inhibitor to intervene in biological processes such as plant flowering [8], fruit maturation [9], and seed dormancy [10], but also directly responds to drought, salt, and low temperature through the ABA signal transduction pathway including mediation of stress-resistant genes expression [11], reducing transpiration [12,13], and inducing stomatal closure [14,15].

Carotenoid cleavage dioxygenases (CCDs) are mainly responsible for the oxidative cleavage of carotenoids in higher plants, which results in the biosynthesis of biologically smaller apocarotenoids [5]. CCDs are also a kind of non-heme iron dioxygenase, which contain the RPE65 (retinal segment epithelial membrane protein) domain responsible for binding Fe^{2+} [16,17]. In higher plants, the carotenoid cleavage dioxygenase (CCD) and 9-cis carotenoid cleavage dioxygenase (NCED) have been identified as a subfamily of CCDs based on the epoxy structure of substrate [18,19]. NCEDs can catalyze the cleavage of 11, 12 double bond of violaxanthin (C_{40}) or neoxanthin (C_{40}) to form the xanthoxin (C_{15}), and this catalytic reaction carried out by NCEDs is considered as the rate-limiting step in ABA biosynthesis [20,21]. A variety of studies have been focused on the regulation mechanism of NCEDs in resistance to stress conditions. In Arabidopsis, NCED members were classified into NCED2, 3, 5, 6, and 9 according to phylogeny and gene function [18]. The transcript levels of *AtNCED3* could be induced by drought stress, and AtNCED3 controls the level of endogenous ABA. Overexpression of *AtNCED3* could improve drought tolerance of Arabidopsis by reducing leaf transpiration rates [22]. AtNCED6 and AtNCED9 were associated with developmental control of ABA biosynthesis in seeds. In addition, both AtNCED2, AtNCED3, and AtNCED5 have essential roles in endogenous ABA accumulation [23]. In rice, OsNCED members have also been classified into OsNCED1, OsNCED2, OsNCED1, OsNCED4, and OsNCED5, and the divergent expression of *OsNCED* members in tissues resulted in different functions [24]. OsNCED1 and OsNCED3 were involved in the drought resistance, and OsNCED2 was associated with the delay of seed germination [25]. Overexpression of *OsNCED5* and *OsNCED3* in Arabidopsis increased the tolerance to drought stress and delayed seed dormancy and changed plant size and leaf morphology [26,27]. *CsNCED* genes from *Crocus sativus* had a closer relationship with ABA accumulation under the drought, salt, and lower temperature [28]. The se results suggested that NCEDs control ABA biosynthesis and then affect the ABA-mediated signal transduction pathway involved in plant stress responses.

In contrast, CCD enzymes do not have the specific cleavage sites of substrates. Some CCDs can cleave carotenoid or apocarotenoid substrates, while others recognize specific carotenoid or apocarotenoid substrates [29]. CCD subfamily of Arabidopsis was divided into CCD1, CCD4, CCD7, and CCD8 [17]. AtCCD1 and AtCCD4 mainly cleaved 9, 10 double bonds leading to catalyze the C_{40} carotenoids and C_{27} terpenoids to form the small molecule volatile substances, such as C_{13} β-ionone and C_{14} geranyl acetone geranyl acetone, which suggested that both AtCCD1 and AtCCD4 have an important influence on the formation of plant aroma [30,31]. AtCCD7 catalyzed β-carotene to form β-ionone and C27 10′-apo-β-carotenal, and AtCCD8 or AtCCD1 could further cleave C_{27} 10′-apo-β-carotenal to generate C_{18} 3′-apo-β-carotenal or β-ionone and one apo-10,10′-carotendial, respectively [32,33]. The observations illustrated that AtCCD7 and AtCCD8 could oxidize β-carotene to produce the strigolactone (SL) precursor related to the important biological processes such as branch formation, lateral root formation, seed germination, and response to drought and salt stresses [34]. The composition of CCD4 members is complex, and only one CCD4 member has been identified in Arabidopsis, while many CCD4 members have been identified in tomato, grape, and saffron [35–37]. The evidences on CCD4 members show that CCD4 members may be involved in forming a flower, peel, and pulp color. The β-cryptoxanthin and zeaxanthin, thought as the substrate of citrus CCD4b1, are cleaved at the 7, 8 double bonds to form unique C_{30} carotenoid associated with the color of orange peel [38]. In addition, the expression level of *Glycine max CCDs* has been significantly changed under salt, drought, low temperature, and high-temperature stresses, indicating that soybean CCDs are involved in its abiotic stress response process [39]. *Brassica oleracea* CCD1 and CCD4 are responsive to drought and salt stresses [40]. The expressions of *Malus domestica CCDs* are significantly affected under salt and drought stress, indicating that MdCCD members are involved in response to the abiotic stresses [41]. Given the above evidence, CCD members have various kinds of biological functions and participant in regulating plant growth and development, abiotic stresses, and color formation. However, so far, there is no genome-wide identification of CCD family genes in poplar.

Studies on genome-wide identification of gene families have focused on their characterizations and functions and provided valuable methods for analyzing gene networks or biological functions. For example, analysis of pin-formed (PIN) gene family in wheat displayed PIN may be involved in various developmental processes and biotic and abiotic stress conditions [42]. Identification of β-ketoacyl CoA synthetase (KCS) in barley exhibited barley KCS members function in regulating physiological and biochemical processes and participant in drought stress [43]. In addition, genome-wide identification and characterization of lncRNAs in *Capsicum annuum* showed that lncRNAs interaction with miRNAs takes part in different abiotic stress by regulating transcription factors (TFs), including tryptophan arginine lysine tyrosine (WRKY), myeloblastosis (MYB), basic leucine zipper domain (bZIP), and so on [44]. In addition, genome-wide analysis of the wheat brassinazole-resistant (BZR) gene family exhibited that BZRs play crucial roles in plant developmental processes and are associated with diverse biotic and abiotic stresses [45]. Genome-wide identification of *Salvia miltiorrhiza* antisense transcripts (NATs) considered as a class of long noncoding RNAs documented that NATs interaction with sense transcripts (STs) potentially plays significant regulatory roles in the biosynthesis of bioactive compounds [46]. Genome-wide investigation of the soybean AT-hook motif nuclear localized (AHL) gene family proved that AHLs mainly react to mediating stress responses [47].

Moreover, the identification of tobacco CCD family revealed that tobacco CCD genes have essential roles in response to different hormones, including ABA, methyl jasmonate (MeJA), indoleacetic acid (IAA), salicylic acid (SA), and abiotic stresses [48]. The above genome-wide identification provides an essential theoretical basis for understanding family gene physiological and biochemical functions in different species. Poplar and willow mainly distributed in the cold zone to the temperate zone of the northern hemisphere belong to *Salicaceae Mirb* [49,50]. Poplar is also a model plant to study the molecular mechanism of growth and development, material properties, stress responses, and other vital traits. In addition to being used as an energy tree for industrial production, poplar can also protect the soil structure, prevent and control soil erosion, and have a significant ecological value. In recent years, with the destruction of the ecological environment and the drastic change of the global climate, the inevitable natural disasters such as drought, salt, and freezing have seriously restricted the natural growth and development of poplar and have caused a decline in poplar production capacity and seriously damaged the balance of the local ecological environment. Considering the physiological functions of CCD members under the various stresses, we identify CCD family members of *Populus trichocarpa* at the whole genome level. We systematically analyze molecular evolution, gene structure, cis-acting elements, and conserved motifs of PtCCD family members. In addition, we evaluate the transcript levels of *PtCCDs* in various tissues and identify the expression patterns of *PtCCDs* under abiotic stress. The above results will lay a foundation for further elucidating the biological functions of PtCCD family members.

2. Results

2.1. Characterization of the CCD Family Members in Poplar

The 30 candidate PtCCD members were obtained from the genome database of *P. trichocarpa* (Phytozome https://phytozome-next.jgi.doe.gov/pz/portal.html (accessed on 28 July 2021)) based on sequence alignment with Arabidopsis CCD members (Table S1). The complete RPE65 domain was used as the standard for screening the above PtCCD members and four members (Potri.006G238500, Potri.006G239301, Potri.018G043201, Potri.018G0

44100) displayed no dominant RPE65 domain were eliminated from the PtCCD members. In addition, Potri.T074400 and Potri.T167700 remained on scaffolds, which were not spliced with the poplar chromosome, so Potri.T074400 and Potri.T167700 were abandoned for analysis. Finally, 23 PtCCD members possessing complete RPE65 domain were retrieved from the genome of poplar (Table 1). The PtCCD members were named based on their orthologous relationship with Arabidopsis and rice CCD family members (Table 1). Sequence alignment illustrated significant differences in PtCCD sequences, but the four histidines (His) sites are highly conserved (Figure S1).

Table 1. The poplar, Arabidopsis, and rice *CCD* accession numbers and gene names.

Arabidopsis thaliana		*Oryza sativa*		*Populus trichocarpa*	
Accessions	Gene name	Accessions	Gene name	Accessions	Gene name
				Potri.001G265400	*PtCCD1a*
AT3G63520	*AtCCD1*	LOC_Os12g44310	*OsCCD1*	Potri.001G265600	*PtCCD1b*
				Potri.001G265900	*PtCCD1c*
				Potri.009G060500	*PtCCD1d*
				Potri.004G190700	*PtCCD4a*
		LOC_Os02g47510	*OsCCD4a*	Potri.005G069100	*PtCCD4b*
AT4G19170	*AtCCD4*			Potri.009G151900	*PtCCD4c*
				Potri.009G152200	*PtCCD4d*
		LOC_Os12g24800	*OsCCD4b*	Potri.009G152300	*PtCCD4e*
				Potri.019G093400	*PtCCD4f*
AT2G44990	*AtCCD7*	LOC_Os04g46470	*OsCCD7*	Potri.014G056800	*PtCCD7*
		LOC_Os01g54270	*OsCCD8a*	Potri.006G239200	*PtCCD8a*
				Potri.006G239400	*PtCCD8b*
		LOC_Os09g15240	*OsCCD8b*	Potri.018G042650	*PtCCD8c*
AT4G32810	*AtCCD8*			Potri.018G042900	*PtCCD8d*
		LOC_Os01g38580	*OsCCD8c*	Potri.018G043000	*PtCCD8e*
				Potri.018G043100	*PtCCD8f*
		LOC_Os08g28240	*OsCCD8d*	Potri.018G043400	*PtCCD8g*
				Potri.018G043500	*PtCCD8h*
AT4G18350	*AtNCED2*	LOC_Os12g42280	*OsNCED2*	Potri.011G084100	*PtNCED2*
AT3G14440	*AtNCED3*	LOC_Os07g05940	*OsNCED3*	Potri.001G393800	*PtNCED3a*
				Potri.011G112400	*PtNCED3b*
AT1G30100	*AtNCED5*				
AT3G24220	*AtNCED6*			Potri.003G176300	*PtNCED6*
AT1G78390	*AtNCED9*	LOC_Os03g44380	*OsNCED9*		

The open reading from (ORF) lengths ranged from 1221 to 1842, except for the incomplete sequence information on Potri.004G190700, Potri.009G152300, Potri.018G042900, and Potri.018G043400. The deduced amino acids of PtCCDs varied from 406 to 613, and the molecular weights (MWs) ranged from 46.28–69.07 KD. A high proportion of PtCCDs theoretical PI was less than 7, indicating that a large portion of them belongs to acidic protein, except PtCCD4a, 4c, 4d, 4e, PtCCD7, and PtNCED2. In addition, the grand average of hydropathicities of PtCCDs was less than 1, which illustrated that PtCCDs are hydrophilic and non-transmembrane proteins. The analysis of instability index showed that PtCCD4c, 4d, 4e, 4f, PtCCD8d, 8e 8f, 8g, 8h, and PtNCED2, 3a, and 3b belong to unstable proteins, while others belong to stable proteins. The signal peptide prediction showed that only PtCCD8a has a classical secretion signal peptide, and the cleavage site is located in positions 16 and 17. Other PtCCD proteins do not have signal peptides. In addition, subcellular localization prediction showed PtCCDs are distributed in chloroplast and cytoplasm.

2.2. Phylogenetic Analysis of PtCCD Family

To better understand the evolutionary relationships among CCD members, 9 Arabidopsis, 11 rice, and 23 poplar CCD members were chosen to construct a phylogenetic tree.

According to the evolutionary relationship of Arabidopsis CCDs and rice CCDs, PtCCD family members were divided into two families and named PtCCD subfamily (PtCCD1, PtCCD4, PtCCD7, and PtCCD8 classes) and PtNCED subfamily (Figure 1). Based on clustering analysis, four CCD members were clustered into the CCD1 class, whereas only one AtCCD1 or OsCCD1 was fell into the CCD1 class. In addition, 6 PtCCD4, 1 AtCCD4, and 2 OsCCD4 members were classified into CCD4 class. In addition, the CCD8 class contained eight members in poplar, four in rice, and only one in Arabidopsis, suggesting that CCD8 members were probably expanded only within some species.

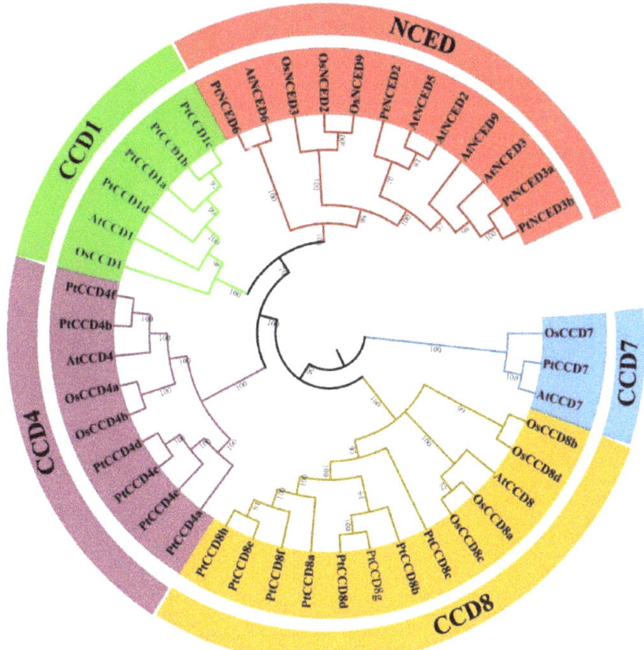

Figure 1. Phylogenetic tree displaying the evolutionary relationships among poplar, Arabidopsis, and rice CCDs. The neighbor-joining (NJ) method tree with 1000 bootstrap replicates was applied to draw a phylogenetic tree with the MEGA7 software. The CCD from poplar, Arabidopsis, and rice were involved in CCD and NCED families, and the CCD family were clustered into CCD1, 4, 7, and 8 classes.

Moreover, NCED5 and NCED9 classes were not found in poplar, but two members are identified as PtNCED3 class. Furthermore, the phylogenetic tree revealed that CCD4 and CCD8 classes have a closer evolutionary relationship, and CCD1 class and NCED subfamily were clustered together, indicating that the NCED type may be evolved from the CCD1 class. The CCD1, CCD4, CCD7, CCD8, and NCED among Arabidopsis, rice, and poplar were assembled in the evolutionary clade, suggesting that the plant CCD gene family is relatively more conservative among different species, and CCD evolution is later than that of herbs and woody plants, monocotyledons, and dicotyledons.

2.3. Gene Structures and Conserved Motifs of CCD Members

To better understand the relationship among the phylogenetic evolution, gene structures, and the conserved motifs of CCD family members, the exon/intron structures and distributions of conserved motifs together with the phylogenetic tree of the CCD family were analyzed. As expected, the *CCD* members possessed similar exons/introns were clustered into the same clade. All *PtNCED* members had no intron, and similar gene structures were identified in *PtCCD4* (except *PtCCD4a* and *PtCCD4e*). Furthermore, the *PtCCD1, 7,* and *8* had many introns and relatively comprehensive exon/intron structures compared with *PtNCED* and *PtCCD4* (Figure S2). Obviously, in addition to *PtCCD* gene structures, the *AtCCD* and *OsCCD* genes shared similar exon/intron structures (Figure 2).

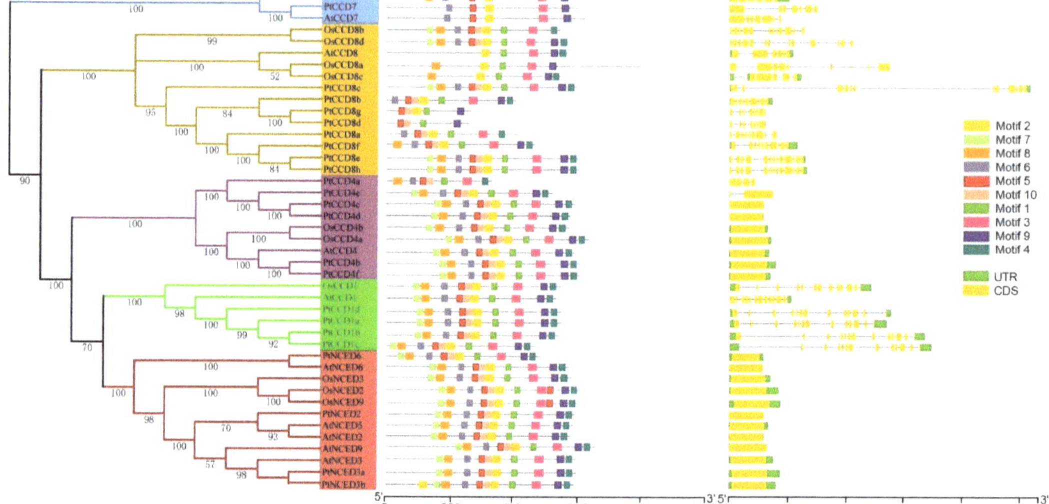

Figure 2. The poplar, Arabidopsis, and rice *CCD* gene structures and conserved motifs. Exons and introns were represented using bars and rectangles, respectively. The conserved motifs were displayed using colorful rectangles.

To explore the features of CCD member motifs, ten distinct motifs were chosen to investigate using the MEME tool. Interestingly, motif distributions of CCDs were similar in the same clade. The CCD1, CCD4, and NCED proteins contained motifs 1–10, suggesting that motifs 1–10 might be involved in their standard functions. In addition, the motif compositions of CCD8 members were divergent, which illustrated that CCD8 members may participate in various physiological processes and could be responsible for different functions. In addition, the number of motifs in CCD7s was lower than that in CCD1, CCD4, and NCED members, which illustrated that CCD7s are speculated to be lacking some specific biological functions compared with CCD1, CCD4, and NCED members (Figure 2). In conclusion, these observations suggested different conserved motifs and gene structures in CCD classes, which further supported the CCD family phylogenetic clustering.

2.4. Expansion and Contraction of PtCCD Genes

The chromosomal distributions of *PtCCDs* were determined and visualized based on the TBtools and poplar genome annotation information. In general, 23 *PtCCD* members were unevenly anchored to Chr01, Chr03–06, Chr09, Chr11, Chr14, Chr18, and Chr19 (Supplemental Figure S3). Chr01, as the largest chromosome, contained 4 *PtCCD* genes, and 6 *PtCCD* genes were mapped to Chr08, while Chr03–05, both 14 and 19, contained only one *PtCCD* gene, respectively. The method of gene family formation usually contains tandem, segmental, and whole-genome duplications. Gene duplication analysis might provide helpful information for *CCD* family gene formation. Here, tandem duplication and segmental duplication are evaluated by the multiple collinearity scan toolkit (MCScanX) method. In general, the tandem duplication origin from homologous genes located on the same chromosome. The gene pairs, including *PtCCD4d* and *PtCCD4e*, and *PtCCD8e* and *PtCCD8f* thought as tandem duplication events were identified in the *PtCCD* gene family. Meanwhile, the *PtCCD4d* and *PtCCD4e*, and *PtCCD8e* and *PtCCD8f* shared similar gene structures and motif distributions. In addition, homologous genes on the different chromosomes were probably the result of segmental duplication. Collinearity analysis of the *PtCCDs* showed that *PtCCD1a* and *PtCCD1d*, *PtCCD4a* and *PtCCD4c*, and *PtNCED3a* and *PtNCED3b* have collinearity relationships, respectively (Figure 3). The *PtCCD1a* and *PtCCD1d*, or *PtNCED3a* and *PtNCED3b* share 91.65% or 91.67% similarity, respectively, which illustrated that gene pairs are formed by segmental duplication. The above observations indicated that expansion of *PtCCD* genes might probably be original from both tandem duplication and segmental duplication. Moreover, *PtCCD8c*, *8d*, *8e*, *8f*, *8g*, and *8h* are located on chromosome 18, *PtCCD1a*, *1b*, and *1c* located on chromosome 1 *PtCCD4c*, *4d*, and *4e* located on chromosome 9 formed various gene clusters. Furthermore, to clarify the role of selection pressure in the evolution of the *PtCCD* genes, TBtools were used to analyze the Ks, Ka, and Ka/Ks of homologous *PtCCD* genes. The results suggested that Ka/Ks values in 2 pairs of tandem duplication and 3 pairs of genome duplication are less than 1, indicating that the *PtCCD* family has undergone strong purification selection during the evolution process.

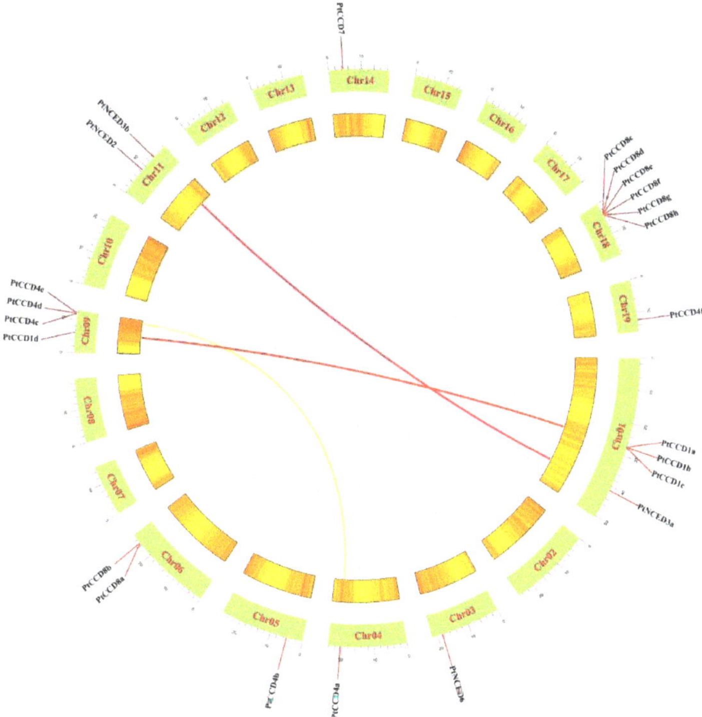

Figure 3. The collinearity analysis of *PtCCDs*. Chromosomes 01–19 were indicated using green rectangles. The homologous *PtCCD* genes in the poplar genome were displayed using red and yellow curves. The poplar gene with collinearity relationship was represented using gray curves.

To explore the orthologous relationship among poplar, Arabidopsis, rice, and willow, the analysis of collinearity relationship was performed using TBtools with MCScanX. The result showed that 5 gene pairs are found between poplar and Arabidopsis, 1 gene pair is found between poplar and rice, and 16 gene pairs are found in poplar and willow, respectively (Figure 4). In the evolutionary relationship of species, poplar and willow were classified to Salicaceae Mirb, and poplar and Arabidopsis were classified to dicotyledonous plants. The collinearity relationship among poplar, Arabidopsis, rice, and willow suggested that poplar has a closer evolutionary relationship with willow.

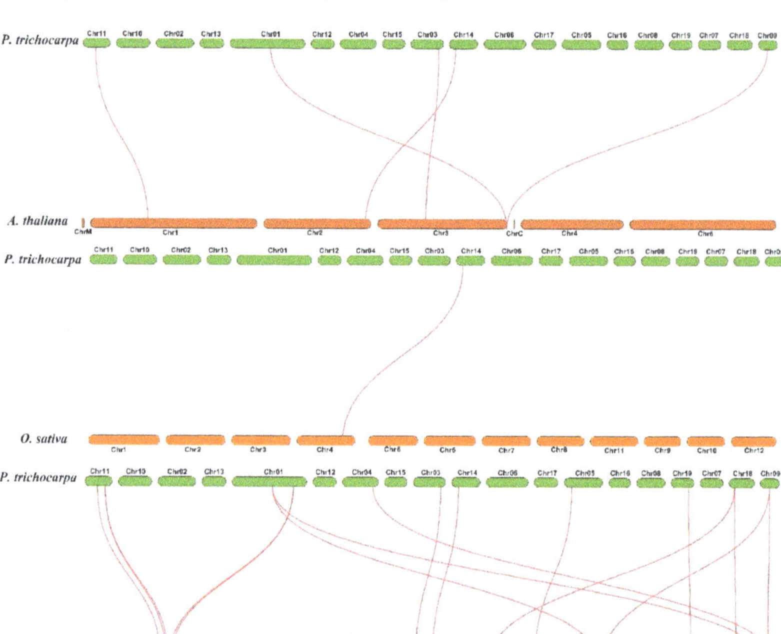

Figure 4. Analysis of collinearity among poplar, Arabidopsis, rice, and willow *CCD* genes. All putative orthologous genes were represented using gray curves, and the orthologous *CCD* genes among the genomes of poplar, Arabidopsis, rice, and willow were displayed using red curves.

2.5. Three-Dimensional (3D) Structures and Cis-Acting Elements Analysis

The results of 3D structure prediction showed that CCD proteins consist of coils, strands, and helixes (Supplemental Figure S3). The coils and strands occupied most CCD structures, while helixes only accounted for a small part of CCD structures. The same class, including CCD1, 4, and 7 and NCED, was visualized to possess the relatively higher similarity, and the divergences lay mainly in the coils and helixes. The 3D structures of CCD8 members were relatively comprehensive and various. The 3D structures of PtCCD8c, PtCCD8d, and PtCCD8g were dominantly divided into two parts marked A and B (Supplemental Figure S4), and the two parts of 3D structures were connected by coils or helixes.

In comparison, the other CCD8 members have no two significant parts of 3D structures. Based on the comparative analysis of CCD8 structures, nearly all 3D structures of CCD8 members could be highly merged in structure A, suggesting that structure A in CCD8 members may be relatively conservative compared to structure B. In addition, the CCD1 members had similar 3D structures with NCED members, which might explain the closer evolutionary relationship between CCD1 and NCED members in a phylogenetic tree.

To understand the putative biological functions of CCD members, PlantCARE was used to analyze the cis-acting elements in the promoters of *CCDs*. The results showed that the *CCD* promoters contain various cis-acting elements, divided into two categories. One is abiotic stress response elements, such as anaerobic inducible element (ARE), stress response element (STRE), MYB drought inducible binding site (MBS), DREB/CBF transcription factor recognition site (DRE), disease resistance, and stress-inducing elements (TC-rich repeats) and low-temperature response elements (LTR). The other category is plant hormone response elements, such as auxin response elements (TGA-element), salicylic acid (SA) inducing elements (TCA-element), methyl jasmonate (MeJA) response element (CGTCA-motif), and ABA response element (ABRE). In addition, the number of light-responsive elements, anaerobic inducing elements, and ABA-responsive elements occupied a large part of cis-acting elements in the *PtCCD* and *AtCCD* promoters, and *OsCCD* promoters contained large numbers of light-responsive elements, MeJA-responsiveness elements, and ABA-responsive elements. Both of *PtCCD* promoters, *PtCCD1b*, *PtCCD4d*, *PtCCD7*, and *PtCCD8f* had a large proportion of MeJA-responsive elements in their promoters, and *PtCCD4b* and *PtNCED2* contained relatively some low-temperature responsive elements. In addition, ABA-responsive elements were widely distributed in *PtCCD4e*, *PtNCED3b*, *PtCCD1a*, *PtNCED3a*, and *PtCCD4e* promoters, while *PtCCD8a*, *PtCCD8b*, and *PtCCD8e* contained fewer cis-acting elements (Supplemental Figure S5). For *AtCCD* promoters, ABA-responsive elements were widely present in *AtNCED5*, *AtNCED9*, and *AtNCED3* promoters. MYB binding sites involved in drought-inducibility existed in *AtNCED9*, *AtCCD4*, *AtCCD8*, *AtNCED3*, and *AtNCED6*. For analysis of *OsCCD* promoters, *OsCCD8c*, *OsCCD8a*, *OsNCED9*, *OsNCED3*, and *OsCCD8d* contained large numbers of ABA-responsive elements. *OsCCD8c*, *OsCCD7*, *OsNCED3*, *OsCCD8b*, *OsCCD4b*, *OsNCED2*, and *OsCCD1* contained the certain number of MeJA-responsive elements. MYB binding sites involved in drought-inducibility were predicted in *OsCCD4a*, *OsCCD4b*, *OsCCD8c*, *OsNCED2*, and *OsNCED9* promoters. Overall, the analysis of cis-acting elements in *CCD* promotes illustrated that *CCDs* transcript levels may be regulated by light, hormone, and abiotic stress, and CCDs were speculated to be involved in diverse stress resistances (Figure 5).

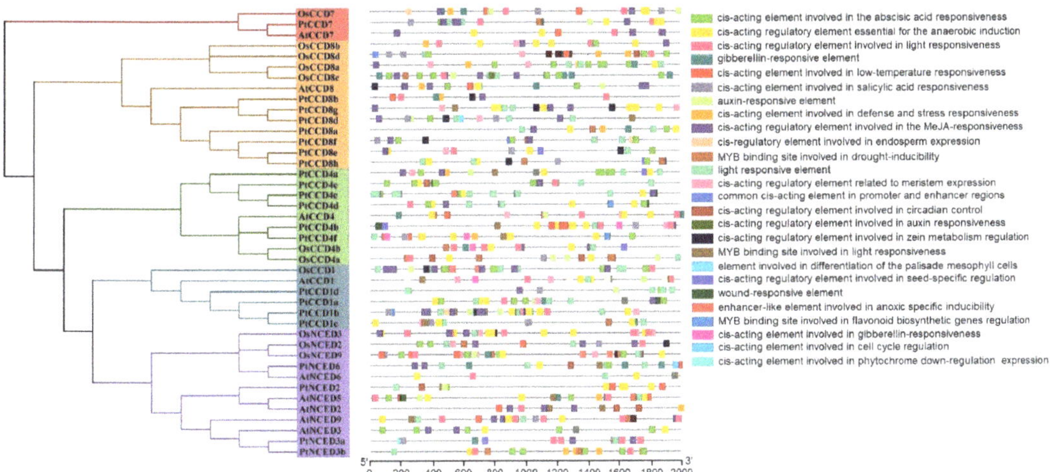

Figure 5. Analysis of poplar and Arabidopsis *CCD* gene promoter cis-elements. Left panel, the phylogenetic tree constructed by NJ method. Right panel, the kinds of *CCD* promoter cis-elements displayed using colorful rectangles.

To analyze the influence of CCDs evolution on their promoter elements, *CCD* cis-acting elements and CCDs phylogenetic tree were compared. The results showed a minor difference in *PtCCDs* cis-acting elements in the same clade. The compositions and distributions of *PtCCDs* cis-acting elements in different evolutionary clades have significant differences. In addition, the other species' *CCD* cis-acting elements in the same clades showed substantial differences, suggesting that orthologous CCDs within the different species may exist the functional differentiation. All above evidence indicated that the categories of *CCD* cis-acting elements are different with CCD evolutionary relationship.

2.6. Interaction Networks of Protein-Protein Assays

Interaction network analysis can find the relationship of protein–protein. The se proteins may be mutually regulated, tightly related in function, or members involved in the same signaling pathway or physiological process. Here, the String database (https://string-db.org/ (accessed on 28 July 2021)) and Cytoscape software were used to identify the interaction network. As shown in Supplemental Figure S6, AtCCD7 was located at the core position in interaction network of AtCCD family members. In addition, the interaction network in PtCCD family members was more sophisticated than that in AtCCD family members. In interaction network of PtCCD family members, PtCCD8g, PtCCD7, and PtCCD4a were predicted to interact together and were in the core position. The above results predicted that PtCCD8g, PtCCD7, and PtCCD4a interaction with other PtCCDs functions together and participants in the same signal transduction and biological process.

Moreover, to further discover the functions of CCDs, interaction assays were performed to construct the relationship between CCD members and other proteins. The result showed that CCDs might interact with cytochrome P (CYP), phytoene synthase (PSY), abscisic acid insensitive (ABI), gibberellic oxidase (GAox), thaumatin-like protein (TCP), and so on, implying CCDs might exert functions by interacting with other genes (Figure 6 and Supplemental Figure S7). The above results predicted that PtCCDs might play an essential role in ABA and GA biosynthesis or ABA and GA signal transduction. Overall, interaction networks could provide crucial references for identifying the regulation mechanism of PtCCDs.

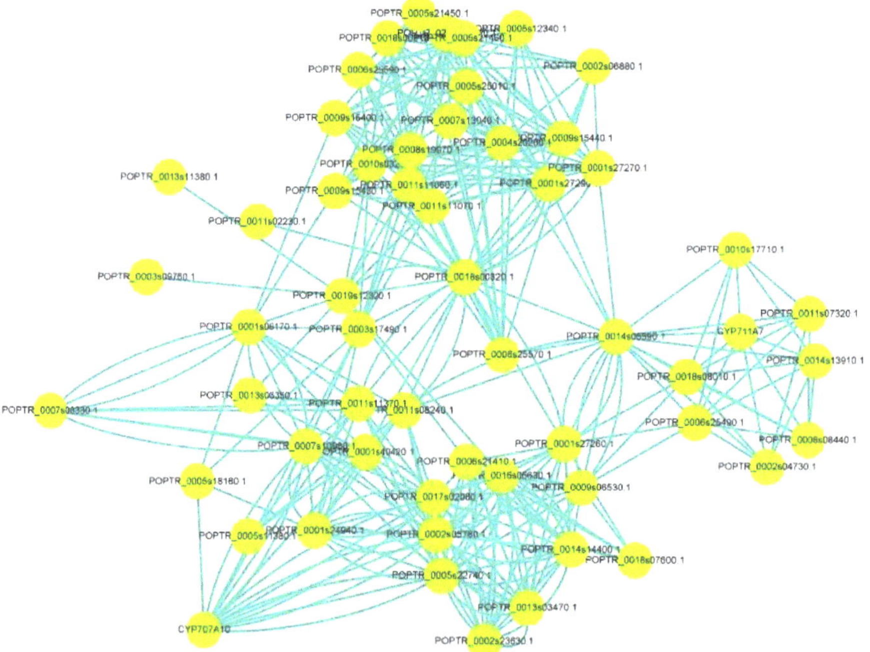

Figure 6. The interaction network for PtCCD members in poplar. The String database (https://string-db.org/ (accessed on 28 July 2021)) was used to predict the interaction relationship between PtCCD members and other proteins, and Cytoscape software was applied to visualize the interaction network.

2.7. Expression Patterns of PtCCDs in Different Tissues

To gain insight into expression profiling of *PtCCDs* in different tissues, the qRT-PCR was used to illustrate gene expression levels in mature and young leaves, the upper and lower region of stems, and roots. The re was a dominant gene expression divergence over different tissues and distinct expression patterns within poplar varieties (Figure 7). For tissue-specific expression patterns of *PtCCDs* in *P. trichocarpa*, the *PtCCD1b* and *1c* were highly expressed in roots. The relatively higher expression accumulations of *PtNCED3a, 3b,* and *6* were presented in a lower region of stems. The mRNA transcript levels of *PtCCD4e* were abundant in the upper region of stems. The higher expression levels of *PtCCD1d, 4b, 4f, 8d, 8e, 8f,* and *8h* were shown in young leaves (Figure 7A). In addition, for expression profiling of *PtCCDs* in 'Nanlin 895' (*P. deltoides* × *P. euramericana*), *PtCCD1b, 8f, 8e,* and *8h* showed higher abundances in roots, the higher expression levels of *PtNCED6* and *PtCCD4e* were accumulated in the lower region of stems, *PtCCD4a, 4b,* and *4f* showed higher mRNA expression in young leaves, and *PtCCD1c* and *1d* showed higher expression levels in the mature leaves (Figure 7B). In addition, for expression patterns of *PtCCDs* in 'Shanxinyang' (*P. davidiana* × *P. bolleana* Loucne), 4 (*PtCCD8d, 8e, 8f,* and *8h*) were illustrated to be highly expressed in roots, 6 (*PtCCD1b, 1c, 1d, 4b,* and *4f*) shared the highest expression in young leaves, and *PtNCED3a* and *PtCCD4a* and *4e* were highly expressed in the lower region of stems (Figure 7C).

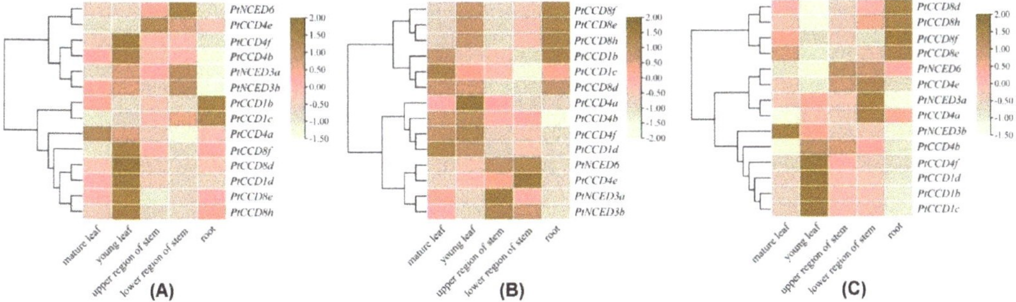

Figure 7. The qRT-PCR analysis of *PtCCDs* expression patterns in different tissues of poplar varieties including *P. trichocarpa* (**A**), 'Nanlin 895' (**B**), and 'Shanxinyang' (**C**). The gene expression levels in mature and young leaves, upper and lower regions of stems and roots were identified using qRT-PCR. Three independent experiments were performed. The data are normalized to poplar *Ptactin* (XM-006370951). *PtCCD* transcript levels were normalized to that in mature leaves; color scale, log2-fold change.

To explore the putative relationship among the *PtCCD* genes, the clustering analysis was performed based on the *PtCCD* expression patterns. It is distinct that *PtCCD* genes in *P. trichocarpa*, 'Nanlin 895', and 'Shanxinyang' can be clustered into different clades, respectively. For example, the *PtCCD8d, 8e, 8f, 8h,* and *1d* clustered into the same clade possessed relatively lower expression levels in young leaves, while the *PtCCD1b* and *1c* clustered into the same clade showed the higher accumulations in roots. In addition, clustering analysis in 'Nanlin 895' indicated that *PtCCD8e, 8f,* and *8h* clustered into the same clade shared similar tissue-specific expression patterns. In addition, *PtCCD* genes transcript patterns in 'Shanxinyang' were divided into three clusters. Cluster 1, 2, and 3 displayed higher expression levels in roots, stems, and leaves, respectively. All above observations revealed that *PtCCD* genes showed tissue-specific expression patterns in poplar, and diverse expression patterns of *PtCCDs* are presented at poplar varieties. The results could illustrate that *PtCCD* genes may be involved in various physiological processes, and the function of *PtCCD* genes possibly experience evolution in response to different environments.

2.8. PtCCD Expression in Response to Abiotic Stress

To illustrate the expression profiling of *PtCCD* genes under abiotic stress, time-course changes were analyzed according to the qRT-PCR. Figure 8 and Supplemental Figure S8 shows *PtCCD* gene expression levels were distinctly affected by abiotic stress, and there are connections and differences in the expression pattern of each *PtCCD* member. Under the ABA treatment, the expression of *PtCCD8d* and *8e* were decreased during the early ABA treatment and recovered at 12 and 24 h, respectively; *PtCCD4b* and *4f* expression was up-regulated with 6h and maintained low expression level during 12–48 h; and other *PtCCD* genes showed relatively higher expression levels within the ABA treatment period, especially *PtNCED3b* and *6*. Under the H_2O_2 treatment, the expression abundances of *PtNCED3a* and *3b* and *PtCCD4b* and *8d* were downregulated during 0–12 h, and the higher expression level was found at 24 h; *PtCCD1c*, *4e*, *8h*, and *PtNCED6* were upregulated and displayed the higher abundances after H_2O_2 stress, but difference existed in reaching the highest expression level. Under the PEG_{6000} treatment, the expression levels of *PtCCD4b* and *4f* were increased firstly and then kept lower expression levels, while the transcript levels of *PtCCD8d*, *8e*, and *8f* displayed the negative results. Expression accumulations of *PtCCD1c*, *1d*, and *4e* and *PtNCED3a*, *3b*, and *6* were dominantly induced during the period of PEG_{6000} treatment. Under the NaCl treatment, the higher accumulations of *PtCCD1b* and *1c* were illustrated during 0–48 h; the expression levels of *PtCCD4f* and *4b* were significantly promoted in early-stage and were dominantly decreased in the later stage. In addition, it is distinct noting that some *PtCCD* genes clustered into the same clade had relatively similar expression profiling. For example, *PtNCED3a* and *3b* had similar mRNA transcript profiling under the H_2O_2 treatment. In addition, the expression levels of *PtCCD8d*, *8e*, and *8f* were generally downregulated in the early stage of PEG_{6000} treatment and then increased (Supplemental Figure S8). The above results showed that poplar *CCD* genes might be involved in response to abiotic and hormone stresses, and different *PtCCD* members had divergent response patterns.

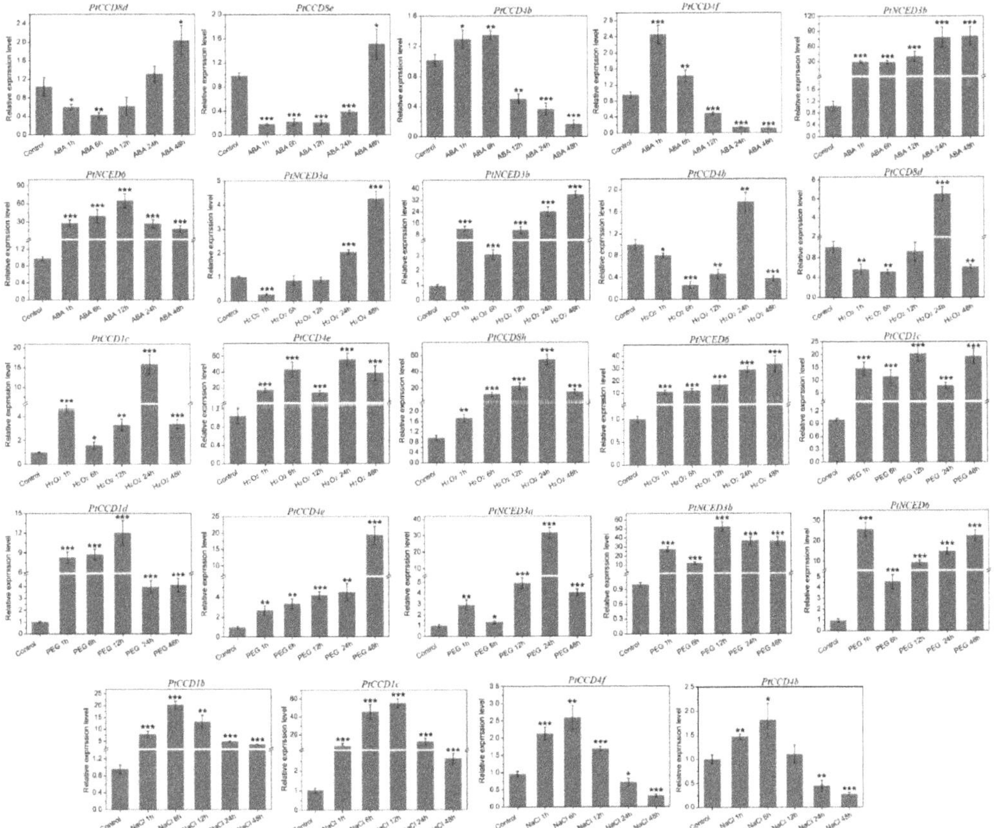

Figure 8. The qRT-PCR analysis of the *PtCCD* expression levels induced by abiotic treatments including ABA, H_2O_2, PEG_{6000}, and NaCl. The vertical bars are representative of SD. Asterisk represents significant difference (*t*-test, * $p < 0.05$, ** $p < 0.01$, and *** $p < 0.001$). Three independent experiments were performed. The data are normalized to poplar *Ptactin* (XM-006370951). *PtCCD* transcript levels were normalized to that in untreated leaves (control).

3. Discussion

The CCD family containing the RPE65 domain participates in the carotenoid metabolic pathway. Almost all eukaryotes contain the CCD family, especially various *CCD* genes in plants, from yeast to human beings. For example, 9, 11, 9, and 19 *CCD* genes were identified in Arabidopsis [17], rice [35], tomato [51], and grape [36], respectively. In this study, 23 *PtCCD* genes were identified from poplar genome, and the compositions of PtCCD1, PtCCD4, PtCCD8, and PtNCED classes differed with other species. For example, each Arabidopsis CCD class contained one *CCD* gene. Rice contained above fourth CCD classes, and CCD8 class contained 8 *OsCCD* genes. While the PtCCD1, 4, and 8 classes contained 4, 6, and 8 *PtCCD* genes, respectively. Compared with Arabidopsis CCD classes, the number of poplar *CCD* genes was significantly increased, suggesting that poplar *CCD* genes might experience gene expansion in the process of evolution. In addition, Arabidopsis NCED subfamily contained *AtNCED2, 3, 5, 6*, and *9*, while the poplar NCED subfamily only contained *NCED3* and *6*. Although the number of *PtNCED* genes was lower than the number of *AtNCED* genes, the NCED subfamily was closely associated with the CCD1 class, and *PtCCD1* genes were dominantly more than *AtCCD1* genes. It has been speculated that parts of PtCCD1 members may make up for the lack of PtNCED function. PtCCD family members are unevenly distributed on 10 poplar chromosomes, of which chromosome 18 contained the relatively more *PtCCD* genes, and only one *PtCCD* gene was identified to be distributed on chromosomes 03, 04, and 05. Moreover, 5 gene pairs of 23 *PtCCD* genes had a homologous evolutionary relationship, including tandem and segmental replication. Among them, *PtCCD4d* and *PtCCD4e*, and *PtCCD8e* and *PtCCD8f* originated from gene tandem replication events, while *PtCCD1a* and *PtCCD1d*, *PtCCD4a*, and *PtCCD4c*, and *PtNCED3a* and *PtNCED3b* originated from segmental replication events, which suggested that gene replication events may result in expansion of *PtCCD* gene family. The Ka/Ks ratios of 5 *PtCCD* pairs were far lower than 1, indicating that they have undergone strong purification selection in the process of evolution. In addition, the harmful non-synonymous substitution disappeared in the evolutionary process, indicating that gene duplication is the main reason for *PtCCD* gene expansion.

In Arabidopsis, AtCCD7 and AtCCD8 could catalyze β-carotene to form caprolactone, the precursor of strigolactone (SL) reported being involved in plant growth and development [52]. In addition, the CsCCD7 and CsCCD8 in saffron were identified to affect the synthesis of SL and control bud sprouting [53]. Similarly, CCD7 and CCD8 in kiwifruit [54], tomato [55], and rice [56] were involved in the regulation of cell senescence, root growth, branch, tiller, and flower organ morphogenesis. In this study, PtCCD7 and PtCCD8 groups showed a closer evolutionary relationship, and gene structures and motif compositions were usually similar in the same clade. Due to the conservation of *CCD* genes, *CCD* genes with similar functions are often clustered in the same clade, which provided an essential basis for studying the role of PtCCD7 and PtCCD8 classes. NCEDs were rate-limiting enzymes of ABA biosynthesis, and ABA as a crucial signal molecule participated in plant growth and development and stress response [57,58]. NCED could be associated with the contents of endogenous ABA and play an essential role in stress tolerance. In Arabidopsis, *AtNCED3* expression was induced by drought stress and participated in response to drought treatment by regulating leaf transpiration rate and controlling the level of endogenous ABA [22]. Also, AtNCED6 and AtNCED9 were involved in ABA biosynthesis during seed development [59]. CsNCED from *Crocus sativus* was closely associated with the content of endogenous ABA under salt, low temperature, and drought stresses [28]. In the present study, PtNCED2, 3, and 6 had similar motif compositions and gene structures with AtNCED2, 3, and 6, respectively, which suggested that PtNCED2, 3, and 6 may be involved in accumulations of endogenous ABA and response to abiotic stress. Compared with the PtCCD subfamily, a part of PtNCED clades was absent from the PtNCED subfamily, which indicated that functions of some PtNCEDs are replaced in the process of poplar evolution. The cleavage of carotenoids catalyzed by CCD4 was related to the coloring of pulp and flower organs. Overexpression of Arabidopsis *AtCCD4* in rice decreased contents of β-carotene and lutein and improved β-violone accumulation [60]. The loss function of *CCD4* led to changes in the color of fruit and flower organs. For example, loss-of-function of *CCD4* resulted in the change of azalea petal color from yellow to white [61] and the change of *Eustoma grandiflorum* petal color from light yellow to white [62]. The previous studies showed differences in the cleavage sites of carotenoids and substrates catalyzed by CCD4. In general, CCD4 cleaved carotenoids at $9'$–$10'$ double bonds, while the cleavage position of CcCCD4b1 in citrus was $7'$–$8'$ double bond, and the product was β-citraurin, β-citaurinene as the unique C_{30} carotenoids [63]. VvCCD4a from grape could catalyze red lycopene to form 6-methyl-5-heptene-2-one. In this study, 6 *PtCCD4s* were identified from poplar genome, and PtCCD4b and 4f had the closer relationship with AtCCD4. All the observations indicated that PtCCD4b and 4f might be involved in poplar pigment formation, and the activity of other PtCCD4 members may be complex. It should be comprehensively analyzed in further studies.

To identify the putative functions of PtCCDs associated with poplar growth and development, the transcript profiling of *PtCCDs* in young and mature leaves, upper and lower region of stems, and roots of poplar varieties were analyzed using qRT-PCR. The divergences in the *PtCCD* expression patterns were observed in poplar varieties, and even the same *PtCCD* was identified to have different expression levels in different poplar varieties. Those results indicated that the same PtCCD might be involved in various physiological processes in diverse poplar varieties. The Arabidopsis and petunia *CCD1* genes were highly expressed in all tested tissues [17,64], while *PtCCD1* genes had high expression levels in leaves of poplar varieties. *PtCCD1* expression patterns were inconsistent with its homologs Arabidopsis and petunia *CCD1* expression patterns, implying that *PtCCD1* genes participate in different physiological processes from Arabidopsis and petunia *CCD1* genes. Although the PtCCD1 class and PtNCED subfamily had the closer evolutionary relationship, the higher expression levels of *PtNCED3a* and *PtNCED6* were found in stems, and *PtCCD1* members highly expressed in leaves of poplar varieties, suggesting that PtNCED family and PtCCD1 class may have divergent functions in stems and leaves. In addition, *PtCCD7* gene was orthologous with *AtCCD7* gene, and previous studies showed *CCD7* has a distinct expression accumulation in roots [65,66]. Interestingly, in this study, the relatively lower expression levels of *PtCCD7* were identified in all tested tissues, which suggested that PtCCD7 in poplar may have different functions with *A. thaliana* CCD7 regarding root development and metabolite synthesis. However, the precise role and related regulation mechanism of PtCCD7 need to be confirmed in future studies. In addition, *PtCCD8* members had been confirmed to have the higher expression levels in *P. trichocarpa* young leaves, while they were identified to have the expression abundances in 'Shanxinyang' and 'Nanlin 895' roots. The se results suggested that PtCCD8 members have divergent functions in the leaves and roots development of poplar varieties.

Since little knowledge focuses on the functions of the poplar CCD genes in regulating abiotic stress responses, the *PtCCDs* expression patterns in response to abiotic stress were illustrated. Previous studies have pointed out that the CCDs play an essential role in ABA synthesis and ABA signal transduction in response to diverse pressures [34,40]. This study identified various kinds of cis-elements involved in hormone-responsive and stress-related elements in the *PtCCD* promoters. *PtCCDs* mRNA levels in leaves induced by abiotic stress were investigated using qRT-PCR. *PtNCED3* and *6* expression abundances improved under the ABA treatment, similar to *AtNCED* gene accumulations caused by ABA treatment [67]. Although the distinct divergence was identified in the tissue-special expression of *PtCCD1* class, the *PtCCD1* members expression levels were also significantly up-regulated by ABA treatment. Compared with the previous conclusion that soybean *CCD7* and *CCD8* gene dominantly respond to ABA treatment [39], the *PtCCD8d* and *8e* expression levels were dominantly down-regulated at the early stage of ABA treatment, and *PtCCD8f* and *8h* expression levels were accumulated at the early stage of ABA treatment. Accordingly, *PtCCD1* members expression was significantly improved after ABA treatment, and even *PtCCD1a* expression abundance underwent a 56-fold increase after a 48 h ABA treatment. The se results suggested that PtCCDs may play an essential role in response to ABA treatment. The precise regulation mechanism of plant stress responses by the ABA signal complex network might need to be explored in further study. Osmotic stress could change the plant physiological processes and affect plant growth and development by decreasing the photosynthetic rate and transpiration [68]. Also, under osmotic stress, the level of membrane lipid peroxidation was increased significantly, which destroyed the plant cell membrane and affected cell integrity [69,70]. Generally, osmotic stress significantly increased the contents of superoxide anion and endogenous H_2O_2 in the plant. Excessive accumulation of superoxide anion would cause irreversible damage to the cell membrane and seriously inhibit the progress of photosynthesis [71]. To explore the putative physiological changes caused by osmotic stress, the expression patterns of *PtCCDs* under the NaCl, PEG_{6000}, and H_2O_2 were analyzed. The expression levels of *PtNCEDs* were considerably upregulated under osmotic stress, and differences only existed in expression trends with time-course. Those results were consistent with the higher expression of *Brassica rapa NCED* under the osmotic stress [40].

In addition, apple *MdCCD8a* expression levels were significantly improved under salt and drought treatments, and the soybean *CCD8* expression was decreased during drought and salt stresses. *PtCCD8s* showed diverse transcript levels under the NaCl and PEG_{6000} treatments. For example, *PtCCD8f* expression experienced an 11-fold increase in response to a 48-h PEG_{6000} treatment, while the peak of *PtCCD8f* expression was identified in 12 h under the H_2O_2 treatment. In addition, under the PEG_{6000} treatment, *PtCCD4b* and *4f* expression levels were up-regulated from 1 to 6 h, but there was distinct down-regulation of their expression levels from 12 to 48 h. However, under the H_2O_2 treatment, *PtCCD4b* expression level was significantly higher than in control, except its expression at 24 h. It was noteworthy that PtCCDs have different response abilities under ABA treatment and osmotic stress, indicating that PtCCDs play vital roles in abiotic stress. This study lays the foundation for identifying the biological functions of PtCCDs and helps to find stress-resistant gene resources.

4. Materials and Methods

4.1. Identification and Classification of Poplar CCD Genes

The RPE65 (PF03055) was achieved from the Pfam database (http://pfam.xfam.org/ (accessed on 28 July 2021)). The sequence information and innovation of poplar, Arabidopsis, rice, and willow were downloaded from the Phytozome database (https://phytozome-next.jgi.doe.gov/ (accessed on 28 July 2021)). The RPE65 is considered as a query to search the putative CCD members from the poplar genome. In addition, the Arabidopsis AtCCDs as a query to search the PtCCD members. The n, the SMART database and NCBI Conserved Domain Search online were applied further to verify the conserved domains in putative PtCCD members. Based on the homologous relationship with AtCCDs, the PtCCD members were named CCD1s, 4s, 7, 8s, and NCEDs.

4.2. Evolutionary Relationship and CCD Sequence Analysis

To illustrate the characterizations and putative functions of PtCCDs, multiple sequence alignment was performed using ClustalX2 software. In addition, according to the neighbor-joining (NJ) method, a phylogenetic tree on PtCCDs, AtCCDs, and OsCCDs evolutionary relationship was constructed. In addition, the poplar, Arabidopsis, rice annotation, and whole-genome information were applied to build the *CCD* gene structures. The MEME was used to identify the CCD motif compositions and distributions. Finally, all those generated files related to *CCD* gene structures and motifs were visualized using TBtools software [72,73].

4.3. Chromosomal Localization and Collinearity Analysis of PtCCDs

According to the *PtCCDs* annotation, the 23 *PtCCD* genes were mapped onto poplar chromosomes. The TBtools with MCScanX was applied to analyze tandem and segmental duplication events of *PtCCDs*, and TBtools with synteny visualization was used to visualize the collinearity relationship. In addition, the TBtools with a simple Ka/Ks calculator was applied to calculate Ka/Ks values between gene pairs. In addition, MCScanX was also used to identify the gene pairs with collinearity relationship among poplar, Arabidopsis, rice, and willow. Similarly, TBtools was applied to visualize syntenic blocks of orthologous genes.

4.4. Analyses of 3D Structures and Cis-Elements

The SWISS-MODEL (https://swissmodel.expasy.org/ (accessed on 28 July 2021)) was used to predict the structures of PtCCDs, and the α-helix, random coil, and strand PtCCDs were represented using Chimera software. The 2000 bp sequences upstream of the translation start sites of *CCD* genes were obtained from poplar, Arabidopsis, and rice genome database, and Plant CARE online tool was applied to predict cis-elements of *CCD* genes.

4.5. Plant Treatments and qRT-PCR Analysis

The leaves, stems, and roots harvested from the *P. trichocarpa*, 'Shanxinyang' (*P. davidiana* × *P. bolleana* Loucne), and 'Nanlin 895' (*P. deltoides* × *P. euramericana*) were used for tissue-specific gene expression analysis. Additionally, the leaves of 'Nanlin 895' grown on MS medium were treated with 2 mM of H_2O_2, 10% PEG_{6000}, 200 mM of NaCl, and 200 µM of ABA, and leaves were collected at 0 (untreated leaves served as a control), 1, 6, 12, 24, and 48 h after each treatment. The abiotic stress treatments were performed with three replicates, with three poplars per replicate. Leaves harvested from treated and untreated poplars were stored at $-80\,°C$.

Total RNA was extracted from various tissues and treated and untreated leaves using an RNA extraction kit (Takara, Japan). The reverse transcriptase (Takara, Japan) was used to synthesize first-strand cDNA of poplar. The UltraSYBR Green I Mixture (CWBIO, China) with 10 µL of Green I Mixture in a 20-µL reaction volume was applied to identify the *PtCCD* gene expression patterns. The Primer3.0 online tool was used to design the primers used in qRT–PCR. The qRT–PCR procedure was as follows: 95 °C for 10 min; 40 cycles of 94 °C for 10 s, 60 °C for 30 s, and 72 °C for 30 s. The relative *PtCCD* expression levels were identified according to the $2^{-\Delta\Delta CT}$ method, with the *Ptactin* (XM-006370951), considered the internal control.

5. Conclusions

In this study, the characterizations and putative functions of 23 poplar CCD members were identified. The se genes were divided into PtCCD1, PtCCD4, PtCCD7, PtCCD8, and PtNCED classes based on the molecular phylogenetic tree. The *PtCCD* gene structures and conserved motifs were also embodied in the evolutionary relationship. The 5 *PtCCD* pairs of homologous genes were identified in poplar genome, and 5 or 16 *CCD* pairs of orthologous genes were illustrated between poplar and Arabidopsis or willow, respectively. In addition, various kinds of stress-responsive cis-elements were identified in the promoters of *PtCCDs*, indicating that PtCCDs have a hand in comprehensive stress resistances. In addition, *PtCCDs* exhibited tissue-special expression patterns in poplar, implying they might be involved in divergent tissue and organ developments. Many *PtCCDs* expression levels were affected by ABA treatment and osmotic stress, suggesting that they may participate in ABA signal transduction or play an essential role in response to osmotic stress. The putative regulation mechanism of PtCCDs in response to abiotic and biotic stresses in poplar was predicted.

On one hand, abiotic and biotic stresses induce higher abundances of PtCCDs in poplar. *PtNCEDs* encode key enzymes for biosynthesis of ABA, ultimately causing the producing ABA signal transduction, resulting in abiotic and biotic stress resistance. On the other hand, other CCDs can catalyze carotenoids to form apocarotenoid substrates in response to stresses (Supplemental Figure S8). The se will be beneficial for exploring PtCCD functions and potential regulatory mechanisms and lay the basis for illustrating corresponding gene networks involved in osmotic stress.

Supplementary Materials: The following are available online at https://www.mdpi.com/article/10 3390/ijms23031418/s1.

Author Contributions: J.Z. designed and funded experiments. H.W. wrote the first draft of the manuscript. J.Z. and A.M. revised the manuscript. H.W., G.L., Y.L., S.L., C.Y., Y.C., and F.Z. experimented. All authors have read and agreed to the published version of the manuscript.

Funding: This work was supported by China Forestry Science and Technology Innovation and Promotion Project (2021TG03), Nantong University Scientific Research Start-up Project for Introducing Talents (135421609106), and the Priority Academic Program Development of Jiangsu Higher Education Institutions.

Institutional Review Board Statement: Not applicable.

Informed Consent Statement: Not applicable.

Data Availability Statement: Not applicable.

Acknowledgments: We gratefully thank Sheng Zhu (Nanjing Forestry University) for providing poplar varieties.

Conflicts of Interest: The authors declare no conflict of interest.

Abbreviations

Carotenoid cleavage dioxygenase: CCD; 9-cis epoxycarotenoid dioxygenases: NCED; abscisic acid: ABA; multiple collinearity scan toolkit: MCScanX; anaerobic inducible element: ARE; stress response element: STRE; MYB drought inducible binding site: MBS; DREB/CBF transcription factor recognition site: DRE; low-temperature response elements: LTR; auxin response elements: TGA-element; salicylic acid: SA; methyl jasmonate: MeJA; ABA response element: ABRE; cytochrome P: CYP; phytoene synthase: PSY; abscisic acid insensitive: ABI; gibberellic oxidase: GAox; thaumatin-like protein: TCP; strigolactone: SL; neighbor-joining: NJ.

References

1. Chen, K.; Li, G.J.; Bressan, R.A.; Song, C.P.; Zhu, J.K.; Zhao, Y. Abscisic acid dynamics, signaling, and functions in plants. *J. Integr. Plant Biol.* **2020**, *62*, 25–54. [CrossRef] [PubMed]
2. Rodriguez-Concepcion, M.; Avalos, J.; Bonet, M.L.; Boronat, A.; Gomez-Gomez, L.; Hornero-Mendez, D.; Limon, M.C.; Meléndez-Martínez, A.J.; Olmedilla-Alonso, B.; Palou, A.; et al. A global perspective on carotenoids: Metabolism, biotechnology, and benefits for nutrition and health. *Prog. Lipid Res.* **2018**, *70*, 62–93. [CrossRef] [PubMed]
3. Woitsch, S.; Romer, S. Expression of xanthophyll biosynthetic genes during light-dependent chloroplast differentiation. *Plant Physiol.* **2003**, *132*, 1508–1517. [CrossRef] [PubMed]
4. Bartley, G.E.; Scolnik, P.A. Plant carotenoids: Pigments for photoprotection, visual attraction, and human health. *Plant Cell.* **1995**, *7*, 1027.
5. Bouvier, F.; Isner, J.C.; Dogbo, O.; Camara, B. Oxidative tailoring of carotenoids: A prospect towards novel functions in plants. *Trends Plant Sci.* **2005**, *10*, 187–194. [CrossRef] [PubMed]
6. Ilahy, R.; Siddiqui, M.W.; Tlili, I.; Montefusco, A.; Piro, G.; Hdider, C.; Lenucci, M.S. When color really matters: Horticultural performance and functional quality of high-lycopene tomatoes. *Crit. Rev. Plant Sci.* **2018**, *37*, 15–53. [CrossRef]
7. Wang, C.; Qiao, A.; Fang, X.; Sun, L.; Gao, P.; Davis, A.R.; Liu, S.; Luan, F. Fine mapping of lycopene content and flesh color related gene and development of molecular marker-assisted selection for flesh color in watermelon (*Citrullus lanatus*). *Front. Plant Sci.* **2019**, *10*, 1240. [CrossRef]
8. Trivellini, A.; Ferrante, A.; Vernieri, P.; Mensuali-Sodi, A.; Serra, G. Effects of promoters and inhibitors of ethylene and aba on flower senescence of *Hibiscus rosa-sinensis* L. *J. Plant Growth Regul.* **2011**, *30*, 175–184. [CrossRef]
9. Chernys, J.T.; Zeevaart, J.A. Characterization of the 9-cis-epoxycarotenoid dioxygenase gene family and the regulation of abscisic acid biosynthesis in avocado. *Plant Physiol.* **2000**, *124*, 343–354. [CrossRef]
10. Shu, K.; Liu, X.D.; Xie, Q.; He, Z.H. Two faces of one seed: Hormonal regulation of dormancy and germination. *Mol. Plant* **2016**, *9*, 34–45. [CrossRef]
11. Seo, M.; Koshiba, T. Complex regulation of ABA biosynthesis in plants. *Trends Plant Sci.* **2002**, *7*, 41–48. [CrossRef]
12. Kuromori, T.; Sugimoto, E.; Shinozaki, K. Arabidopsis mutants of AtABCG22, an ABC transporter gene, increase water transpiration and drought susceptibility. *Plant J.* **2011**, *67*, 885–894. [CrossRef]
13. McAdam, S.A.; Brodribb, T.J.; Banks, J.A.; Hedrich, R.; Atallah, N.M.; Cai, C.; Geringer, M.A.; Lind, C.; Nichols, D.S.; Stachowski, K.; et al. Abscisic acid controlled sex before transpiration in vascular plants. *Proc. Natl. Acad. Sci. USA* **2016**, *113*, 12862–12867. [CrossRef]
14. Islam, M.M.; Ye, W.; Matsushima, D.; Munemasa, S.; Okuma, E.; Nakamura, Y.; Biswas, S.; Mano, J.I.; Murata, Y. Reactive carbonyl species mediate ABA signaling in guard cells. *Plant Cell Physiol.* **2016**, *57*, 2552–2563. [CrossRef]
15. Li, X.; Zhao, J.; Sun, Y.; Li, Y. *Arabidopsis thaliana* CRK41 negatively regulates salt tolerance via H_2O_2 and ABA cross-linked networks. *Environ. Exp. Bot.* **2020**, *179*, 104210.
16. Kloer, D.P.; Ruch, S.; Al-Babili, S.; Beyer, P.; Schulz, G.E. The structure of a retinal-forming carotenoid oxygenase. *Science* **2005**, *308*, 267–269. [CrossRef] [PubMed]
17. Auldridge, M.E.; Block, A.; Vogel, J.T.; Dabney-Smith, C.; Mila, I.; Bouzayen, M.; Magallanes-Lundback, M.; DellaPenna, D.; McCarty, D.R.; Klee, H.J. Characterization of three members of the Arabidopsis carotenoid cleavage dioxygenase family demonstrates the divergent roles of this multifunctional enzyme family. *Plant J.* **2006**, *45*, 982–993. [CrossRef]
18. Tan, B.C.; Joseph, L.M.; Deng, W.T.; Liu, L.; Li, Q.B.; Cline, K.; McCarty, D.R. Molecular characterization of the Arabidopsis 9-cis epoxycarotenoid dioxygenase gene family. *Plant J.* **2003**, *35*, 44–56. [CrossRef]
19. Wang, P.; Lu, S.; Zhang, X.; Hyden, B.; Qin, L.; Liu, L.; Bai, Y.; Han, Y.; Wen, Z.; Xu, J.; et al. Double NCED isozymes control ABA biosynthesis for ripening and senescent regulation in peach fruits. *Plant Sci.* **2021**, *304*, 110739. [CrossRef] [PubMed]

20. Tan, B.C.; Schwartz, S.H.; Zeevaart, J.A.; McCarty, D.R. Genetic control of abscisic acid biosynthesis in maize. *Proc. Natl. Acad. Sci. USA* **1997**, *94*, 12235–12240. [CrossRef]
21. Schwartz, S.H.; Tan, B.C.; Gage, D.A.; Zeevaart, J.A.; McCarty, D.R. Specific oxidative cleavage of carotenoids by VP14 of maize. *Science* **1997**, *276*, 1872–1874.
22. Iuchi, S.; Kobayashi, M.; Taji, T.; Naramoto, M.; Seki, M.; Kato, T.; Tabata, S.; Kakubari, Y.; Yamaguchi-Shinozaki, K.; Shinozaki, K. Regulation of drought tolerance by gene manipulation of 9-cis-epoxycarotenoid dioxygenase, a key enzyme in abscisic acid biosynthesis in Arabidopsis. *Plant J.* **2001**, *27*, 325–333. [CrossRef]
23. Fan, J.; Hill, L.; Crooks, C.; Doerner, P.; Lamb, C. Abscisic acid has a key role in modulating diverse plant-pathogen interactions. *Plant Physiol.* **2009**, *150*, 1750–1761. [CrossRef]
24. Changan, S.S.; Ali, K.; Kumar, V.; Garg, N.K.; Tyagi, A. Abscisic acid biosynthesis under water stress: Anomalous behavior of the 9-cis-epoxycarotenoid dioxygenase1 (NCED1) gene in rice. *Biol. Plantarum* **2018**, *62*, 663–670. [CrossRef]
25. Song, S.; Dai, X.; Zhang, W.H. A rice F-box gene, Os Fbx352, is involved in glucose-delayed seed germination in rice. *J. Exp. Bot.* **2012**, *63*, 5559–5568. [CrossRef]
26. Hwang, S.G.; Chen, H.C.; Huang, W.Y.; Chu, Y.C.; Shii, C.T.; Cheng, W.H. Ectopic expression of rice *OsNCED3* in Arabidopsis increases ABA level and alters leaf morphology. *Plant Sci.* **2010**, *178*, 12–22. [CrossRef]
27. Huang, Y.; Jiao, Y.; Xie, N.; Guo, Y.; Zhang, F.; Xiang, Z.; Wang, R.; Wang, F.; Gao, Q.; Tian, L.; et al. OsNCED5, a 9-cis-epoxycarotenoid dioxygenase gene, regulates salt and water stress tolerance and leaf senescence in rice. *Plant Sci.* **2019**, *287*, 110188. [CrossRef] [PubMed]
28. Ahrazem, O.; Rubio-Moraga, A.; Trapero, A.; Gómez-Gómez, L. Developmental and stress regulation of gene expression for a 9-cis-epoxycarotenoid dioxygenase, CstNCED, isolated from *Crocus sativus* stigmas. *J. Exp. Bot.* **2012**, *63*, 681–694. [CrossRef] [PubMed]
29. Huang, F.C.; Molnár, P.; Schwab, W. Cloning and functional characterization of carotenoid cleavage dioxygenase 4 genes. *J. Exp. Bot.* **2009**, *60*, 3011–3022. [CrossRef]
30. Baldermann, S.; Kato, M.; Kurosawa, M.; Kurobayashi, Y.; Fujita, A.; Fleischmann, P.; Watanabe, N. Functional characterization of a carotenoid cleavage dioxygenase 1 and its relation to the carotenoid accumulation and volatile emission during the floral development of *Osmanthus fragrans* Lour. *J. Exp. Bot.* **2010**, *61*, 2967–2977. [CrossRef] [PubMed]
31. Yahyaa, M.; Bar, E.; Dubey, N.K.; Meir, A.; Davidovich-Rikanati, R.; Hirschberg, J.; Aly, R.; Tholl, D.; Simon, P.W.; Tadmor, Y.; et al. Formation of norisoprenoid flavor compounds in carrot (*Daucus carota* L.) roots: Characterization of a cyclic-specific carotenoid cleavage dioxygenase 1 gene. *J. Agric. food Chem.* **2013**, *61*, 12244–12252. [CrossRef] [PubMed]
32. Ilg, A.; Beyer, P.; Al-Babili, S. Characterization of the rice carotenoid cleavage dioxygenase 1 reveals a novel route for geranial biosynthesis. *FEBS J.* **2009**, *276*, 736–747. [CrossRef]
33. Schwartz, S.H.; Qin, X.; Loewen, M.C. The biochemical characterization of two carotenoid cleavage enzymes from Arabidopsis indicates that a carotenoid-derived compound inhibits lateral branching. *J. Biol. Chem.* **2004**, *279*, 46940–46945. [CrossRef]
34. Alder, A.; Jamil, M.; Marzorati, M.; Bruno, M.; Vermathen, M.; Bigler, P.; Ghisla, S.; Bouwmeester, H.; Beyer, P.; Al-Babili, S. The path from β-carotene to carlactone, a strigolactone-like plant hormone. *Science* **2012**, *335*, 1348–1351. [CrossRef]
35. Vallabhaneni, R.; Bradbury, L.M.; Wurtzel, E.T. The carotenoid dioxygenase gene family in maize, sorghum, and rice. *Arch. Biochem. Biophys.* **2010**, *504*, 104–111. [CrossRef]
36. Lashbrooke, J.G.; Young, P.R.; Dockrall, S.J.; Vasanth, K.; Vivier, M.A. Functional characterisation of three members of the *Vitis vinifera* L. carotenoid cleavage dioxygenase gene family. *BMC Plant Biol.* **2013**, *13*, 156. [CrossRef]
37. Ohmiya, A.; Kishimoto, S.; Aida, R.; Yoshioka, S.; Sumitomo, K. Carotenoid cleavage dioxygenase (CmCCD4a) contributes to white color formation in chrysanthemum petals. *Plant Physiol.* **2006**, *142*, 1193–1201. [CrossRef] [PubMed]
38. Rodrigo, M.J.; Alquézar, B.; Alós, E.; Medina, V.; Carmona, L.; Bruno, M.; Al-Babili, S.; Zacarías, L. A novel carotenoid cleavage activity involved in the biosynthesis of Citrus fruit-specific apocarotenoid pigments. *J. Exp. Bot.* **2013**, *64*, 4461–4478. [CrossRef]
39. Wang, R.K.; Wang, C.E.; Fei, Y.Y.; Gai, J.Y.; Zhao, T.J. Genome-wide identification and transcription analysis of soybean carotenoid oxygenase genes during abiotic stress treatments. *Mol. Biol. Rep.* **2013**, *40*, 4737–4745. [CrossRef]
40. Kim, Y.; Hwang, I.; Jung, H.J.; Park, J.I.; Kang, J.G.; Nou, I.S. Genome-wide classification and abiotic stress-responsive expression profiling of carotenoid oxygenase genes in *Brassica rapa* and *Brassica oleracea*. *J. Plant Growth Regul.* **2016**, *35*, 202–214. [CrossRef]
41. Chen, H.; Zuo, X.; Shao, H.; Fan, S.; Ma, J.; Zhang, D.; Zhao, C.; Yan, X.; Liu, X.; Han, M. Genome-wide analysis of carotenoid cleavage oxygenase genes and their responses to various phytohormones and abiotic stresses in apple (*Malus domestica*). *Plant Physiol. Bioch.* **2018**, *123*, 81–93. [CrossRef]
42. Kumar, M.; Kherawat, B.S.; Dey, P.; Saha, D.; Singh, A.; Bhatia, S.K.; Ghodake, G.S.; Kadam, A.A.; Kim, H.U.; Chung, S.M.; et al. Genome-Wide Identification and Characterization of PIN-FORMED (PIN) Gene Family Reveals Role in Developmental and Various Stress Conditions in *Triticum aestivum* L. *Int. J. Mol. Sci.* **2021**, *22*, 7396. [CrossRef] [PubMed]
43. Tong, T.; Fang, Y.X.; Zhang, Z.; Zheng, J.; Zhang, X.; Li, J.; Niu, C.; Xue, D.; Zhang, X. Genome-wide identification and expression pattern analysis of the KCS gene family in barley. *Plant Growth Regul.* **2021**, *93*, 89–103. [CrossRef]
44. Baruah, P.M.; Krishnatreya, D.B.; Bordoloi, K.S.; Gill, S.S.; Agarwala, N. Genome wide identification and characterization of abiotic stress responsive lncRNAs in *Capsicum annuum*. *Plant Physiol. Biochem.* **2021**, *162*, 221–236. [CrossRef]

45. Kesawat, M.S.; Kherawat, B.S.; Singh, A.; Dey, P.; Kabi, M.; Debnath, D.; Saha, D.; Khandual, A.; Rout, S.; Ali, A.; et al. Genome-wide identification and characterization of the brassinazole-resistant (BZR) gene family and its expression in the various developmental stage and stress conditions in wheat (*Triticum aestivum* L.). *Int. J. Mol. Sci.* **2021**, *22*, 8743.

46. Jiang, M.; Chen, H.; Liu, J.; Du, Q.; Lu, S.; Liu, C. Genome-wide identification and functional characterization of natural antisense transcripts in Salvia miltiorrhiza. *Sci. Rep.* **2021**, *11*, 4769. [CrossRef] [PubMed]

47. Wang, M.; Chen, B.; Zhou, W.; Xie, L.; Wang, L.; Zhang, Y.; Zhang, Q. Genome-wide identification and expression analysis of the AT-hook Motif Nuclear Localized gene family in soybean. *BMC Genom.* **2021**, *22*, 361. [CrossRef]

48. Zhou, Q.; Li, Q.; Li, P.; Zhang, S.; Liu, C.; Jin, J.; Cao, P.; Yang, Y. Carotenoid cleavage dioxygenases: Identification, expression, and evolutionary analysis of this gene family in tobacco. *Int. J. Mol. Sci.* **2019**, *20*, 5796. [CrossRef]

49. Tuskan, G.A.; Difazio, S.; Jansson, S.; Bohlmann, J.; Grigoriev, I.; Hellsten, U.; Putnam, N.; Ralph, S.; Rombauts, S.; Salamov, A.; et al. The genome of black cottonwood, *Populus trichocarpa* (Torr. & Gray). *Science* **2006**, *313*, 1596–1604. [PubMed]

50. Zhang, J.; Yuan, H.; Li, Y.; Chen, Y.; Liu, G.; Ye, M.; Yu, C.; Lian, B.; Zhong, F.; Jiang, Y.; et al. Genome sequencing and phylogenetic analysis of allotetraploid *Salix matsudana* Koidz. *Hortic. Res.* **2020**, *7*, 201. [CrossRef]

51. Wei, Y.; Wan, H.; Wu, Z.; Wang, R.; Ruan, M.; Ye, Q.; Li, Z.; Zhou, G.; Yao, Z.; Yang, Y. A comprehensive analysis of carotenoid cleavage dioxygenases genes in *Solanum lycopersicum*. *Plant Mol. Biol. Rep.* **2016**, *34*, 512–523. [CrossRef]

52. Bruno, M.; Vermathen, M.; Alder, A.; Wüst, F.; Schaub, P.; van der Steen, R.; Beyer, P.; Ghisla, S.; Al-Babili, S. Insights into the formation of carlactone from in-depth analysis of the CCD 8-catalyzed reactions. *FEBS Lett.* **2017**, *591*, 792–800. [CrossRef] [PubMed]

53. Rubio-Moraga, A.; Ahrazem, O.; Pérez-Clemente, R.M.; Gómez-Cadenas, A.; Yoneyama, K.; López-Ráez, J.A.; Molina, R.V.; Gómez-Gómez, L. Apical dominance in saffron and the involvement of the branching enzymes CCD7 and CCD8 in the control of bud sprouting. *BMC Plant Biol.* **2014**, *14*, 171. [CrossRef]

54. Ledger, S.E.; Janssen, B.J.; Karunairetnam, S.; Wang, T.; Snowden, K.C. Modified CAROTENOID CLEAVAGE DIOXYGENASE8 expression correlates with altered branching in kiwifruit (*Actinidia chinensis*). *New Phytol.* **2010**, *188*, 803–813. [CrossRef]

55. Vogel, J.T.; Walter, M.H.; Giavalisco, P.; Lytovchenko, A.; Kohlen, W.; Charnikhova, T.; Simkin, A.J.; Goulet, C.; Strack, D.; Bouwmeester, H.J.; et al. SlCCD7 controls strigolactone biosynthesis, shoot branching and mycorrhiza-induced apocarotenoid formation in tomato. *Plant J.* **2010**, *61*, 300–311. [CrossRef]

56. Kulkarni, K.P.; Vishwakarma, C.; Sahoo, S.P.; Lima, J.M.; Nath, M.; Dokku, P.; Gacche, R.N.; Mohapatra, T.; Robin, S.; Sarla, N.; et al. A substitution mutation in *OsCCD7* cosegregates with dwarf and increased tillering phenotype in rice. *J. Genet.* **2014**, *93*, 389–401. [CrossRef] [PubMed]

57. Pei, X.; Wang, X.; Fu, G.; Chen, B.; Nazir, M.F.; Pan, Z.; He, S.; Du, X. Identification and functional analysis of 9-cis-epoxy carotenoid dioxygenase (NCED) homologs in *G. hirsutum*. *Int. J. Biol. Macromol.* **2021**, *182*, 298–310. [CrossRef]

58. Zhang, J.; Zhang, P.; Huo, X.; Gao, Y.; Chen, Y.; Song, Z.; Wang, F.; Zhang, J. Comparative phenotypic and transcriptomic analysis reveals key responses of upland cotton to salinity stress during postgermination. *Front. Plant Sci.* **2021**, *12*, 606. [CrossRef] [PubMed]

59. Lefebvre, V.; North, H.; Frey, A.; Sotta, B.; Seo, M.; Okamoto, M.; Nambara, E.; Marion-Poll, A. Functional analysis of Arabidopsis *NCED6* and *NCED9* genes indicates that ABA synthesized in the endosperm is involved in the induction of seed dormancy. *Plant J.* **2006**, *45*, 309–319. [CrossRef] [PubMed]

60. Song, M.H.; Lim, S.H.; Kim, J.K.; Jung, E.S.; John, K.M.; You, M.K.; Ahn, S.N.; Lee, C.H.; Ha, S.H. In planta cleavage of carotenoids by Arabidopsis carotenoid cleavage dioxygenase 4 in transgenic rice plants. *Plant Biotechnol. Rep.* **2016**, *10*, 291–300. [CrossRef]

61. Ureshino, K.; Nakayama, M.; Miyajima, I. Contribution made by the carotenoid cleavage dioxygenase 4 gene to yellow colour fade in azalea petals. *Euphytica* **2016**, *207*, 401–417. [CrossRef]

62. Liu, H.; Kishimoto, S.; Yamamizo, C.; Fukuta, N.; Ohmiya, A. Carotenoid accumulations and carotenogenic gene expressions in the petals of *Eustoma grandiflorum*. *Plant Breed.* **2013**, *132*, 417–422. [CrossRef]

63. Zheng, X.; Zhu, K.; Sun, Q.; Zhang, W.; Wang, X.; Cao, H.; Tan, M.; Xie, Z.; Zeng, Y.; Ye, J.; et al. Natural variation in CCD4 promoter underpins species-specific evolution of red coloration in citrus peel. *Mol. Plant* **2019**, *12*, 1294–1307. [CrossRef]

64. Simkin, A.J.; Underwood, B.A.; Auldridge, M.; Loucas, H.M.; Shibuya, K.; Schmelz, E.; Clark, D.G.; Klee, H.J. Circadian regulation of the PhCCD1 carotenoid cleavage dioxygenase controls emission of β-ionone, a fragrance volatile of petunia flowers. *Plant Physiol.* **2004**, *136*, 3504–3514. [CrossRef]

65. Booker, J.; Sieberer, T.; Wright, W.; Williamson, L.; Willett, B.; Stirnberg, P.; Turnbull, C.; Srinivasan, M.; Goddard, P.; Leyser, O. MAX1 encodes a cytochrome P450 family member that acts downstream of MAX3/4 to produce a carotenoid-derived branch-inhibiting hormone. *Dev. Cell* **2005**, *8*, 443–449. [CrossRef]

66. Delaux, P.M.; Xie, X.; Timme, R.E.; Puech-Pages, V.; Dunand, C.; Lecompte, E.; Delwiche, C.F.; Yoneyama, K.; Bécard, G.; Séjalon-Delmas, N. Origin of strigolactones in the green lineage. *New Phytol.* **2012**, *195*, 857–871. [CrossRef]

67. Iuchi, S.; Kobayashi, M.; Yamaguchi-Shinozaki, K.; Shinozaki, K. A stress-inducible gene for 9-cis-epoxycarotenoid dioxygenase involved in abscisic acid biosynthesis under water stress in drought-tolerant cowpea. *Plant Physiol.* **2000**, *123*, 553–562. [CrossRef] [PubMed]

68. Lawlor, D.W.; Cornic, G. Photosynthetic carbon assimilation and associated metabolism in relation to water deficits in higher plants. *Plant Cell Environ.* **2002**, *25*, 275–294. [CrossRef]

69. Candan, N.; Tarhan, L. Tolerance or sensitivity responses of *Mentha pulegium* to osmotic and waterlogging stress in terms of antioxidant defense systems and membrane lipid peroxidation. *Environ. Exp. Bot.* **2012**, *75*, 83–88. [CrossRef]
70. Zhang, C.; Shi, S. Physiological and proteomic responses of contrasting alfalfa (*Medicago sativa* L.) varieties to PEG-induced osmotic stress. *Front. Plant Sci.* **2018**, *9*, 242. [CrossRef] [PubMed]
71. Li, Q.; Lv, L.R.; Teng, Y.J.; Si, L.B.; Ma, T.; Yang, Y.L. Apoplastic hydrogen peroxide and superoxide anion exhibited different regulatory functions in salt-induced oxidative stress in wheat leaves. *Biol. Plantarum* **2018**, *62*, 750–762. [CrossRef]
72. Chen, C.; Chen, H.; Zhang, Y.; Thomas, H.R.; Frank, M.H.; He, Y.; Xia, R. TBtools: An integrative toolkit developed for interactive analyses of big biological data. *Mol. Plant* **2020**, *13*, 1194–1202. [CrossRef] [PubMed]
73. Zhang, J.; zheng Shi, S.; Jiang, Y.; Zhong, F.; Liu, G.; Yu, C.; Lian, B.; Chen, Y. Genome-wide investigation of the AP2/ERF superfamily and their expression under salt stress in Chinese willow (*Salix matsudana*). *PeerJ* **2021**, *9*, e11076. [CrossRef] [PubMed]

International Journal of
Molecular Sciences

Article

NAC Transcription Factor PwNAC11 Activates *ERD1* by Interaction with ABF3 and DREB2A to Enhance Drought Tolerance in Transgenic *Arabidopsis*

Mingxin Yu [†], Junling Liu [†], Bingshuai Du, Mengjuan Zhang, Aibin Wang and Lingyun Zhang *

Key Laboratory of Forest Silviculture and Conservation of the Ministry of Education, The College of Forestry, Beijing Forestry University, Beijing 100083, China; ymxbjfu@163.com (M.Y.); liujunling022@163.com (J.L.); dubingshuai624@163.com (B.D.); liujunling02@126.com (M.Z.); wangaibin126@126.com (A.W.)
* Correspondence: lyzhang@bjfu.edu.cn; Tel.: +86-10-6233-6044
† These authors contributed equally to this work.

Abstract: NAC (NAM, ATAF1/2, and CUC2) transcription factors are ubiquitously distributed in eukaryotes and play significant roles in stress response. However, the functional verifications of NACs in *Picea (P.) wilsonii* remain largely uncharacterized. Here, we identified the NAC transcription factor PwNAC11 as a mediator of drought stress, which was significantly upregulated in *P. wilsonii* under drought and abscisic acid (ABA) treatments. Yeast two-hybrid assays showed that both the full length and C-terminal of PwNAC11 had transcriptional activation activity and PwNAC11 protein cannot form a homodimer by itself. Subcellular observation demonstrated that PwNAC11 protein was located in nucleus. The overexpression of *PwNAC11* in *Arabidopsis* obviously improved the tolerance to drought stress but delayed flowering time under nonstress conditions. The steady-state level of antioxidant enzymes' activities and light energy conversion efficiency were significantly increased in PwNAC11 transgenic lines under dehydration compared to wild plants. *PwNAC11* transgenic lines showed hypersensitivity to ABA and PwNAC11 activated the expression of the downstream gene *ERD1* by binding to ABA-responsive elements (ABREs) instead of drought-responsive elements (DREs). Genetic evidence demonstrated that PwNAC11 physically interacted with an ABA-induced protein—ABRE Binding Factor3 (ABF3)—and promoted the activation of *ERD1* promoter, which implied an ABA-dependent signaling cascade controlled by PwNAC11. In addition, qRT-PCR and yeast assays showed that an ABA-independent gene—DREB2A—was also probably involved in PwNAC11-mediated drought stress response. Taken together, our results provide the evidence that PwNAC11 plays a dominant role in plants positively responding to early drought stress and ABF3 and DREB2A synergistically regulate the expression of *ERD1*.

Keywords: *Picea wilsonii*; transcription factor; PwNAC11; drought stress; ABA signaling

Citation: Yu, M.; Liu, J.; Du, B.; Zhang, M.; Wang, A.; Zhang, L. NAC Transcription Factor PwNAC11 Activates *ERD1* by Interaction with ABF3 and DREB2A to Enhance Drought Tolerance in Transgenic *Arabidopsis*. Int. J. Mol. Sci. **2021**, 22, 6952. https://doi.org/10.3390/ijms22136952

Academic Editor: Víctor Quesada

Received: 31 May 2021
Accepted: 22 June 2021
Published: 28 June 2021

1. Introduction

Water deficit is one of the most disruptive abiotic stresses influencing plant growth and development [1,2]. As environments of plants have gradually deteriorated as a result of global warming, it is of great significance to understand the molecular mechanisms of plants themselves responding to abiotic stress, especially drought stress, for further cultivating new varieties with strong adaptability. In order to fight against abiotic stress, plants have evolved a series of sophisticated but effective strategies to protect themselves from the environmental disadvantages, including stress escape and stress tolerance [3,4]. Stress escape mainly occurs under mild stress conditions, which allows plants to accelerate the growth and flowering processes in order to avoid prematurely dying of water deprivation. Stress tolerance is a regulatory mechanism by which plants can endure long-term stress and maintain their vitality under such environmental pressure [5].

Particularly, transcription factors (TFs) play a vital role in stress regulatory mechanisms by selectively binding to *cis*-elements in the promoter of downstream genes [6]. NACs are one of the most widely reported families of transcription factors in plants, which can be divided into NAM, ATAF and CUC subfamilies according to their interspecific DNA-binding domains. The N-terminal domain of NAC TF can be divided into five conserved subdomains (A, B, C, D, and E). Among them, the A domain is related to the formation of dimers, the C and D domains are highly conserved with binding affinity to promoters, but the B and E domains are relatively non-conserved. By comparison, the C-terminal exhibits more structural variability and is considered as the transcription activation domain [7]. Ooka et al. conducted a comprehensive analysis of 180 NAC family proteins in *Arabidopsis* and rice for the first time [8]. According to the similarity of the NAC domain sequences between *Arabidopsis* and rice, the 180 proteins were divided into two large groups and 18 subgroups. A growing body of evidence has shown that NAC transcription factors are widely involved in plant response to abiotic/biotic stress [9], senescence [10], fruit ripening [11] and hormonal regulation [12], etc.

Over the years, a variety of NAC TFs in different plant species have been reported to play positive roles in response to adverse environments. For example, three NAC transcription factors—ANAC019, ANAC055 and ANAC072—were substantially induced by drought, salt and ABA treatments, the overexpression of which improved the drought tolerance of transgenic *Arabidopsis* [13]. Similarly, the overexpression of pepper *CaNAC46* in *Arabidopsis* significantly inhibited ROS accumulation and the transgenic lines are less susceptible to salt stress [14]. Similarly, *TaSNAC8-6A* overexpression in wheat exhibited increased drought stress tolerance and this process was proven to be controlled by the auxin- and drought response pathways by transcriptomic analysis [15]. The NAC family member JUB1 was proven to be a regulator of drought stress, and the overexpression of tomato *SlJUB1* or *Arabidopsis AtJUB1* could increase drought tolerance in transgenic tomato by directly binding to the promoters of *SlDELLA*, *SlDREB2* and *SlDREB1* [16], suggesting that the regulation mechanisms of the NAC transcription factors involved in abiotic stress are relatively specific among different species. In woody plants, BpNAC012 derived from *Betula platyphylla* activated the core CGTG/A elements to promote the expression of stress-responsive genes, which further enhanced osmotic and salt stress tolerance in *BpNAC012* transgenic birch [17]. In addition, many of the NAC TFs were demonstrated to regulate abscisic acid (ABA)-induced stress response process. *ONAC022* showed transcriptional abundance after ABA and drought treatments in rice and *NAC022* OE lines improved sensitivity to exogenous ABA [18]. ANAC096 could directly interact with ABF2 and ABF4 to activate the expression of *RD29A*, thus participating in dehydration and osmotic stresses regulation in an ABA-dependent pathway in *Arabidopsis* [19]. Although the functions of the NAC TF family have been investigated in various plant species, it has been difficulties of in-depth study for revealing the complicated and efficient molecular mechanisms inside.

ABA, an important hormone upstream of transcription factors, acts as a messenger in the biological processes of stomatal conductance, seed dormancy and germination, leaf senescence, and stress response [20–22]. Once suffering biotic or abiotic stress, the content of ABA shows a notable increase, and endogenous ABA will be transported to the ground part through the transmission of xylem to reduce stomatal conductance, further reducing water loss and activating the corresponding receptors to initiate signal transduction in response to stress [23]. ABA synthesis in plants is mainly formed through the degradation of carotenoids: several crucial enzymatic proteins, such as NCED, ABA2 and AAO, are involved in this process [24]. The transduction of the ABA signal mainly consists of three key ABA receptors, PP2C, PYP/PYL/RCAR, and SNRK2 kinase [25]. In the presence of ABA, SnRK2 could phosphorylate downstream ABA-responsive element (ABRE)-binding factors (ABFs). Meanwhile, some transcription factors such as NAC, bZIP, MYB and WRKY, are reported to activate the expression of ABA-dependent genes through binding to cis-acting elements on promoters of downstream genes containing specific ABRE motifs. In addition, other cis-acting regulatory elements, such as dehydration responsive elements

(DREs) are also demonstrated to respond to external stresses in an ABA-independent pathway [26,27].

Although the roles of NAC transcription factors involved in abiotic stress response have been extensively explored, the knowledge regarding NAC TFs participating in the regulation of coniferous forests responding to abiotic stress is still limited. *Picea wilsonii*, an endemic species of coniferous forests, which is best characterized by tolerance to abiotic stress and environment adaptability, is widespread in northern China. *Picea wilsonii* also has ornamental value in landscaping based on its impressing appearance and lush green crown [28]. Given its advantageous characteristics, it is of great significance to screen and reveal the molecular mechanisms of its abiotic resistance. Previously, we identified two NAC transcription factors—PwNAC2 and PwNAC30—from *P. wilsonii*, which acted as positive and negative regulators in response to abiotic stress, respectively [28,29], suggesting the complexity and functional diversity of NAC TF members in coniferous forests. Here, based on the high-throughput RNA-sequencing (RNA-seq) of the expression profiles of *P. wilsonii* in the absence or presence of drought treatments [30], we identify PwNAC11 as one of the most differentially expressed genes (DEGs) on a transcriptional level and further elucidate the function and molecular mechanism underlying the ABA-mediated drought stress response. We found that overexpression of *PwNAC11* in *Arabidopsis* obviously improved tolerance to drought stress by binding to the promoter of *ERD1*. During this process, the interaction of PwNAC11 with ABF3 synergistically activated the expression of the *ERD1* promoter, resulting in increased resistance to drought stress. The biochemical and physiological evidence showed that ABA positively promoted the regulation of the PwNAC11-mediated abiotic stress response. Our study provides the basis for a new understanding of NAC TF in genetic breeding and the improvement of coniferous forest germplasm resources.

2. Results

2.1. Bioinformatics Analysis of PwNAC11

PwNAC11 was cloned from cDNA library of *P. wilsonii*. It has an open reading frame with a length of 954 bp and encodes a protein with 317 amino acid. Protein multiple sequence alignment and BLAST analysis showed that the PwNAC11 protein had a typical NAM (no apical meristem) domain without transmembrane domain (TMD) (Figure 1A). Phylogenetic analysis indicated that PwNAC11 was close to the homology of *Picea sitchensis* (Figure 1B).

To determine whether PwNAC11 potentially functions as a transcription factor, we first examined the subcellular localization of the PwNAC11 protein. The results showed that the fluorescent signals for the empty vector were widely detected in the nucleus, cell membrane, and cytoplasm, but the PwNAC11-GFP fusion protein was only detected in the nucleus, indicating that PwNAC11 is a nucleus-located transcription factor (Figure 1C). The yeast two-hybrid assay was conducted to detect whether PwNAC11 had transcriptional activating activity. We found that only the yeast cells containing pGBKT7-PwNAC11, pGBKT7-PwNAC11-C, and pGBKT7-ANAC092 grew well on SD/Trp-His-Ade selective medium, suggesting that both full length and C-terminal of PwNAC11 have transcriptional activating activity (Figure 1D). Moreover, yeast strains AH109 co-transformed with pGBKT7-PwNAC11 + pGADT7-PwNAC11 and pGBKT7-PwNAC11ΔC + pGADT7-PwNAC11 cannot grow on SD/Trp-Leu-His-Ade selective medium, indicating that PwNAC11 cannot form a homodimer by itself (Figure 1E).

To elucidate the response of PwNAC11 to drought stress, qRT-PCR was performed with 4-week-old *P. wilsonii* seedlings. The results showed that the expression of *PwNAC11* was dramatically induced at 3 h by PEG or ABA treatment, indicating that PwNAC11 is probably involved in the process of plants' early responses to adverse environments (Figure 1F).

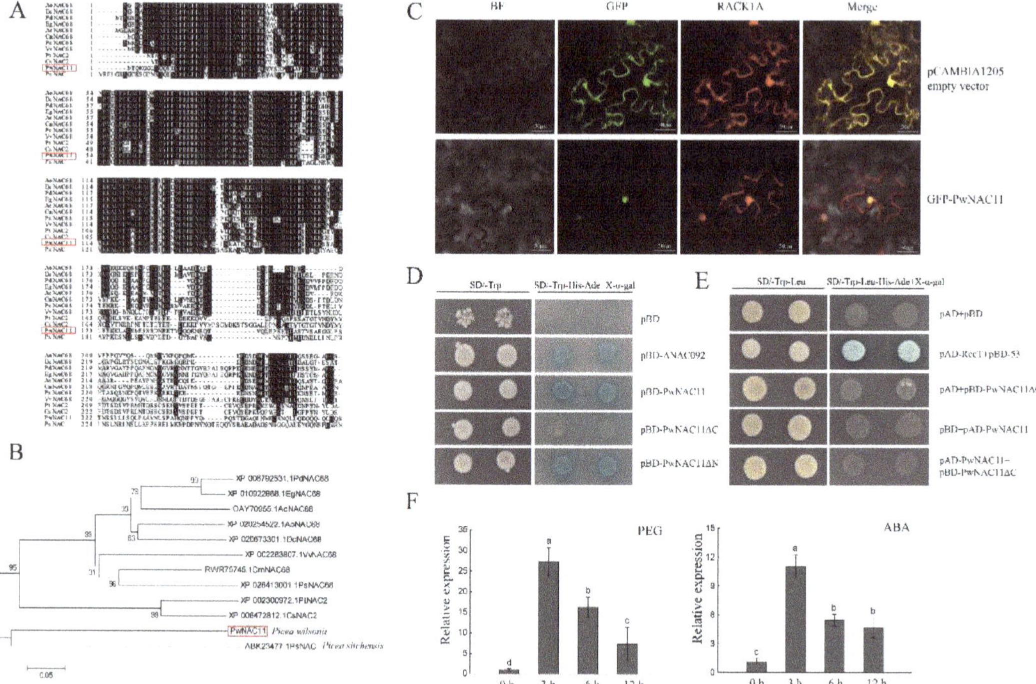

Figure 1. (**A**) Protein multiple sequence alignment of PwNAC11. (**B**) Phylogenetic analysis of PwNAC11. Numbers on each branch represent the confidence of 1000 repetitions. (**C**) Subcellular localization of PwNAC11. RACK1A-RFP is used as a marker gene which locates in nucleus, cytoplasm and cell membrane. The bar in each picture is 10 μm. (**D**) Yeast trans-acting activity assay. pBD-ANAC092 was used as positive control. pBD-PwNAC11ΔN refers to the PwNAC11 protein without N terminal (171–318) and pBD-PwNAC11ΔC refers to the PwNAC11 protein without C terminal (1–153). (**E**) Yeast two-hybrid assay for PwNAC11 forming homodimers. pAD-RecT+pBD-53 was used as positive control and pAD + pBD was used for negative control. (**F**) Expression profiles of *PwNAC11* in *P. wilsonii* under drought and ABA treatments. Different letters indicate the significant differences at $p < 0.05$.

2.2. Overexpression of PwNAC11 Enhances Drought Tolerance in Transgenic A. thaliana

In order to investigate the function of PwNAC11 in response to drought stress, we heterogeneously expressed *PwNAC11* in *Arabidopsis* under the control of CaMV 35S promoter. Two T3 lines with *PwNAC11* highly expressed, named OE2 and OE3, were selected for further functional analysis (Figure S1). The seeds of wild type (WT), empty vector (VC) and overexpression (OE2 and OE3) lines were sown on MS medium with selected concentrations of mannitol. It was found that there was no significant difference in the germination rates among these lines under normal conditions. Under the simulated drought conditions, the growth of the WT and VC groups were severely inhibited, whereas the germination rates of the seeds of the two OE lines were less affected (Figure 2A). For example, the germination percentage of OE lines were over 80% under mannitol conditions of 200 mM compared to less than 60% of the WT and VC lines (Figure 2F,H). Moreover, the *PwNAC11* transgenic lines showed stronger germination rates. Under 100 mM mannitol treatment, the germination rates of OE lines were 1.6 times higher than control groups on the 4th day. Meanwhile, in the presence of 300 mM mannitol, the germination rate of OE3 line was nearly 20% on the 3rd day, but the seeds of control groups showed almost no germination at the same time, implying that the OE lines had stronger seed vitality at the early stage of germination under drought treatment. Similar results were found in the root length experiment, especially under the treatment of 200 mM mannitol, in which the average root length of the OE lines was 10 mm longer than that of WT and VC groups (Figure 2B,C).

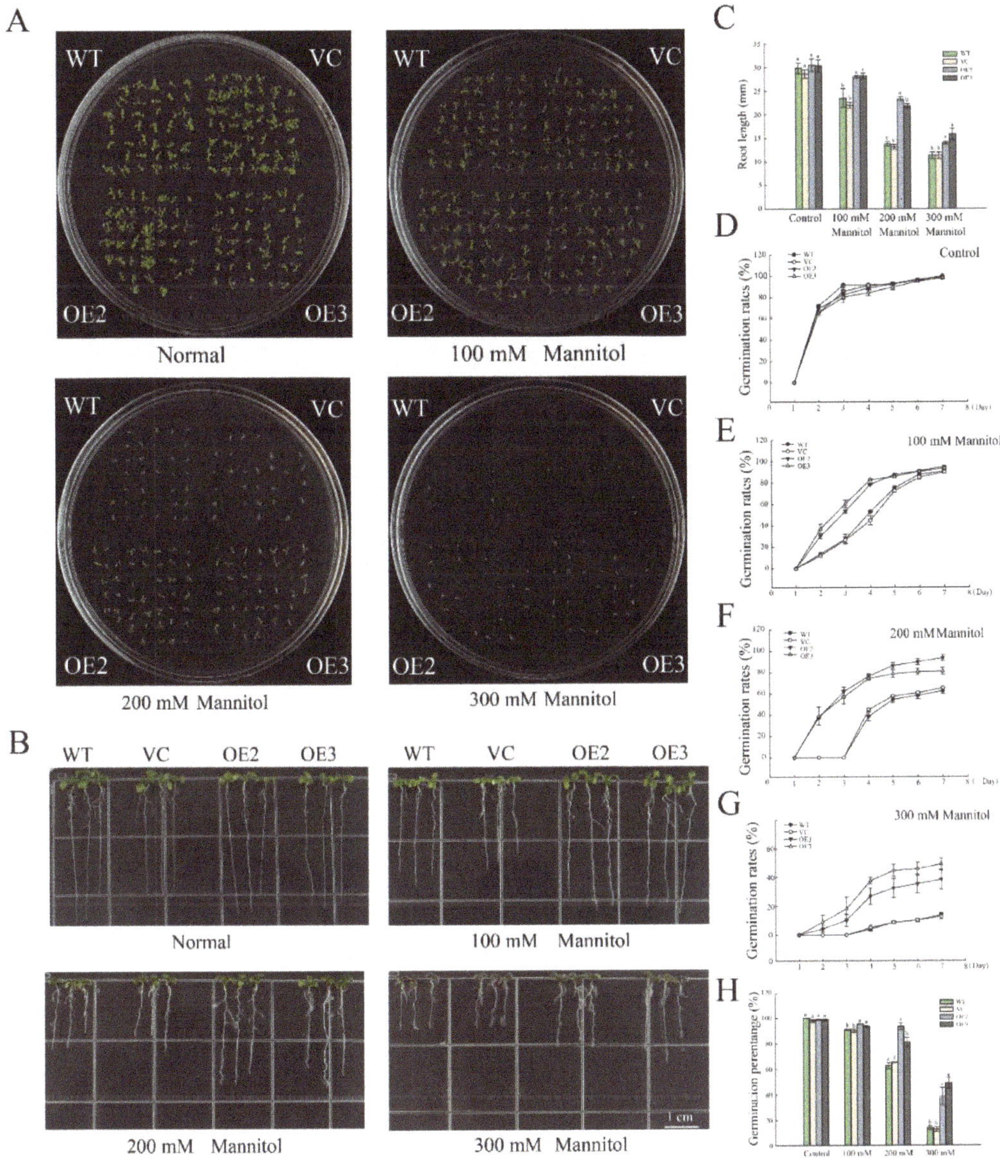

Figure 2. PwNAC11 overexpression promotes seed germination under simulated drought conditions. (**A**) Germination assay under different concentrations of mannitol treatments. (**B**) Root elongation assays under different concentrations of mannitol treatments. (**C**) Quantification of root length under different concentrations of mannitol treatments. Different letters indicate significant differences at p-value < 0.05. (**D–G**) Germination rates of different lines under different concentrations of mannitol treatments within 7 days. (**H**) Germination percentage of each line after 7-day simulated drought treatment. Different letters indicate significant differences at p-value < 0.05.

To analyze the role of PwNAC11 in drought stress during adult stage, 4-week-old seedlings were exposed to drought treatment. We found that the transgenic lines only wilted slightly and still maintained a normal growth state with a final survival percentage of 74% after drought treatment and followed by re-watering (Figure 3C), but the leaves of most of the WT and VC plants were wilted seriously or even dead, which were unable to recover after re-watering (Figure 3A). We also counted the relative water content (RWC) of each line. The RWC of the OE lines was approximately 50%, while it was only 25% for WT and VC lines (Figure 3B).

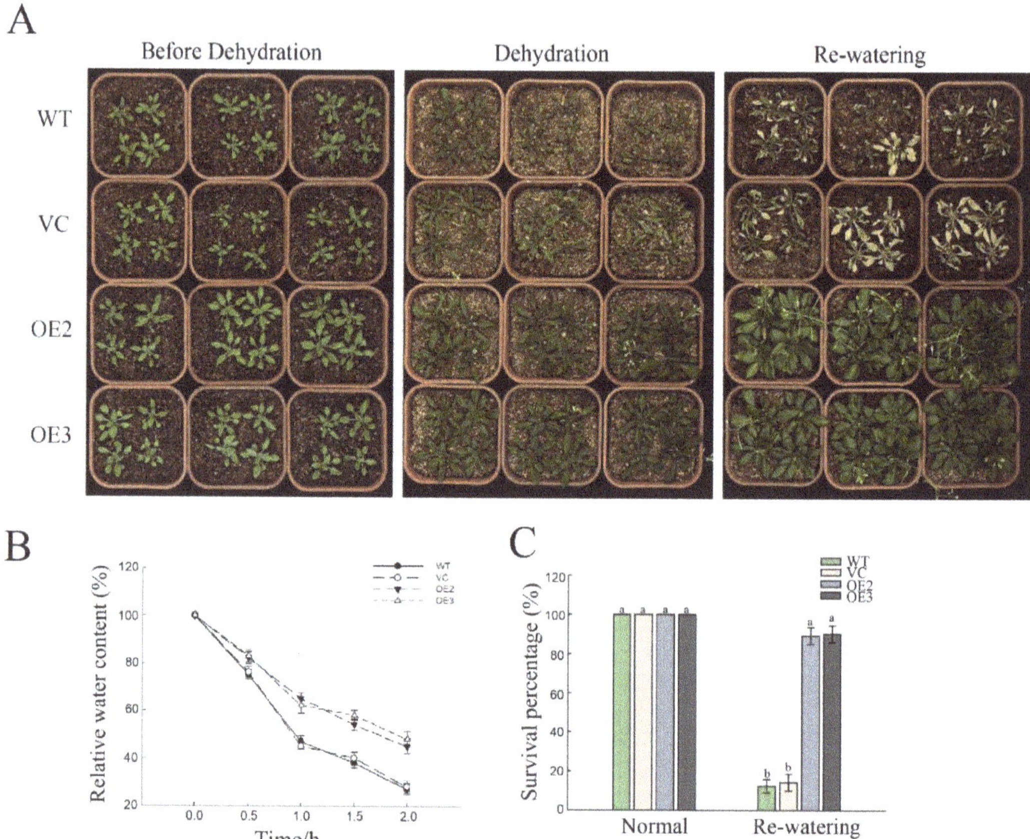

Figure 3. Overexpression of *PwNAC11* improves tolerance to drought stress. (**A**) Phenotypes of different lines under drought conditions. Dehydration lasted 14 days, followed by re-watering for 3 days. (**B**) Relative Water Content (RWC) after drought stress. Isolated leaves were placed at room temperature and dehydrated for 2 h. (**C**) Survival percentage before drought treatment and after re-watering. Different letters indicate the significant differences at $p < 0.05$.

2.3. PwNAC11 Is Involved in the Regulation of ROS Accumulation and Photosynthetic Efficiency under Drought Conditions

Drought stress has been reported to accelerate the accumulation of ROS. In order to investigate whether PwNAC11 promotes the degradation of superfluous hydrogen peroxide and superoxide in the leaves, we conducted the DAB and NBT staining. It was found that the leaves of each line were hardly stained under normal conditions. Nevertheless, once submitted to drought stress, the staining intensity of the leaves for WT and VC groups was substantially deeper than those of OE lines (Figure 4A,B), suggesting more ROS accumulated in WT and VC groups compared with OE lines. Further antioxidant enzymes activity assay indicated the activities of SOD and CAT increased apparently in two OE lines during drought treatments compared to normal conditions (Figure 4C,D). The severity of lipid oxidation was reflected in the MDA content. Upon drought treatment, the content of MDA in the OE lines was below 80 nmol/mg, which was 37% lower than that in WT and VC lines (Figure 4E). These results indicate that overexpression of *PwNAC11* significantly improved the scavenging abilities of ROS and contributed to the protection from peroxidation damage under drought stress. In addition, the decomposition rate of chlorophyll was reduced in transgenic lines (Figure 4F), suggesting that PwNAC11 probably participates in the regulation of the photosynthetic system.

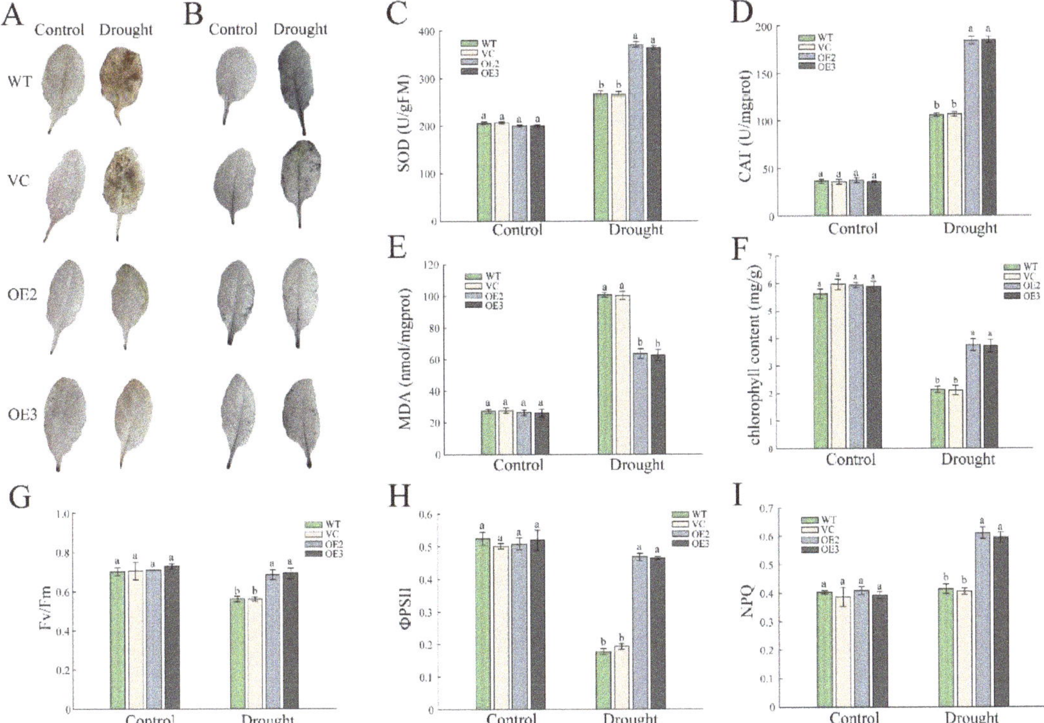

Figure 4. Overexpression of PwNAC11 enhances ROS scavenging ability in transgenic *Arabidopsis*. (**A**) Histochemical detection of hydrogen peroxide using DAB staining. (**B**) Histochemical detection of superoxide using NBT staining. (**C**) The activity of SOD. (**D**) The activity of CAT. (**E**) The content of MDA. (**F**) The content of chlorophyll. (**G**) Measurement of Fv/Fm under normal and drought conditions. (**H**) Measurement of ΦPSII under normal and drought conditions. (**I**) Measurement of NPQ under normal and drought conditions. Different letters indicate significant differences at p-value < 0.05.

Drought stress may also reduce the photosynthetic rate, leading to the decline in the efficiency of photosynthesis (Fv/Fm) and quantum yield of PSII electron transport (ΦPSII). In order to verify whether PwNAC11 is involved in the protection of plant photosynthesis under drought stress, several photosynthetic parameters were subsequently measured after 14 days of drought stress. It was observed that Fv/Fm fell by 22% in WT and VC after exposure to drought stress, but it was almost unchanged in the two OE lines (Figure 4G). Similarly, the ΦPSII decreased significantly by 65.4% in WT and only by 10% in OE lines (Figure 4H). On the contrary, nonphotochemical quenching (NPQ) increased dramatically in the OE lines under drought treatment and showed no significant difference in the WT and VC groups in the absence or presence of drought treatment (Figure 4I), suggesting that PwNAC11 confers increased photoprotection in OE lines.

2.4. Overexpression of PwNAC11 Activates the Expression of Stress-Responsive Genes

Our study found that overexpression of *PwNAC11* improved plant resistance to drought stress and can be greatly induced by ABA. To further quantify the expression levels of stress-responsive and ABA-induced genes of different lines, we selected such genes as *ANAC019*, *ANAC055*, *DREB2A*, *ERD1*, *ATHB-7* and *ABF3* for qRT-PCR analysis based on the reports on the regulation network of homologous genes of *PwNAC11* in *Arabidopsis* (Figure 5). Under simulated drought conditions, most of these genes showed higher expression levels in OE lines compared with the WT and VC groups. *ANAC055* and *ERD1* were even upregulated in normal conditions compared with WT and VC groups (Figure 5C,E). For *ERD1*, the expression was 12 times higher in OE lines than that in the other groups under PEG treatment, and *ATHB-7* showed more than a twenty-fold increase when subjected to 12 h PEG treatment (Figure 5F). These results suggested that PwNAC11 promoted the expression of stress-related and ABA-responsive genes, and thus, improved the stress tolerance in the plants.

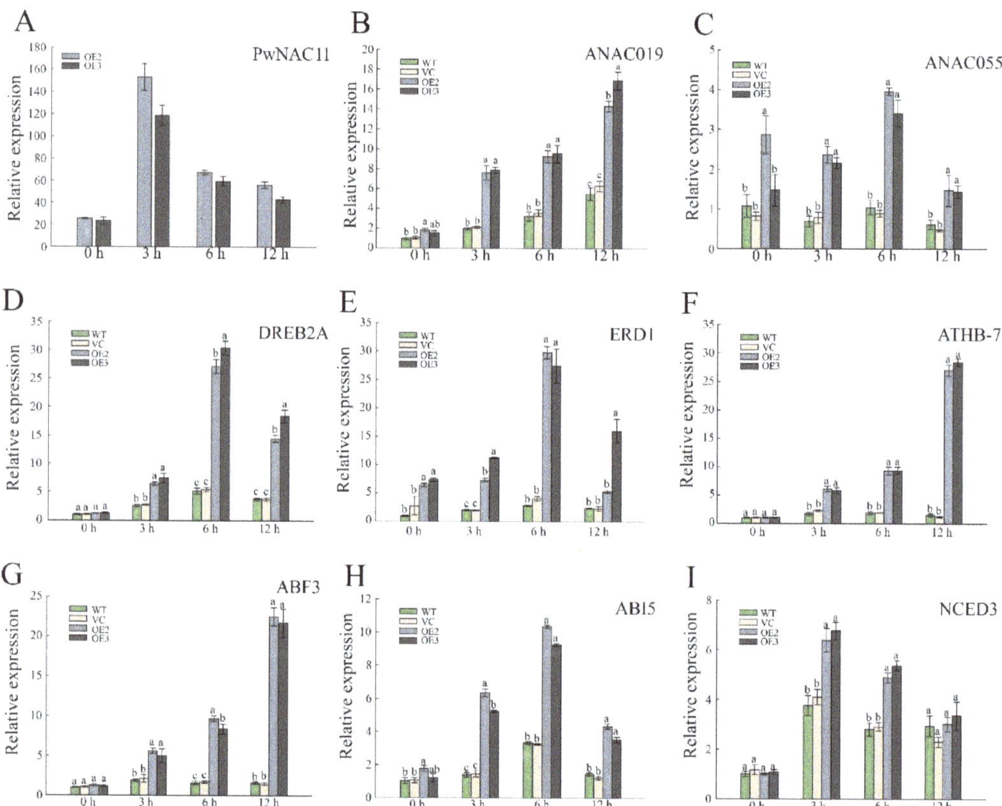

Figure 5. Expression profiles of some stress-responsive genes and ABA-responsive genes after PEG treatment. (**A–I**) Expression profiles of these genes in OE, WT, and VC lines under simulated drought conditions. The data are represented as means ± standard error from three independent replications. Different letters indicate significant differences at *p*-value < 0.05.

2.5. Overexpression of PwNAC11 Increases ABA Sensitivity and Promotes ABA-Induced Stomatal Closure in A. thaliana

Since the expression of *PwNAC11* was upregulated under exogenous ABA treatment in *P. wilsonii*, we speculate that PwNAC11 is responsive to the ABA signal. To prove this hypothesis, we conducted another phenotypic experiment by exposing the seeds of each line to different concentrations of exogenous ABA to test the response of PwNAC11 to the ABA signal. We found that the germination percentage of OE2 and OE3 decreased to 70% under 1 μM ABA treatment, while it was maintained at nearly 80% in the WT and VC lines (Figure 6A,C). The results of the root length experiment also proved that the root length of the OE lines was almost 5 mm shorter than that of the WT and VC lines, whether under 0.5 μM or 1 μM ABA treatment (Figure 6B,D).

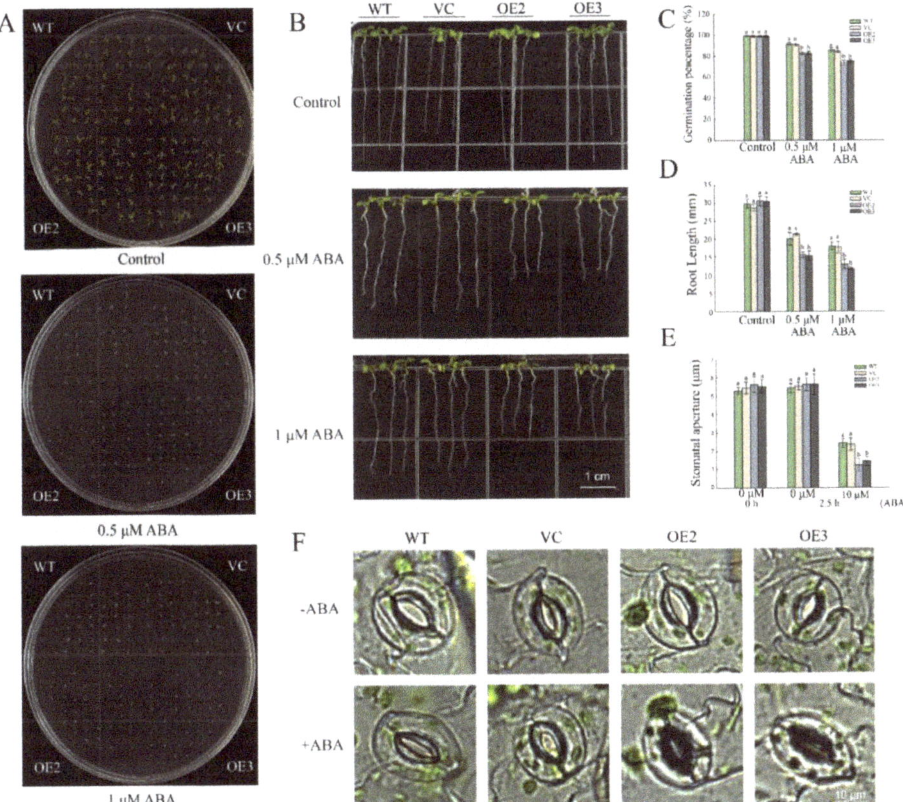

Figure 6. The stomatal aperture of *PwNAC11* overexpression lines was positively modulated by ABA. (**A**) Germination assays for different lines under different concentrations of ABA treatments. (**B**) Root length assays for different lines under different concentrations of ABA treatments. (**C**) Quantification of germination percentage under ABA treatments. (**D**) Quantification of root length under ABA treatments. (**E**) Quantitative comparisons of stomatal aperture calculated by ImageJ software. (**F**) The observation of stomatal aperture of OE, WT and VC lines photographed by biomicroscope (DM2500, Leica). Different letters indicate the significant differences at *p* < 0.05.

Recent studies indicated that increased ABA content could promote stomatal closure to improve the drought tolerance of plants [31]. We further observed and calculated the level of close stomata by exogenously applying ABA in different lines. The results showed that there was no distinct discrepancy under normal light conditions. Once adding 10 μM ABA, however, an acceleration of stomatal closure occurred in all plants, of which the OE lines exhibited a more obvious response to exogenous ABA treatment (Figure 6E,F). These results indicated that *PwNAC11* overexpression lines were more sensitive to exogenous ABA.

2.6. PwNAC11 Interacts with ABF3 and DREB2A

According to the prediction of the STRING and related reports of *PwNAC11* homologous genes in *Arabidopsis*, five proteins were selected for potential interaction verification (Figure 7A). The transcriptional activity results showed that only pGBKT7-ANAC019 could grow well on selective medium and activate the reporter gene expression, which implied that ANAC019 has transcriptional activity, but not for other chosen proteins (Figure S2). Subsequent yeast two-hybrid (Y2H) assays revealed that the yeast cells containing pGADT7-PwNAC11 + pGBKT7-ABF3 and pGADT7-PwNAC11 + pGBKT7-DREB2A grew well on SD/Trp-Leu-His-Ura selective medium (Figure 7B), suggesting that PwNAC11 can interact with ABF3 and DREB2A in yeast. Bimolecular fluorescence complementation (BiFC) assay further demonstrated that the YFP signals were observed in the nuclei of tobacco leaves co-expressing PwNAC11 and ABF3 or DREB2A (Figure 7C), which confirmed that PwNAC11 physically interacted with ABF3 or DREB2A in vivo.

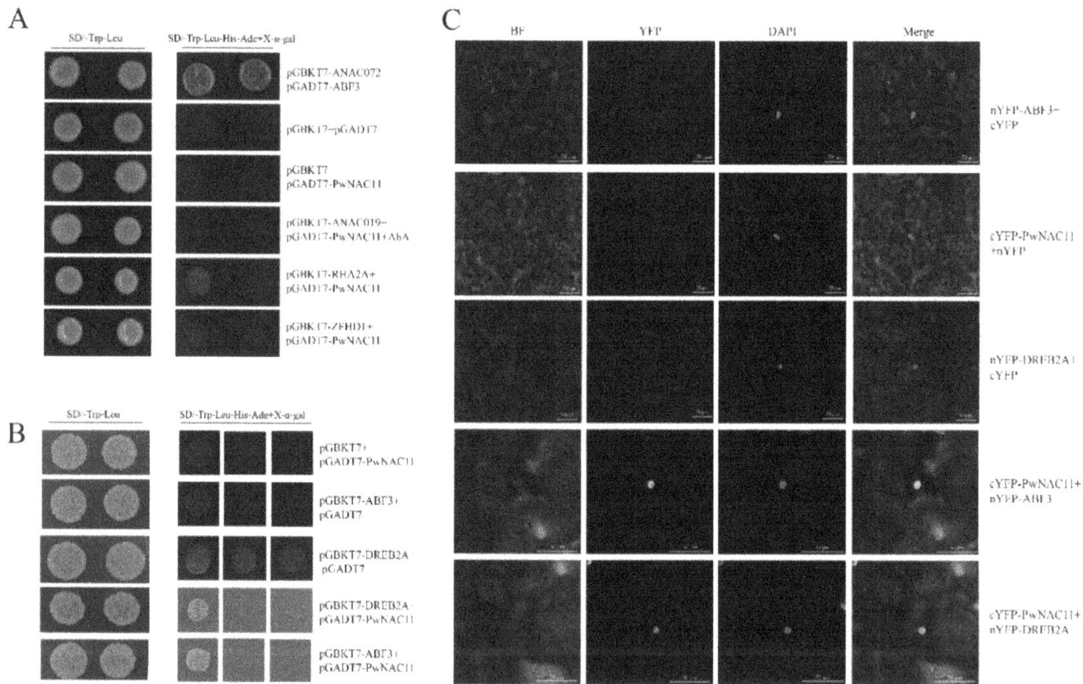

Figure 7. PwNAC11 interacts with ABF3 and DREB2A. (**A**) Yeast two-hybrid assays for PwNAC11 with ANAC019, RHA2A and ZFHD1. The yeast strains containing pGADT7-ABF3 and pGBKT7-ANAC072 fusion plasmids were used as positive control. Empty pGADT7 and pGBKT7 were used as negative control. (**B**) Yeast two-hybrid assays for PwNAC11 with ABF3 and DREB2A. (**C**) BiFC assays for PwNAC11 and ABF3/DREB2A interactions. Bars represent 50 μm. The C-terminal or N-terminal of YFP was used as negative control.

2.7. PwNAC11 and ABF3 Combine with the Promoter Region of ERD1

Considering that *ERD1* could be activated by *ANAC019/ANAC072* in *Arabidopsis* [32], and the expression of *ERD1* was prominently upregulated in *PwNAC11* OE lines, we subsequently conducted the yeast one-hybrid (Y1H) assays to confirm the regulation of

PwNAC11 on *ERD1* promoter region. In contrast with the negative control, both pGADT7-PwNAC11 + pAbAi-*ERD1*pro and pGADT7-ABF3 + pAbAi-*ERD1*pro yeast strains could grow on SD/Ura-Leu + AbA selective medium (Figure S3), suggesting both PwNAC11 and ABF3 can combine with the *ERD1* promoter (Figure 8A,B). As ABRE and DRE motifs were enriched in the promoter region of *ERD1*, we performed another Y1H assay to test whether PwNAC11 and ABF3 could bind to the specific cis-elements of *ERD1* promoter. The results showed that both PwNAC11 and ABF3 could specifically combine with ABRE motifs, but neither PwNAC11 nor ABF3 could bind to the DRE motifs (Figure 8C). These results implied that PwNAC11 and ABF3 synergistically regulate the *ERD1* expression by binding to the ABRE motifs. Furthermore, we performed dual-luciferase assays in transformed tobacco. As is shown in Figure 8D, the expression of PwNAC11 promoted the activation of reporter gene by 2.67-fold than empty vector in tobacco, while ABF3 limitedly activated the LUC expression. The simultaneous expression of PwNAC11 and ABF3 together enhanced the activation of *ERD1* promoter than PwNAC11 alone, which were 3.04 times higher than in the control group (Figure 8D). These results proved that PwNAC11 functions as a positive regulator of the expression of *ERD1*, and co-expression of PwNAC11 and ABF3 further improves this activation by binding to the ABRE motifs. Moreover, co-expression of PwNAC11 and DREB2A also activated the expression of *ERD1* promoter (Figure 8E), suggesting that the regulation of PwNAC11 on *ERD1* was also involved in an ABA-independent signaling pathway.

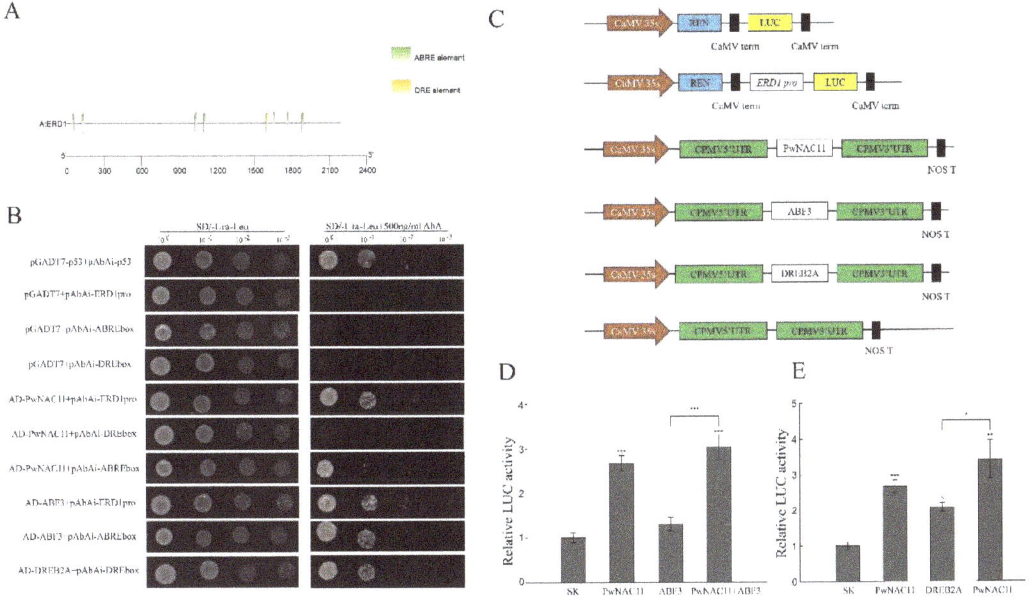

Figure 8. PwNAC11 and ABF3 cooperatively activates *ERD1* transcription. (**A**) Diagram of the *ERD1* promoter. The seven ABREs detected in *ERD1* promoter are indicated in green and one DRE is marked in yellow. (**B**) Yeast one-hybrid assays for the interaction of PwNAC11 with *ERD1* promoter, ABRE and DRE element. Aureobasidin A (AbA) of 500ng/mL was used to inhibit autoactivation. pGADT7-p53 and pAbAi-p53 were used as positive control. Empty PGADT7 was used as negative control. (**C**) Sketch map of the double-effector and reporter plasmids in dual-luciferase reporter assay. (**D**) Dual-luciferase assays for PwNAC11 and ABF3 interactions. (**E**) Dual-luciferase assays for PwNAC11 and DREB2A. The data are represented as means ± standard error from three independent replications. The grey lines indicate the comparisons and asterisks indicate significant differences compared with SK (* $p < 0.05$; ** $p < 0.01$; *** $p < 0.001$).

2.8. PwNAC11 Overexpression Delays Flowering

To test whether overexpression of *PwNAC11* affected the growth and development of plants, some relative parameters were calculated. Under nonstress conditions, the rosette leaf morphology, plant height and pod length of different lines showed almost no differences (Figure S4A,B,D,E). However, the OE lines showed a 4–5-day delayed flowering phenotype compared with the WT and VC groups (Figure S4C). Meanwhile, we also examined the expression of some flowering-related genes by qRT-PCR and found that *LFY*, *FY* and *AP1* were significantly downregulated in 20-day-old transgenic *Arabidopsis* (Figure S4G–I). These results indicated that the delayed flowering phenotype was associated with the suppressed expression of flowering-related genes in *PwNAC11* overexpression lines.

3. Discussion

3.1. PwNAC11 Positively Regulates Drought Resistance in Transgenic Arabidopsis

For most woody plants, it is still a great challenge to conducting functional verification in a homologous way via establishing transformation platforms [29]. Alternatively, we can still explore the gene function by heterogeneously transforming model plants. NAC transcription factors are reported to be involved in abiotic stress responses in various plant species, including model plants, crops and woody plants. Our previous work revealed that *PwNAC2*, cloned from *P. wilsonii*, could interact with PwRFCP1 to positively regulate abiotic stress response and enhance drought and salt resistance in transgenic *Arabidopsis* [28]. Another NAC, TF PwNAC30, acted as a negative regulator in plant responses to abiotic stress. Overexpression of *PwNAC30* apparently reduced the survival rates of transgenic *Arabidopsis* when subjected to drought and salt treatment [29], suggesting the distinct role and multiple function of different members of the NAC family from *P. wilsonii* in response to abiotic stress. Here, based on the high-throughput RNA-seq of the expression profiles of *P. wilsonii* in the absence or presence of drought treatments [30], we identified PwNAC11 as one of the most differentially expressed genes (DEGs) on a transcriptional level, which is a homologue of *Arabidopsis* NAC transcription factors ANAC019 and ANAC072 (RD26). Protein multiple sequence alignment and BLAST analysis showed that the conserved N-terminal of PwNAC11 protein had a typical NAM domain and its C-terminal was comparatively variable. Both the full length and C-terminal regions of PwNAC11 have transcriptional activation activity (Figure 1). Subcellular localization demonstrated that PwNAC11 was a nuclear-located transcriptional activator, which was consistent with most NAC transcription factors [33]. The transcript level of *PwNAC11* displayed a great increase at 3 h when *P. wilsonii* seedlings were exposed to drought stress, indicating its potential role in early response to abiotic stress. Furthermore, overexpression of *PwNAC11* in *Arabidopsis* considerably improved drought stress tolerance both for seedlings and adult plants. Under PEG treatment, the growth of the tested seedlings was inhibited to some extent, but the OE lines showed significantly higher germination rates, stronger vitality and longer root length compared with the WT and VC lines. Moreover, the survival rates of the OE lines were more than 60% higher than the WT and VC groups at the adult stage after dehydration followed by re-watering. Additionally, the RWC of the adult OE plants was nearly 20% higher than the control group (Figure 3). These results demonstrated that PwNAC11 functions as a positive regulator in response to drought stress throughout plants' growth and development stages.

It is well known that drought can cause the accumulation of large amounts of ROS (mainly H_2O_2 and O_2^-) in chloroplasts and mitochondria [34]. The excessive accumulation of ROS can cause serious peroxidation damage to plant cell membranes. Therefore, ROS eliminating efficiency is an essential indicator of plant resistance to drought stress. In our study, the NBT and DAB staining for H_2O_2 and O_2^- assays proved that ROS was slightly accumulated in the OE lines but accumulated extensively in WT and VC lines (Figure 4). Further measurement of enzymatic activities indicated that the activity of CAT, POD and SOD were upregulated in the OE lines compared to the lesser accumulation of MDA. These results indicated that PwNAC11 inhibits membrane lipid peroxidation via activating the ROS-scavenging system and decreasing the accumulation of ROS in resistance to drought stress.

Chlorophyll fluorescence is another important parameter reflecting the function of Photosystem II (PSII) and the transfer efficiency of electrons from PSII to PSI [35]. The Fv/Fm ratio and ΦPSII value represent the potential maximum light energy conversion efficiency and actual quantum yield of PSII, respectively [36,37]. However, the delicate photosynthetic system is frequently affected by the external environment, especially drought stress, which can decrease the Fv/Fm ratio and ΦPSII value. In our study, despite these indicators of different samples showing a declining trend under drought conditions, the *PwNAC11* OE lines displayed higher Fv/Fm, ΦPSII and chlorophyll content compared with the WT and VC groups. Additionally, significantly, the values of ΦPSII in transgenic lines remained almost unchanged whether in the presence or absence of drought treatment, suggesting the minimizing damage to the photosynthesis apparatus of PSII in *PwNAC11* OE lines and *PwNAC11* can stabilize photosynthetic energy utilization under drought conditions. Meanwhile, the NPQ, an indicator of plant to disperse excess light energy in the form of heat to protect itself under adverse conditions, showed no significant differences under normal and dehydration conditions in control groups, while *PwNAC11* OE lines showed increased NPQ values after drought treatment. These results verified the complete network of photoprotective mechanism in OE lines [38]. Recent studies revealed that the inhibition of light energy conversion could produce more excess excitation energy and promote the accumulation of ROS in plant [39]. In grape, Sun et al. found that the inhibition of Alternative oxidase (AOX) pathway increased the accumulation of H_2O_2, thus resulting in the photoinhibition under heat conditions [40]. Overexpression of Arabidopsis *AtCBF3* enabled transgenic potatoes to possess a favorable ROS removing ability, which decreased the degree of photoinhibition and enhanced heat resistance [41]. Therefore, we conclude that the efficient ROS-scavenging system in *PwNAC11* OE lines may partly contribute to the stable light energy conversion efficiency under drought conditions.

Many NAC TFs involved in stress response were reported to have an effect on plant development or flowering. For example, EsNAC1, an NAC TF isolated from *E. salsugineum*, was proven as a positive regulator of salt and oxidative stress but inhibited the process of vegetative growth [42]. Additionally, MLNAC5, a stress-related transcription factor from *Miscanthus*, improved drought and cold tolerance in transgenic *Arabidopsis* but inhibited the growth of the plant and accelerated the process of leaf senescence [43]. The overexpression of sugarcane NAC transcription factor *ScNAC23* in *Arabidopsis* accelerated flowering compared with WT plants [44]. Our previous study also revealed that another *P. wilsonii* transcription factor, PwNAC2, enhances drought and salt tolerance in plants and can interact with PwRFCP1 to co-regulate flowering via regulating the expression of *SOC1*, *FLC* and *FT* [28]. These results indicate that the different NAC transcription factors function differentially, and some of them can balance and coordinate the regulation of plant growth and flowering and coping with adverse stress. In this study, a late flowering phenotype was discovered in 35S: *PwNAC11* OE lines. Further investigation demonstrated that PwNAC11 delayed flowering time by inhibiting the expression of genes such as *LYF*, *FT* and *AP1*. Interestingly, the growth retardation disappeared when the plants were subjected to drought conditions (Figure 3A). Previous research showed that under non-stress conditions the growth inhibition phenotype was partly affected by the constitutive expression of 35S promoter [45]. Under normal conditions, for example, ectopic expression of *35S: AtDREB1A* in *Salvia miltiorrhiza* inhibited plant growth but the overexpression of stress-inducible *RD29A: AtDREB1A* caused no growth suppression. Therefore, we cannot exclude the effect of 35S promoter on flowering time in *PwNAC11* transgenic *Arabidopsis*. To further determine whether the stunted growth phenotype was mainly caused by functional *PwNAC11* expression or 35S promoter, knock-out studies are required to verify this hypothesis and decipher the role of PwNAC11 on flowering regulation in *P. wilsonii* in the future. In addition, considering most transcription factors are simultaneously involved in plant growth, development and response to stress, the stress-induced promoters or self-gene promoters are probably recommended for plant gene transformation to reduce the impact on the development process of the plant itself.

3.2. PwNAC11 Improves Drought Tolerance in an ABA-Dependent Manner

Plants' responses to drought stress are usually controlled by hormonal signals, in which ABA acts as a mediator to regulate a range of processes in stress-signaling pathways [46]. CmBBX19, a zinc finger protein, suppressed drought resistance in chrysanthemum in an ABA-dependent way [47]. At a translational level, alternative splicing of rice NAC transcription factor ONAC054 led to premature termination of translation in response to exogenous ABA, thus producing more ABA-responsive ONAC054β variants to positively modulate the homeostasis [48]. In our study, expression of *PwNAC11* was greatly upregulated especially at 3 h after ABA treatment. *PwNAC11* overexpression lines conferred hypersensitivity to ABA at an earlier phase than the WT and VC groups. Meanwhile, some ABA synthesis genes, such as *NCED3* and *ABI5*, showed transcriptional abundance after exposure to PEG. These results implied that PwNAC11 is likely to participate in water-deficit regulation in an ABA-dependent pathway. In addition, previous investigations revealed that most TFs were involved in the regulation of stomata closure under the control of ABA signals. The heterologous expression of *AlNAC1* in *Arabidopsis* increased the endogenous ABA level, thus resulting in an enhanced percentage of close stomata and reduced water loss [49]. CsMYB61, a MYB TF from *Citrus sinensis*, was implicated in the process of stomatal closure. Overexpression of *CsMYB61* driven by a stomata-specific promoter conferred a repressed opening of stomata pores, especially under the treatment of exogenous ABA [50]. Similarly, *AtMYB61* transgenic *Arabidopsis* displayed a higher percentage of closed stomata under drought treatment [51]. Here, we found that PwNAC11 acted in a similar way to regulate ABA-induced stomatal closure. Overexpression of *PwNAC11* transcript in *Arabidopsis* led to a significant increase in the number of close stomata (Figure 6). It was reported that stomatal closure is generally triggered by drought-induced hormone ABA as well as the application of exogenous ABA [52,53]. Our results also demonstrate that PwNAC11 functions as an ABA signal factor by promoting stomatal closure and ABA synthesis to directly or indirectly improve plants' tolerance to drought stress.

The ABF TFs are considered as pivotal recipients of ABA signaling and function downstream of various ABA-mediated stress response, especially drought and salt stresses [54,55]. *ABI5*, an important ABA synthesis gene, is governed by AtABF3 in *Arabidopsis*, which could enhance salt stress tolerance by activating the ABA signal pathway [56]. In *Populus euphratica*, PeABF3 could activate the expression of *PeADF5* to enhance drought tolerance and promote ABA-induced stomatal closure [57]. Although different ABFs exhibited high levels of homology, they could also work antagonistically in definite conditions to generate a negative feedback effect. Pear ABF protein PpyABF3 can activate *PpyDAM3* expression by binding to the ABRE motif on *PpyDAM3* promoter, while the binding affinity of PpyABF3 to *PpyDAM3* promoter was disturbed by interacting with PpyABF2, thus revealing the antagonistic regulatory network of ABFs towards downstream genes [58]. In the present study, we found that PwNAC11 could physically interact with AtABF3 by Y2H and BiFC assay, which is consistent with another homologous gene, *ANAC072*, in *Arabidopsis*. ANAC072 participated in the ABA-responsive signaling pathway by interacting with ABF3 [59]. The data showed the conservation of function and further proves that NAC TFs could perform similar functions in both woody plants and *Arabidopsis*. However, ANAC072 was reported to promote leaf senescence and accelerate chlorophyll degradation, which was not observed in *PwNAC11* transgenic lines.

3.3. The Regulatory Networks of ABA-Induced Drought Resistance Mediated by PwNAC11

TF acts as a transcriptional activator/repressor by specifically binding to cis-acting elements in the promoters of target genes. Accordingly, it will enable us to uncover the unknown regulatory mechanisms by identifying the relative target genes. *ERD1* encoded a homologous protein of ATP binding subunit in *Escherichia coli* [13], and the expression of *ERD1* was much upregulated by abiotic stress or dark-induced senescence. In *Arabidopsis*, a zinc finger homeodomain TF (ZFHD1) could enhance the expression of *ERD1* by binding to its promoter region, which contributed to an obvious improvement of drought stress endurance in the *ZFHD1* overexpression lines [60]. Likewise, we found that PwNAC11 bound to the promoter of *ERD1* and activated *ERD1* expression. RT-qPCR analysis also showed that the expression of *PwNAC11* reached its peak at 3 h after PEG treatment, while at 6 h for *ERD1*. We next performed a Y1H assay to prove that PwNAC11 could specifically bind to ABRE motifs instead of DRE motifs. These results provide genetic evidence that PwNAC11 functions as a transcriptional activator upstream of *ERD1*, leading to enhanced dehydration resistance in an ABA-dependent manner. In addition, our results also confirmed the interaction between PwNAC11 and ABF3 (Figure 7), both of which could bind to the ABRE elements (Figure 8), but whether PwNAC11 and ABF3 simultaneously bind to the same ABRE element on ERD1 promoter, and whether the interaction of PwNAC11 and ABF3 interferes with the binding affinity to target genes remains to be further investigated. Nevertheless, the expression of PwNAC11 promoted the activation of the reporter gene by 2.67-fold greater than empty vector in tobacco, while ABF3 limitedly activated LUC expression. However, the simultaneous expression of PwNAC11 and ABF3 together enhanced the activation of the *ERD1* promoter more than PwNAC11 alone (Figure 8D). These results showed that PwNAC11 positively regulates plant response to early drought stress and the interaction of PwNAC11 and ABF3 further improves this activation by binding to the ABRE motifs. Generally, ABF3 can interact physically with diverse components to promote or inhibit the expression of downstream targets, especially stress-responsive genes. For example, in *Arabidopsis*, *AtSOC1* was ubiquitously expressed in different vegetative tissues and accelerated flowering. *AtABF3* could form a complex with NF-Y TF and activated the expression of *AtSOC1*, and this phenotype was totally abolished in *nf-yc3 yc4 yc9* mutants [61]. Similarly, the physical interaction between CmBBX19 and CmABF3 in *Chrysanthemum* withheld the activation of *CmRAB18*, resulting in decreased tolerance to drought stress [47].

However, our results by no means suggest the minimal role of ABA-independent pathway present in PwNAC11-mediated drought stress response, since DRE-BINDING PROTEIN 2A (DREB2A) also physically interacted with PwNAC11 by Y2H and BiFC assay in this study (Figure 7). It was noteworthy that dehydration-responsive element (DRE) also exists in the promoter of *ERD1*, which is recognized as an essential cis-element for ABA-independent regulation in response to dehydration and heat stresses [62]. In woody plants, the expression of *BpDREB2*, a transcription factor of DREB protein family in *Broussonetia papyrifera*, was remarkably increased under salt and dehydration conditions, but no obvious change was observed under ABA treatment. Importantly, overexpression of *BpDREB2* in *Arabidopsis* improved its salt and freezing tolerance [63]. Similar results were found in *Jatropha curcas*. The expression patterns of *JcDREB* showed that it was induced by cold, drought and salt stresses, not by ABA signal [64]. However, some reports have also suggested that DREB transcription factors can be mediated in both ABA-dependent and ABA-independent pathways in response to abiotic stress. In *Leymus chinensis*, the transcript of *LcDREB3a* accumulated in response to ABA treatments and the protein was shown to bind to DRE motifs, indicating that *LcDREB3a* was involved in both ABA-dependent and ABA-independent signal transduction in the process of abiotic stress response. Here, we also found that PwNAC11 remarkably improved the transcription activity of *ERD1* promoter by interacting with DREB2A rather than directly binding to the DRE motif. We thus speculate that the drought response of *ERD1* is also modulated in an indirect ABA-independent approach, in which PwNAC11 may also participate. This assertion is also supported by the apparent up-regulation of *DREB2A* in transgenic *Arabidopsis*, whose expression levels are less dependent on ABA signals [65,66].

In addition, combined with the data in our study that PwNAC11 was dramatically induced at 3 h after PEG treatment, while ABF3 highly expressed at 12 h and DREB2A at 6 h, we speculated that PwNAC11 plays a dominant role in positively regulating plants' responses to early drought stress, and ABF3 or DREB2A were subsequently induced and synergistically regulate the *ERD1* expression. Based on these results and the function of homologous genes, we sketched the regulatory mechanism and linkage of ABA signals and the drought response pathway mediated by PwNAC11 (Figure 9). First, PwNAC11 functions to improve the expression of ABF3 or DREB2A at transcriptional-level control under drought stress, then the protein interaction between PwNAC11 and ABF3 strengthens the transcription of target genes by activating ABRE-containing promoters or indirectly activating DRE-containing promoters by interaction with DREB2A. The existence of such synergistic pathways could help plants to respond to environmental stress efficiently and rapidly, but it is necessary to further explore how precise networks PwNAC11 participate or contribute to the drought stress response via ABA-dependent or -independent pathways, which will provide further information on the drought tolerance mechanism of *Picea wilsonii*.

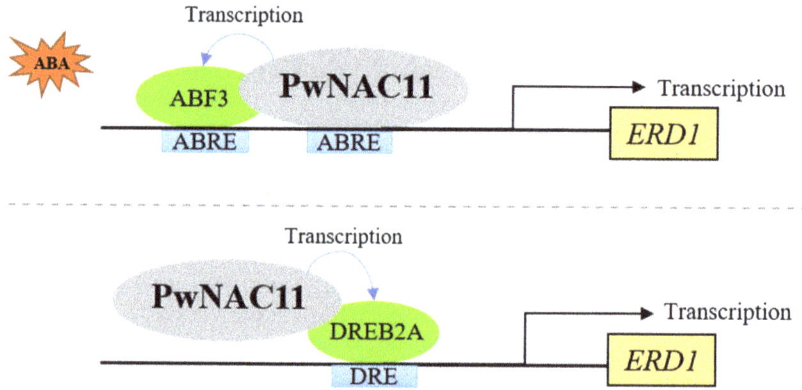

Figure 9. Proposed model of ABA-dependent and ABA-independent abiotic stress responses via PwNAC11.

4. Materials and Methods

4.1. Plant Materials and Gene Expression Analysis

For RT-qPCR assay, the 4-week-old *P. wilsonii* seedlings grown on nutrient soil were watered with 20% polyethylene glycol (PEG) or treated with 100 µM ABA solution and the 2-week-old *Arabidopsis* seedlings were watered with 10% PEG solution. The whole plants were then harvested at 0, 3, 6 and 12 h after treatment for RNA extraction. Total RNA was extracted from *P. wilsonii* or *Arabidopsis* with plant RNA extraction kit (Kangwei Century Biotechnology Co., Ltd., Beijing China) and first strand cDNA synthesis kit (ABM, Beijing, China), followed by RT-qPCR analysis. Step One plus Real-time PCR system (ABI, Vernon, CA, USA) was used to perform RT-qPCR reactions with SYBR Green Master Mix enzymes (ABI, Vernon, CA, USA). The 2−∆∆CT method was employed to calculate expression levels and all experiments were repeated 3 times for biological replications and 3 times for technical replications. *PwEF-1α* in *P. wilsonii* and *ACTIN* in *Arabidopsis* were selected as housekeeping genes in qRT-PCR assay, respectively.

Before sowing in the MS medium, all seeds were sterilized by 5% NaClO solution for 20 min. After 3 days dark treatment, the seeds were then placed in a 21 °C incubator under a 16-h light/8-h dark photoperiod at 70% relative humidity. At 14-days-old, plants were transferred into soil.

4.2. Bioinformatic Analysis

The sequences of PwNAC11 homologous proteins were obtained from NCBI database (https://www.ncbi.nlm.nih.gov/ (accessed on 1st November 2018)). Multiple sequence alignments of 12 NAC amino acid sequences were analyzed using Clustal X 1.83. The neighbor-joining method was subsequently used to construct the phylogenetic tree via MEGA-X (Mega Limited, Auckland, New Zealand) program.

4.3. Plant Phenotype Experiments under Drought and ABA Treatments

Seeds of *Arabidopsis* have Col-0 genetic background. To generate transgenic plants, full length of *PwNAC11* CDS was cloned into pCAMBIA1205 vector, which was driven by 35S promotor [28]. The agrobacterium stains GV3101 containing PwNAC11-pCAMBIA1205 were subsequently transformed into *Arabidopsis* using floral dipping. Stable homozygous lines were selected by hygromycin resistance and all T3 seedings could survive on selective medium, in which two dependent lines named OE-2 and OE-3, with relatively higher expression levels, were chosen as experimental samples.

The WT and VC lines were set as controls and OE-2 and OE-3 lines were chosen for phenotype analysis. For the seed germination experiment, the disinfected seeds were sown on MS solid medium (100 seeds were sown for each line) containing various concentration of ABA (0, 0.5, 1.0 μM) and mannitol (0, 100, 200, 300 mM), respectively. After placing in a refrigerator at 4 °C for 3 days, the mediums were then transferred into an incubator as described above. The germination number was recorded once a day and the germination criterion was the exposure of radicle. For the root length experiment, the seeds of each line were sown in MS medium for germination and then transferred to MS medium supplemented with different concentrations of mannitol and ABA. For seedlings treated with drought stress, all of the lines were sown on MS medium for 2 weeks and then transferred to soil under normal conditions for another 14 days. For drought stress, the 4-week-old seedlings were dehydrated for 14 days and then re-watered for 3 days. Phenotypic differences were observed after 14 days of drought treatment and re-watering. Each experiment was repeated three times.

4.4. Subcellular Localization Assay

The CDS of *PwNAC11* was constructed on pCAMBIA1205 vector, which was controlled by CaMV35S promotor. The RACK1A-RFP plasmid, a nuclear-located marker [29], and recombinant vector were transfected into *Agrobacterium tumefaciens* GV3101 cells, respectively, and then co-infiltrated into *Nicotiana benthamiana* leaves, which were then placed in the incubator at 25 °C (day)/23 °C (night) for 48–72 h. The TCS SP8 fluorescence microscope (Leica, Wetzlar, Germany) was used to detect GFP fluorescence. Three infiltrated leaves from two *N. benthamiana* were regarded as three biological repeats for each experiment.

4.5. Trans-Acting Activity and YEAST Two-Hybrid (Y2H) Assay

The complete *PwNAC11* CDS and truncated N-terminal (pGBKT7-PwNAC11-1-153) or C-terminal (pGBKT7-PwNAC11-171-318) sequences of *PwNAC31* were cloned into pGBKT7 vector, respectively. The assay was performed as previously described with modification [29]. Briefly, the plasmids were separately transformed into yeast AH109 cells with empty pGBKT7 vector as negative control. The yeast strains were firstly painted on SD/-Trp medium and then transformed into SD/-Trp-His-Ade+X-α-gal selective medium.

For Y2H assay, similar transcriptional activity assays were conducted. The CDS of *ABF3*, *ZFHD1*, *RHA2A* or *ANAC019* were cloned to pGBKT7 vector as the prey and recombinant vectors were then selected on SD/-Trp-His-Ade+X-α-gal medium. Full-length CDS of PwNAC11 was correspondingly constructed on pGADT7 vector as the bait. The recombinant AD and BD vectors were simultaneously inserted into yeast AH109 cells and selected on SD/-Trp-Ura-His-Ade+X-α-gal medium.

4.6. Yeast One-Hybrid (Y1H) Assay

Y1H assay was accomplished with the Matchmaker Gold Yeast One-Hybrid System Kit (TaKaRa, Beijing, China). Briefly, about 2 kb *ERD1* promoter region and short tandem repeats containing core ABRE (ACGTG) and DRE (GTCGGC) motifs were constructed on pAbAi vector. The pGADT7-PwNAC11 or pGADT7-ABF3 vectors were transformed into competent cells of yeast containing the *ERD1* pro sequences. The recombinant vectors were selected on SD/-Ura-Leu+ AbA medium.

4.7. Dual-Luciferase Assay

Dual-luciferase assay was performed as previously described [67]. Briefly, the *ERD1* promoter was cloned into pGreenII 0800-LUC and the CDS of PwNAC11, ABF3 or DREB2A was cloned into pGreenII 62-SK vector. The transfected *Nicotiana benthamiana* leaves were detected by Dual Luciferase Reporter Gene Assay Kit (Beyotime, Shanghai, China). The firefly luciferase (fLUC) and renilla luciferase (rLUC) were detected by Multiscan Spectrum (Inginite M Plex, TECAN). The empty pGreenII 0800-LUC vector was used as control and each experiment was repeated three times.

4.8. Bimolecular Fluorescence Complementation Assay (BiFC)

The complete CDS of *PwNAC11* was cloned into pSPYCE vector to generate PwNAC11-cYFP fusion protein and coding region of *ABF3* was similarly cloned into pSPYNE vector. The recombinant vectors were transformed into agrobacterium strains GV3101 and then co-infiltrated into *Nicotiana benthamiana* leaves as mentioned above. The GFP signal was observed through Leica TCS SP8 fluorescence microscope.

4.9. Physiological Measurement

The Diaminobenzidine (DAB) and Nitro Blue Tetrazolium (NBT) staining assays were conducted as previously described [68]. NBT Chromogen and Metal Enhanced DAB Substrate Kit were employed in this experiment (Solarbio, Beijing, China). Briefly, the leaves of different lines were immersed in the staining solution for 2 h in darkness. Then, the staining solution was replaced by 95% alcohol and the leaves were boiled until decolorization. The catalase, antioxidant enzymes' activities and proline content were measured by CAT, SOD, POD and Pro detection kits (Jiancheng Bioengineering Institute, Nanjing, China). The chlorophyll content was quantified by spectrometer as previously described [69]. The chlorophyll and fluorescence parameters of *Arabidopsis* leaves under drought stress were measured by handy portable fluorometer (PAM-2500, WALZ, Germany), and the fluorescence parameters were calculated as previously described [35]. Some parameters were calculated as follows: NPQ = Fm/Fm' − 1, Fv/Fm = (Fm − Fo)/Fm, ΦPSII = (Fm' − F)/Fm', in which Fo is the minimum recorded fluorescence intensity in the dark-adapted states. Fm and Fm' are the maximal recorded fluorescence intensity in the dark and light-adapted states, respectively.

4.10. Stomatal Aperture Measurements

Stomatal apertures were measured as described previously [70]. Leaves of *Arabidopsis* were pre-incubated in MES-KCl buffer for 2.5 h under normal light conditions, followed by MES-KCl buffer treatment alone or with 10 μM ABA for another 3 h. Stomatal pores of OE, WT and VC lines were photographed by light microscope DM2500 (Leica, Wetzlar, Germany).

4.11. Statistical Analysis

SPSS software version 18.0 (SPSS Corp, Chicago, IL, USA) was used for statistical analysis. Duncan's multiple range test was performed for statistical difference analysis with $p < 0.05$ designated as significant difference.

Supplementary Materials: Supplementary Materials can be found at https://www.mdpi.com/article/10.3390/ijms22136952/s1.

Author Contributions: L.Z. conceived the project and L.Z. and M.Y. designed the experiments; J.L. collected the samples; M.Y., J.L., B.D., A.W. and M.Z. performed the experiments; M.Y. conducted the data analysis and drafted the manuscript; L.Z. revised the manuscript. All authors have read and agreed to the published version of the manuscript.

Funding: This research was funded by a grant from the Agricultural Ministry of China, grant number 2016ZX08009-003.

Institutional Review Board Statement: Not applicable.

Informed Consent Statement: Not applicable.

Data Availability Statement: No new data were created or analyzed in this study. Data sharing is not applicable to this article.

Acknowledgments: We thank Dapeng Zhang (College of Life Sciences, Tsinghua University) for providing the pGreenII 62-SK and pGreenII 0800-LUC vector.

Conflicts of Interest: The authors declare that they have no conflict of interest.

References

1. Guo, Y.Y.; Yu, H.Y.; Kong, D.S.; Yan, F.; Zhang, Y.J. Effects of drought stress on growth and chlorophyll fluorescence of *Lycium ruthenicum Murr.* seedlings. *Photosynthetica* **2016**, *54*, 524–531. [CrossRef]
2. Zhao, P.; Hou, S.; Guo, X.; Jia, J.; Yang, W.; Liu, Z.; Chen, S.; Li, X.; Qi, D.; Liu, G.; et al. A MYB-related transcription factor from sheepgrass, LcMYB2, promotes seed germination and root growth under drought stress. *BMC Plant Biol.* **2019**, *19*, 954. [CrossRef]
3. Du, H.; Huang, F.; Wu, N.; Li, X.; Hu, H.; Xiong, L. Integrative Regulation of Drought Escape through ABA-Dependent and -Independent Pathways in Rice. *Mol. Plant* **2018**, *11*, 584–597. [CrossRef] [PubMed]
4. Yu, Y.; Wang, L.; Chen, J.; Liu, Z.; Park, C.-M.; Xiang, F. WRKY71 Acts Antagonistically Against Salt-Delayed Flowering in Arabidopsis thaliana. *Plant Cell Physiol.* **2017**, *59*, 414–422. [CrossRef]
5. Blum, A. Effective use of water (EUW) and not water-use efficiency (WUE) is the target of crop yield improvement under drought stress. *Field Crop. Res.* **2009**, *112*, 119–123. [CrossRef]
6. Olsen, A.N.; Ernst, H.A.; Leggio, L.L.; Skriver, K. NAC transcription factors: Structurally distinct, functionally diverse. *Trends Plant Sci.* **2005**, *10*, 79–87. [CrossRef]
7. Puranik, S.; Sahu, P.P.; Srivastava, P.S.; Prasad, M. NAC proteins: Regulation and role in stress tolerance. *Trends Plant Sci.* **2012**, *17*, 369–381. [CrossRef] [PubMed]
8. Ooka, H.; Satoh, K.; Doi, K.; Nagata, T.; Otomo, Y.; Murakami, K.; Matsubara, K.; Osato, N.; Kawai, J.; Carninci, P.; et al. Comprehensive Analysis of NAC Family Genes in Oryza sativa and Arabidopsis thaliana. *DNA Res.* **2003**, *10*, 239–247. [CrossRef] [PubMed]
9. Raja, V.; Majeed, U.; Kang, H.; Andrabi, K.I.; John, R. Abiotic stress: Interplay between ROS, hormones and MAPKs. *Environ. Exp. Bot.* **2017**, *137*, 142–157. [CrossRef]
10. Li, S.; Gao, J.; Yao, L.; Ren, G.; Zhu, X.; Gao, S.; Qiu, K.; Zhou, X.; Kuai, B. The role of ANAC072 in the regulation of chlorophyll degradation during age- and dark-induced leaf senescence. *Plant Cell Rep.* **2016**, *35*, 1729–1741. [CrossRef] [PubMed]
11. Zhu, M.; Chen, G.; Zhou, S.; Tu, Y.; Wang, Y.; Dong, T.; Hu, Z. A New Tomato NAC (NAM/ATAF1/2/CUC2) Transcription Factor, SlNAC4, Functions as a Positive Regulator of Fruit Ripening and Carotenoid Accumulation. *Plant Cell Physiol.* **2013**, *55*, 119–135. [CrossRef]
12. Mao, C.; Lu, S.; Lv, B.; Zhang, B.; Shen, J.; He, J.; Luo, L.; Xi, D.; Chen, X.; Ming, F. A Rice NAC Transcription Factor Promotes Leaf Senescence via ABA Biosynthesis. *Plant Physiol.* **2017**, *174*, 1747–1763. [CrossRef]
13. Tran, L.-S.P.; Nakashima, K.; Sakuma, Y.; Simpson, S.D.; Fujita, Y.; Maruyama, K.; Fujita, M.; Seki, M.; Shinozaki, K.; Yamaguchi-Shinozaki, K. Isolation and Functional Analysis of Arabidopsis Stress-Inducible NAC Transcription Factors That Bind to a Drought-Responsive cis-Element in the early responsive to dehydration stress 1 Promoter. *Plant Cell* **2004**, *16*, 2481–2498. [CrossRef]
14. Ma, J.; Wang, L.-Y.; Dai, J.-X.; Wang, Y.; Lin, D. The NAC-type transcription factor CaNAC46 regulates the salt and drought tolerance of transgenic Arabidopsis thaliana. *BMC Plant Biol.* **2021**, *21*, 11. [CrossRef] [PubMed]
15. Mao, H.; Li, S.; Wang, Z.; Cheng, X.; Li, F.; Mei, F.; Chen, N.; Kang, Z. Regulatory changes in TaSNAC8-6A are associated with drought tolerance in wheat seedlings. *Plant Biotechnol. J.* **2019**, *18*, 1078–1092. [CrossRef] [PubMed]
16. Thirumalaikumar, V.P.; Devkar, V.; Mehterov, N.; Ali, S.; Ozgur, R.; Turkan, I.; Mueller-Roeber, B.; Balazadeh, S. NAC transcription factor JUNGBRUNNEN1 enhances drought tolerance in tomato. *Plant Biotechnol. J.* **2017**, *16*, 354–366. [CrossRef] [PubMed]
17. Hu, P.; Zhang, K.; Yang, C. BpNAC012 Positively Regulates Abiotic Stress Responses and Secondary Wall Biosynthesis. *Plant Physiol.* **2019**, *179*, 700–717. [CrossRef] [PubMed]

18. Ehong, Y.; Ezhang, H.; Ehuang, L.; Eli, D.; Esong, F. Overexpression of a Stress-Responsive NAC Transcription Factor Gene ONAC022 Improves Drought and Salt Tolerance in Rice. *Front. Plant Sci.* **2016**, *7*, 4. [CrossRef]

19. Xu, Z.-Y.; Kim, S.Y.; Hyeon, D.Y.; Kim, D.H.; Dong, T.; Park, Y.; Jin, J.B.; Joo, S.-H.; Hong, J.C.; Hwang, D.; et al. The Arabidopsis NAC Transcription Factor ANAC096 Cooperates with bZIP-Type Transcription Factors in Dehydration and Osmotic Stress Responses. *Plant Cell* **2013**, *25*, 4708–4724. [CrossRef]

20. Dodd, I.C.; Egea, G.; Watts, C.W.; Whalley, W.R.; Cegarra, G.E. Root water potential integrates discrete soil physical properties to influence ABA signalling during partial rootzone drying. *J. Exp. Bot.* **2010**, *61*, 3543–3551. [CrossRef]

21. Tuteja, N. Abscisic Acid and Abiotic Stress Signaling. *Plant Signal. Behav.* **2007**, *2*, 135–138. [CrossRef] [PubMed]

22. Xue-Xuan, X.; Hong-Bo, S.; Yuan-Yuan, M.; Gang, X.; Jun-Na, S.; Dong-Gang, G.; Cheng-Jiang, R. Biotechnological implications from abscisic acid (ABA) roles in cold stress and leaf senescence as an important signal for improving plant sustainable survival under abiotic-stressed conditions. *Crit. Rev. Biotechnol.* **2010**, *30*, 222–230. [CrossRef] [PubMed]

23. Sato, H.; Takasaki, H.; Takahashi, F.; Suzuki, T.; Iuchi, S.; Mitsuda, N.; Ohme-Takagi, M.; Ikeda, M.; Seo, M.; Yamaguchi-Shinozaki, K.; et al. Arabidopsis thaliana NGATHA1 transcription factor induces ABA biosynthesis by activating NCED3 gene during dehydration stress. *Proc. Natl. Acad. Sci. USA* **2018**, *115*, E11178–E11187. [CrossRef] [PubMed]

24. Nambara, E.; Marion-Poll, A. Abscisic acid biosynthesis and catabolism. *Annu. Rev. Plant Biol.* **2005**, *56*, 165–185. [CrossRef]

25. Nishimura, N.; Sarkeshik, A.; Nito, K.; Park, S.; Wang, A.; Carvalho, P.C.; Lee, S.; Caddell, D.F.; Cutler, S.R.; Chory, J.; et al. PYR/PYL/RCAR family members are major in-vivo ABI1 protein phosphatase 2C-interacting proteins in Arabidopsis. *Plant J.* **2010**, *61*, 290–299. [CrossRef]

26. Udvardi, M.K.; Kakar, K.; Wandrey, M.; Montanari, O.; Murray, J.; Andriankaja, A.; Zhang, J.-Y.; Benedito, V.; Hofer, J.; Chueng, F.; et al. Legume Transcription Factors: Global Regulators of Plant Development and Response to the Environment. *Plant Physiol.* **2007**, *144*, 538–549. [CrossRef]

27. Narusaka, Y.; Nakashima, K.; Shinwari, Z.K.; Sakuma, Y.; Furihata, T.; Abe, H.; Narusaka, M.; Shinozaki, K.; Yamaguchi-Shinozaki, K. Interaction between two cis-acting elements, ABRE and DRE, in ABA-dependent expression of Arabidopsis rd29A gene in response to dehydration and high-salinity stresses. *Plant J.* **2003**, *34*, 137–148. [CrossRef] [PubMed]

28. Zhang, H.; Cui, X.; Guo, Y.; Luo, C.; Zhang, L. Picea wilsonii transcription factor NAC2 enhanced plant tolerance to abiotic stress and participated in RFCP1-regulated flowering time. *Plant Mol. Biol.* **2018**, *98*, 471–493. [CrossRef]

29. Liang, K.-H.; Wang, A.-B.; Yuan, Y.-H.; Miao, Y.-H.; Zhang, L.-Y. Picea wilsonii NAC Transcription Factor PwNAC30 Negatively Regulates Abiotic Stress Tolerance in Transgenic Arabidopsis. *Plant Mol. Biol. Rep.* **2020**, *38*, 554–571. [CrossRef]

30. Guo, Y.; Zhang, H.; Yuan, Y.; Cui, X.; Zhang, L. Identification and characterization of NAC genes in response to abiotic stress conditions in Picea wilsonii using transcriptome sequencing. *Biotechnol. Biotechnol. Equip.* **2020**, *34*, 93–103. [CrossRef]

31. Li, S.; Zhang, J.; Liu, L.; Wang, Z.; Li, Y.; Guo, L.; Li, Y.; Zhang, X.; Ren, S.; Zhao, B.; et al. SlTLFP8 reduces water loss to improve water-use efficiency by modulating cell size and stomatal density via endoreduplication. *Plant Cell Environ.* **2020**, *43*, 2666–2679. [CrossRef]

32. Fujita, M.; Fujita, Y.; Maruyama, K.; Seki, M.; Hiratsu, K.; Ohme-Takagi, M.; Tran, L.-S.P.; Yamaguchi-Shinozaki, K.; Shinozaki, K. A dehydration-induced NAC protein, RD26, is involved in a novel ABA-dependent stress-signaling pathway. *Plant J.* **2004**, *39*, 863–876. [CrossRef]

33. Nuruzzaman, M.; Sharoni, A.M.; Kikuchi, S. Roles of NAC transcription factors in the regulation of biotic and abiotic stress responses in plants. *Front. Microbiol.* **2013**, *4*, 248. [CrossRef]

34. Baxter, A.; Mittler, R.; Suzuki, N. ROS as key players in plant stress signalling. *J. Exp. Bot.* **2014**, *65*, 1229–1240. [CrossRef] [PubMed]

35. Huang, W.; Yang, Y.-J.; Hu, H.; Cao, K.-F.; Zhang, S.-B. Sustained Diurnal Stimulation of Cyclic Electron Flow in Two Tropical Tree Species Erythrophleum guineense and Khaya ivorensis. *Front. Plant Sci.* **2016**, *7*, 1068. [CrossRef]

36. Williamson, R.; Field, J.G.; Shillington, F.A.; Jarre, A.; Potgieter, A. A Bayesian approach for estimating vertical chlorophyll profiles from satellite remote sensing: Proof-of-concept. *ICES J. Mar. Sci.* **2010**, *68*, 792–799. [CrossRef]

37. Shen, C.; Zhang, Y.; Li, Q.; Liu, S.; He, F.; An, Y.; Zhou, Y.; Liu, C.; Yin, W.; Xia, X. PdGNC confers drought tolerance by mediating stomatal closure resulting from NO and H_2O_2 production via the direct regulation of PdHXK1 expression in Populus. *New Phytol.* **2021**, *230*, 1868–1882. [CrossRef]

38. Acebron, K.; Matsubara, S.; Jedmowski, C.; Emin, D.; Muller, O.; Rascher, U. Diurnal dynamics of nonphotochemical quenching in Arabidopsis npq mutants assessed by solar-induced fluorescence and reflectance measurements in the field. *New Phytol.* **2021**, *229*, 2104–2119. [CrossRef] [PubMed]

39. Liang, K.; Wang, A.; Sun, Y.; Yu, M.; Zhang, L. Identification and Expression of NAC Transcription Factors of *Vaccinium corymbosum* L. in Response to Drought Stress. *Forests* **2019**, *10*, 1088. [CrossRef]

40. Sun, Y.; Liu, X.; Zhai, H.; Gao, H.; Yao, Y.; Du, Y. Responses of photosystem II photochemistry and the alternative oxidase pathway to heat stress in grape leaves. *Acta Physiol. Plant* **2016**, *38*, 232. [CrossRef]

41. Dou, H.; Xv, K.; Meng, Q.; Li, G.; Yang, X. Potato plants ectopically expressing *Arabidopsis thaliana* CBF3 exhibit enhanced tolerance to high-temperature stress. *Plant Cell Environ.* **2014**, *38*, 61–72. [CrossRef]

42. Liu, C.; Sun, Q.; Zhao, L.; Li, Z.; Peng, Z.; Zhang, J. Heterologous Expression of the Transcription Factor EsNAC1 in Arabidopsis Enhances Abiotic Stress Resistance and Retards Growth by Regulating the Expression of Different Target Genes. *Front. Plant Sci.* **2018**, *9*, 1495. [CrossRef] [PubMed]

43. Yang, X.; Wang, X.; Ji, L.; Yi, Z.; Fu, C.; Ran, J.; Hu, R.; Zhou, G. Overexpression of a Miscanthus lutarioriparius NAC gene MlNAC5 confers enhanced drought and cold tolerance in Arabidopsis. *Plant Cell Rep.* **2015**, *34*, 943–958. [CrossRef] [PubMed]

44. Fang, J.; Chai, Z.; Yao, W.; Chen, B.; Zhang, M. Interactions between ScNAC23 and ScGAI regulate GA-mediated flowering and senescence in sugarcane. *Plant Sci.* **2020**, *304*, 110806. [CrossRef]

45. Kovalchuk, N.; Jia, W.; Eini, O.; Morran, S.; Pyvovarenko, T.; Fletcher, S.; Bazanova, N.; Harris, J.; Beck-Oldach, K.; Shavrukov, Y.; et al. Optimization of TaDREB3 gene expression in transgenic barley using cold-inducible promoters. *Plant Biotechnol. J.* **2013**, *11*, 659–670. [CrossRef] [PubMed]

46. He, F.; Wang, H.-L.; Li, H.-G.; Su, Y.; Li, S.; Yang, Y.; Feng, C.-H.; Yin, W.; Xia, X. PeCHYR1, a ubiquitin E3 ligase from Populus euphratica, enhances drought tolerance via ABA-induced stomatal closure by ROS production in Populus. *Plant Biotechnol. J.* **2018**, *16*, 1514–1528. [CrossRef]

47. Xu, Y.; Zhao, X.; Aiwaili, P.; Mu, X.; Zhao, M.; Zhao, J.; Cheng, L.; Ma, C.; Gao, J.; Hong, B. A zinc finger protein BBX19 interacts with ABF3 to affect drought tolerance negatively in chrysanthemum. *Plant J.* **2020**, *103*, 1783–1795. [CrossRef]

48. Sakuraba, Y.; Kim, D.; Han, S.-H.; Kim, S.-H.; Piao, W.; Yanagisawa, S.; An, G.; Paek, N.-C. Multilayered Regulation of Membrane-Bound ONAC054 is essential for Abscisic Acid-Induced Leaf Senescence in Rice. *Plant Cell* **2020**, *32*, 630–649. [CrossRef]

49. Joshi, P.S.; Agarwal, P.; Agarwal, P.K. Overexpression of AlNAC1 from recretohalophyte *Aeluropus lagopoides* alleviates drought stress in transgenic tobacco. *Environ. Exp. Bot.* **2021**, *181*, 104277. [CrossRef]

50. Romero-Romero, J.L.; Inostroza-Blancheteau, C.; Orellana, D.; Aquea, F.; Reyes-Díaz, M.; Gil, P.M.; Matte, J.P.; Arce-Johnson, P. Stomata regulation by tissue-specific expression of the Citrus sinensis MYB61 transcription factor improves water-use efficiency in Arabidopsis. *Plant Physiol. Biochem.* **2018**, *130*, 54–60. [CrossRef] [PubMed]

51. Romano, J.M.; Dubos, C.; Prouse, M.B.; Wilkins, O.; Hong, H.; Poole, M.; Kang, K.-Y.; Li, E.; Douglas, C.J.; Western, T.L.; et al. At MYB61, an R2R3-MYB transcription factor, functions as a pleiotropic regulator via a small gene network. *New Phytol.* **2012**, *195*, 774–786. [CrossRef]

52. Grondin, A.; Rodrigues, O.; Verdoucq, L.; Merlot, S.; Leonhardt, N.; Maurel, C. Aquaporins Contribute to ABA-Triggered Stomatal Closure through OST1-Mediated Phosphorylation. *Plant Cell* **2015**, *27*, 1945–1954. [CrossRef] [PubMed]

53. Ju, Y.-L.; Yue, X.-F.; Min, Z.; Wang, X.-H.; Fang, Y.-L.; Zhang, J.-X. VvNAC17, a novel stress-responsive grapevine (*Vitis vinifera* L.) NAC transcription factor, increases sensitivity to abscisic acid and enhances salinity, freezing, and drought tolerance in transgenic Arabidopsis. *Plant Physiol. Biochem.* **2020**, *146*, 98–111. [CrossRef]

54. Yoshida, T.; Fujita, Y.; Maruyama, K.; Mogami, J.; Todaka, D.; Shinozaki, K.; Yamaguchi-Shinozaki, K. Four A rabidopsis AREB/ABF transcription factors function predominantly in gene expression downstream of SnRK2 kinases in abscisic acid signalling in response to osmotic stress. *Plant Cell Environ.* **2014**, *38*, 35–49. [CrossRef]

55. Yoshida, T.; Fujita, Y.; Sayama, H.; Kidokoro, S.; Maruyama, K.; Mizoi, J.; Shinozaki, K.; Yamaguchi-Shinozaki, K. AREB1, AREB2, and ABF3 are master transcription factors that cooperatively regulate ABRE-dependent ABA signaling involved in drought stress tolerance and require ABA for full activation. *Plant J.* **2010**, *61*, 672–685. [CrossRef] [PubMed]

56. Chang, H.-C.; Tsai, M.-C.; Wu, S.-S.; Chang, I.-F. Regulation of ABI5 expression by ABF3 during salt stress responses in Arabidopsis thaliana. *Bot. Stud.* **2019**, *60*, 1–14. [CrossRef]

57. Yang, Y.; Li, H.-G.; Wang, J.; Wang, H.-L.; He, F.; Su, Y.; Zhang, Y.; Feng, C.-H.; Niu, M.; Li, Z.; et al. ABF3 enhances drought tolerance via promoting ABA-induced stomatal closure by directly regulating ADF5 in *Populus euphratica*. *J. Exp. Bot.* **2020**, *71*, 7270–7285. [CrossRef]

58. Yang, Q.; Yang, B.; Li, J.; Wang, Y.; Tao, R.; Yang, F.; Wu, X.; Yan, X.; Ahmad, M.; Shen, J.; et al. ABA-responsive ABRE-binding factor3 activates DAM3 expression to promote bud dormancy in Asian pear. *Plant Cell Environ.* **2020**, *43*, 1360–1375. [CrossRef]

59. Li, X.; Li, X.; Li, M.; Yan, Y.; Liu, X.; Li, L. Dual Function of NAC072 in ABF3-Mediated ABA-Responsive Gene Regulation in Arabidopsis. *Front. Plant Sci.* **2016**, *7*, 1075. [CrossRef]

60. Tran, L.-S.P.; Nakashima, K.; Sakuma, Y.; Osakabe, Y.; Qin, F.; Simpson, S.D.; Maruyama, K.; Fujita, Y.; Shinozaki, K.; Yamaguchi-Shinozaki, K. Co-expression of the stress-inducible zinc finger homeodomain ZFHD1 and NAC transcription factors enhances expression of the ERD1 gene in Arabidopsis. *Plant J.* **2006**, *49*, 46–63. [CrossRef] [PubMed]

61. Hwang, K.; Susila, H.; Nasim, Z.; Jung, J.-Y.; Ahn, J.H. Arabidopsis ABF3 and ABF4 Transcription Factors Act with the NF-YC Complex to Regulate SOC1 Expression and Mediate Drought-Accelerated Flowering. *Mol. Plant* **2019**, *12*, 489–505. [CrossRef] [PubMed]

62. Yamaguchi-Shinozaki, K. A novel cis-acting element in an Arabidopsis gene is involved in responsiveness to drought, low-temperature, or high-salt stress. *Plant Cell* **1994**, *6*, 251–264. [CrossRef] [PubMed]

63. Sun, J.; Peng, X.; Fan, W.; Tang, M.; Liu, J.; Shen, S. Functional analysis of BpDREB2 gene involved in salt and drought response from a woody plant *Broussonetia papyrifera*. *Gene* **2014**, *535*, 140–149. [CrossRef]

64. Tang, M.; Liu, X.; Deng, H.; Shen, S. Over-expression of JcDREB, a putative AP2/EREBP domain-containing transcription factor gene in woody biodiesel plant Jatropha curcas, enhances salt and freezing tolerance in transgenic Arabidopsis thaliana. *Plant Sci.* **2011**, *181*, 623–631. [CrossRef] [PubMed]

65. Kim, J.-S.; Mizoi, J.; Yoshida, T.; Fujita, Y.; Nakajima, J.; Ohori, T.; Todaka, D.; Nakashima, K.; Hirayama, T.; Shinozaki, K.; et al. An ABRE Promoter Sequence is Involved in Osmotic Stress-Responsive Expression of the DREB2A Gene, Which Encodes a Transcription Factor Regulating Drought-Inducible Genes in Arabidopsis. *Plant Cell Physiol.* **2011**, *52*, 2136–2146. [CrossRef] [PubMed]

66. Nakashima, K.; Fujita, Y.; Katsura, K.; Maruyama, K.; Narusaka, Y.; Seki, M.; Shinozaki, K.; Yamaguchi-Shinozaki, K. Transcriptional Regulation of ABI3- and ABA-responsive Genes Including RD29B and RD29A in Seeds, Germinating Embryos, and Seedlings of Arabidopsis. *Plant Mol. Biol.* **2006**, *60*, 51–68. [CrossRef]

67. Min, M.K.; Kim, R.; Hong, W.-J.; Jung, K.-H.; Lee, J.-Y.; Kim, B.-G. OsPP2C09 Is a Bifunctional Regulator in Both ABA-Dependent and Independent Abiotic Stress Signaling Pathways. *Int. J. Mol. Sci.* **2021**, *22*, 393. [CrossRef]

68. Mergby, D.; Hanin, M.; Saidi, M.N. The durum wheat NAC transcription factor TtNAC2A enhances drought stress tolerance in Arabidopsis. *Environ. Exp. Bot.* **2021**, *186*, 104439. [CrossRef]

69. Ren, G.; An, K.; Liao, Y.; Zhou, X.; Cao, Y.; Zhao, H.; Ge, X.; Kuai, B. Identification of a Novel Chloroplast Protein AtNYE1 Regulating Chlorophyll Degradation during Leaf Senescence in Arabidopsis. *Plant Physiol.* **2007**, *144*, 1429–1441. [CrossRef]

70. Lv, S.; Zhang, Y.; Pan, L.; Liu, Z.; Yang, N.; Pan, L.; Wu, J.; Wang, J.; Yang, J.; Lv, Y.; et al. Strigolactone-triggered stomatal closure requires hydrogen peroxide synthesis and nitric oxide production in an abscisic acid-independent manner. *New Phytol.* **2018**, *217*, 290–304. [CrossRef]

International Journal of
Molecular Sciences

MDPI

Article

Overexpression of ABI5 Binding Proteins Suppresses Inhibition of Germination Due to Overaccumulation of DELLA Proteins

Ruth R. Finkelstein * and Tim J. Lynch

Department of Molecular, Cellular and Developmental Biology, University of California at Santa Barbara, Santa Barbara, CA 93106, USA; lynch@lifesci.ucsb.edu
* Correspondence: finkelst@ucsb.edu; Tel.: +1-805-893-4800

Abstract: Abscisic acid (ABA) and gibberellic acid (GA) antagonistically regulate many aspects of plant growth, including seed dormancy and germination. The effects of these hormones are mediated by a complex network of positive and negative regulators of transcription. The DELLA family of proteins repress GA response, and can promote an ABA response via interactions with numerous regulators, including the ABA-insensitive (ABI) transcription factors. The AFP family of ABI5 binding proteins are repressors of the ABA response. This study tested the hypothesis that the AFPs also interact antagonistically with DELLA proteins. Members of these protein families interacted weakly in yeast two-hybrid and bimolecular fluorescence complementation studies. Overexpression of AFPs in *sleepy1*, a mutant that over-accumulates DELLA proteins, suppressed DELLA-induced overaccumulation of storage proteins, hyperdormancy and hypersensitivity to ABA, but did not alter the dwarf phenotype of the mutant. The interaction appeared to reflect additive effects of the AFPs and DELLAs, consistent with action in convergent pathways.

Keywords: abscisic acid; gibberellic acid; arabidopsis; ABI5; ABI5-binding proteins (AFPs); DELLA proteins; SLEEPY1; germination; dormancy; storage proteins

Citation: Finkelstein, R.R.; Lynch, T.J. Overexpression of ABI5 Binding Proteins Suppresses Inhibition of Germination Due to Overaccumulation of DELLA Proteins. *Int. J. Mol. Sci.* **2022**, 23, 5537. https://doi.org/10.3390/ijms23105537

Academic Editors: Víctor Quesada and Ricardo Aroca

Received: 26 December 2021
Accepted: 9 May 2022
Published: 16 May 2022

Publisher's Note: MDPI stays neutral with regard to jurisdictional claims in published maps and institutional affiliations.

1. Introduction

Seed germination is regulated by diverse environmental signals, including light, temperature, and water availability (reviewed in [1]). Many of these signals affect the balance between germination inhibition by abscisic acid (ABA) and promotion by gibberellic acid (GA) [2–4]. Following many years of studies with exogenously applied hormones, genetic evidence for intrinsic control by this balance was provided by the isolation of non-germinating dwarf mutants that were deficient in GA synthesis [5], and germinating revertants that had additional mutations disrupting ABA synthesis [6]. In addition to control by the relative concentrations of ABA and GA, the abundance and activity of numerous signaling intermediates affect the sensitivity to these hormones.

The ABA-insensitive loci *ABI1*, *ABI2*, *ABI3*, *ABI4*, and *ABI5*, major regulators in ABA response at this stage, were identified by screens for ABA-resistant germination (reviewed in [7]). The initial *abi1-1* and *abi2-1* mutants had dominant negative mutations in members of a clade of protein phosphatases (PP2Cs) later shown to be a central part of the ABA core signaling pathway. Subsequent screens for ABA-hypersensitive germination (*AHG*) identified loss of function mutations in additional members of this clade, reinforcing the interpretation that these were negative regulators of ABA response. In contrast, *ABI3*, *ABI4*, and *ABI5* encode transcription factors that both activate ABA-induced genes and repress ABA-downregulated genes [8–11].

Screens for defects in GA response focused on plants with defects in regulating elongation, either GA insensitive dwarfs (*gai* and *gid*) [12,13] or displayed constitutive response to GA (*spindly*) [14]. The initial *gai* mutant also had a dominant negative mutation [12], reflecting the role of GAI (also known as REPRESSOR OF GA(RGA)2) as an inhibitor of the GA response. Closely related proteins include RGA1, RGA-LIKE(RGL)1, RGL2 and RGL3, and loss of function in these loci results in constitutive GA response [15]; all are members of the DELLA class of transcriptional regulators. GA response is mediated by proteasomal degradation of the DELLA repressors. Additional screens for revertants of ABA-resistance due to the *abi1-1* mutation identified *sleepy1 (sly1)* [16], an F-box protein responsible for ubiquitination of DELLAs in the presence of GA and a functional GA receptor (GID) [17,18]. However, extended after-ripening or increased GID1 receptor expression can alleviate the extreme dormancy of *sly1* mutants by a non-proteolytic signaling mechanism [19].

Numerous direct and regulatory interactions between the ABI transcription factors, DELLA proteins, and additional transcription factors regulating germination have been reported. ABI3, ABI5, and DELLAs interact directly, apparently forming a complex in seeds exposed to high temperatures [20]. This complex activates the expression of *SOMNUS (SOM)*, which encodes a CCCH-type zinc finger protein that inhibits germination by increasing ABA biosynthetic gene expression and decreasing GA biosynthetic gene expression. Under far-red light conditions, the bHLH transcription factor PHYTOCHROME INTERACTING FACTOR 1(PIF1) accumulates, also resulting in positive regulation of ABI5, DELLA proteins, and SOM, inducing expression of yet another inhibitor of germination: MOTHER-OF-FT-AND-TFL1 (MFT) [21].

Although RGA1, RGA2, and RGL2 are expressed at similar levels in dry and germinating seeds [22,23], RGL2 is reported to be the major regulator of seed germination [24,25], and its repressive function is dependent on ABI5 activity [26]. Mechanistically, this involves interactions between RGL2 and three NUCLEAR FACTOR-Y C (NF-YC) homologues (NF-YC3, NF-YC4, and NF-YC9) that synergistically bind and activate the ABI5 promoter [27]. Conversely, INDUCER OF CBF EXPRESSION1 (ICE1) interacts with both ABI5 and DELLAs to antagonize their function, thereby permitting seed germination [28].

Additional proteins implicated in antagonistically regulating the function of ABI5 and related bZIP proteins are the ABI5-binding Proteins (AFPs) [29,30]. Within this family, AFP1 and AFP2 have the strongest effects on germination. The AFPs have also been reported to regulate salt and osmotic stress through interactions with SnRK1 kinases and flowering via effects on expression of *CONSTANS (CO), FLOWERING LOCUS T(FT)*, and *SUPPRESSOR OF OVEREXPRESSION OF CO(SOC)1* [31]. Proposed biochemical functions include inhibiting ABA-dependent gene expression through chromatin modifications mediated at least in part by direct interactions with TOPLESS and histone deactylase subunits [32,33] and promoting ABI5 degradation [29]. However, AFP2 overexpression promotes ABA-resistant germination that precedes ABI5 degradation [34]. The AFPs are predicted to include multiple intrinsically disordered domains, suggesting the possibility that the diversity of their interactions may reflect a role in scaffolding complexes of regulators. Given the extensive interactions already documented for ABI5, the AFPs, and the DELLAs, we sought to determine whether the AFPs also interact with DELLA proteins. We found weak direct interactions between a subset of each family in two protein–protein interaction assays, and additive antagonistic effects of overexpression/overaccumulation of these protein classes in genetic assays.

2. Results

2.1. Direct Interactions between AFPs and DELLA Proteins

We initially tested for possible direct interactions between AFPs and DELLA proteins by yeast two hybrid and bimolecular fluorescence complementation (split YFP) assays, including ABI5 as a positive control (Figures 1 and S1). We focused on the DELLA proteins with similarly significant expression in dry or imbibing seeds, i.e., all except RGL1 (Figure S2). For the yeast two hybrid assays, the DELLAs were presented as GAL4 activation domain (AD)-fusions, the AFPs were fused to the GAL4 binding domain (BD), and ABI5 was present in both AD- and BD-fusions. Both AFP1 and AFP2 interacted strongly with RGA1 and RGA2/GAI, and AFP1 also interacted to a lesser extent with RGL2 and

RGL3. As documented previously [20,30], ABI5 interacted directly with the AFPs and the DELLAs, especially RGA1 and RGA2. The DELLA protein interactions with the AFPs were further mapped to the GRAS domain of RGA2, but this was much weaker than the interaction with the full-length protein. Although ABI5 appeared to interact with both the GRAS and DELLA domains of RGA2, these interactions were again weaker than those with full-length RGA2.

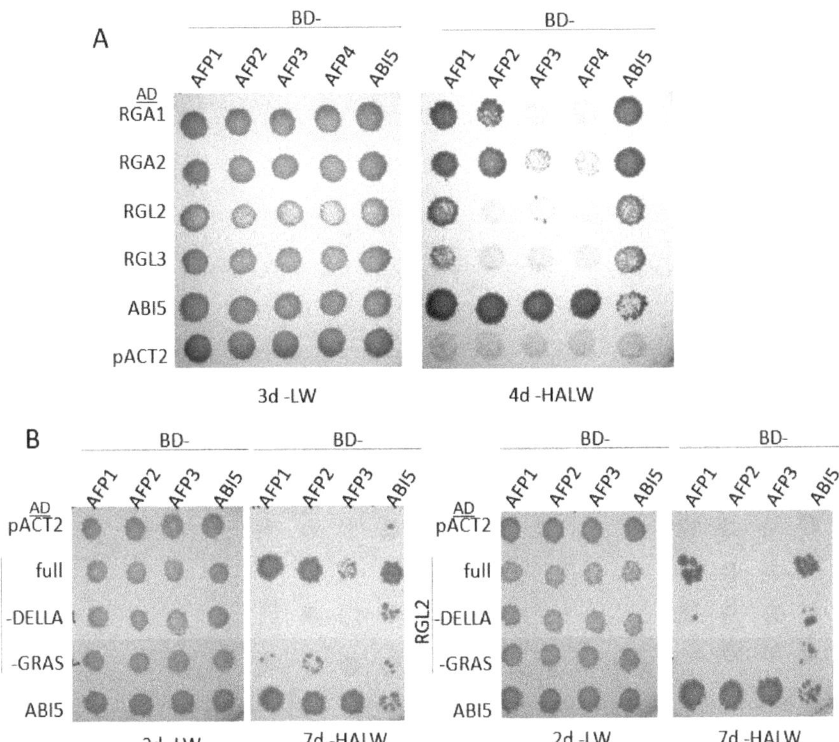

Figure 1. Yeast two-hybrid interactions between AFPs and DELLAs. (**A**) Interactions between the GAL4 AD alone (pACT2) or fusions to the DELLAs or ABI5 and GAL4 BD- fusions to AFPs or ABI5. (**B**) Mapping domains of RGA2 or RGL2 required for interaction with the AFPs or ABI5. Growth on -LW selects for diploids carrying both AD- and BD-fusions. Growth on -HALW requires interaction to produce GAL4-dependent activation of the HIS and ADE reporters.

The split YFP assays also repeated strong interactions between ABI5 and the AFPs, but weaker interactions between DELLA fusions and either ABI5 or the AFPs (Figure S1). Surprisingly, although RGL2 did not interact with AFP2 in the yeast two-hybrid assays, this combination interacted quite strongly in the split YFP assay.

2.2. Genetic Interactions between AFPs and DELLA Proteins

Our previous studies showed that AFP2 overexpression results in extreme ABA resistance of seeds and consequently a loss of dormancy. In contrast, *sleepy (sly1)* mutants lack the F-box protein that ubiquitinates DELLA proteins, such that the DELLAs are hyperstable in these mutants, GA response is repressed and the *sly1* seeds are hyperdormant. To determine whether the apparent physical interactions between AFP2 and DELLAs result in epistatic or additive effects on germination potential, we overexpressed AFP2 in the *sly1-2* background. The *YFP-AFP2* transgene was crossed into *sly1* from a previously characterized line in the wild-type background [32]. Although AFP1 overexpression was less effective than AFP2 for promoting extreme ABA resistance [32], we also tested the effects of AFP1 overexpression in the *sly1* mutant because the strongest interactions observed in the yeast two-hybrid assay were between AFP1 and the DELLAs. The *YFP-AFP1* overexpression transgene from *sly1* line #3 was backcrossed into the wild-type background. As in the wild-type background, all *sly1* lines with good seedling viability were hemizygous for the transgene, such that 25% of these seed populations lacked the transgene. Overexpression of either AFP was sufficient to reduce dormancy and permit some ABA-resistant germination in this background (Figure 2A). In both backgrounds, *YFP-AFP2* overexpression promoted cotyledon expansion and greening of most germinated seeds (Figure 2B,C). However, the fraction capable of germinating, especially in the presence of ABA, was greatly reduced in the *sly1* background; a similar percentage germination was achieved for *YFP-AFP2* seeds in the wild-type background exposed to 50 μM ABA as for those in the *sly1* background on hormone-free medium. In contrast, the *YFP-AFP1* overexpression transgene from *sly1* line #3 permitted germination on media containing at least five-fold higher ABA concentrations in the wild-type background (similar responses to 50 μM vs. 10 μM ABA in wild-type vs. *sly1* backgrounds), but had more limited effects on promoting ABA-resistant greening. An independent *sly1, YFP-AFP1/+* line (#4) had stronger effects in promoting both germination and post-germinative growth in the presence of ABA, but still less than the weaker transgene in the wild-type background.

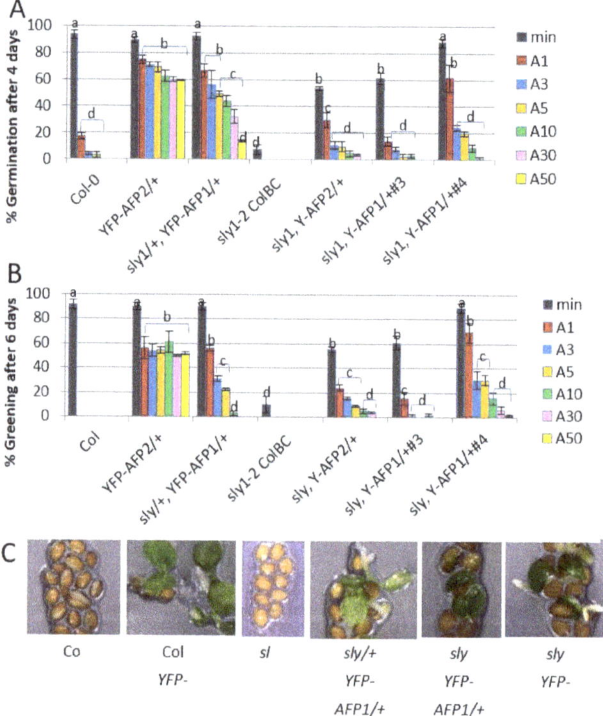

Figure 2. Effects of overexpressing YFP-AFP fusions in wild-type (Col-0) vs. *sly1* mutants on ABA sensitivity of seeds. (**A**) Germination of the indicated genotypes after 4 days on media with no ABA (min) or 1–50 μM ABA (A1–A50). (**B**) Greening of the germinated seedlings after 6 days on the media described in (**A**). Data displayed is the average of at least triplicate assays for each genotype and treatment ± S.E. Bars with different letters represent statistically different values using Tukey's HSD post hoc test (*p* < 0.01). (**C**) Representative images of seeds/seedlings after 6 days on 5 μM ABA.

During seed development, the ABI transcription factors mediate ABA-promoted accumulation of storage proteins and lipids. Recently, RGL3 was shown to be the major DELLA protein expressed in maturing seeds, where it promotes the expression of the cruciferin and At2S families of storage proteins [35]. Consistent with this, we found a slightly increased accumulation of these proteins in the *sly1* mutant background (Figure 3, lane 7). Transgenic lines overexpressing either AFP reduced accumulation of the storage proteins (Figure 3, lanes 2 and 3), with those in the *sly1* background again showing an intermediate phenotype (Figure 3, lanes 4–6).

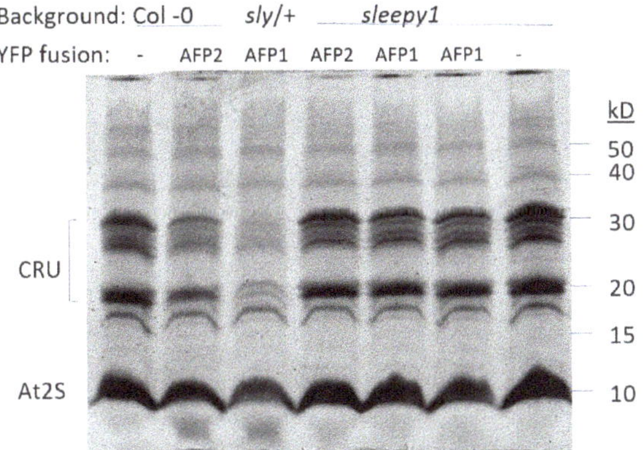

Figure 3. Storage protein accumulation in the indicated genotypes. Total seed protein extracts were separated on 15% SDS-PAGE and stained with Coomassie blue. Major bands represent cruciferin and At2S storage protein families.

In contrast to the effects on seeds, both classes of *sly1*, *35S-YFP-AFP* lines displayed the dwarf stature, slow growth, loss of apical dominance, and poor fertility of the *sly1* parental line, although YFP-AFP2 overexpression slightly suppressed lateral branching in the *sly1* background (Figures 4 and S3). Fertility was delayed the most in the primary inflorescence, with lateral shoots producing much fewer sterile flowers. Although more variable, the YFP-AFP1 transgenes enhanced the sterility in the primary inflorescence of the *sly1* parental line but did not affect fertility in a wild-type background.

Figure 4. Effects of overexpressing YFP-AFP fusions in wild-type (Col-0) vs. *sly1* mutants on vegetative growth. (**A**,**B**) Five-week old plants, (**C**) seven-week old plants.

2.3. AFP, DELLA and ABI5 Protein Accumulation

ABA-resistant germination can result from decreased levels of inhibitors such as ABI5 or DELLA proteins or increased levels of proteins that antagonize their function. To determine whether the altered extent of germination was associated with changes in the levels of these proteins, we used immunoblots to compare their accumulation in seeds and following stratification in the presence or absence of added 1 µM ABA, a concentration sufficient to delay but not block germination (Figure 5). In populations segregating a YFP-AFP transgene in the *sly1* background, ABA-resistant germination was limited to individuals accumulating the fusion protein (Figure S4A).

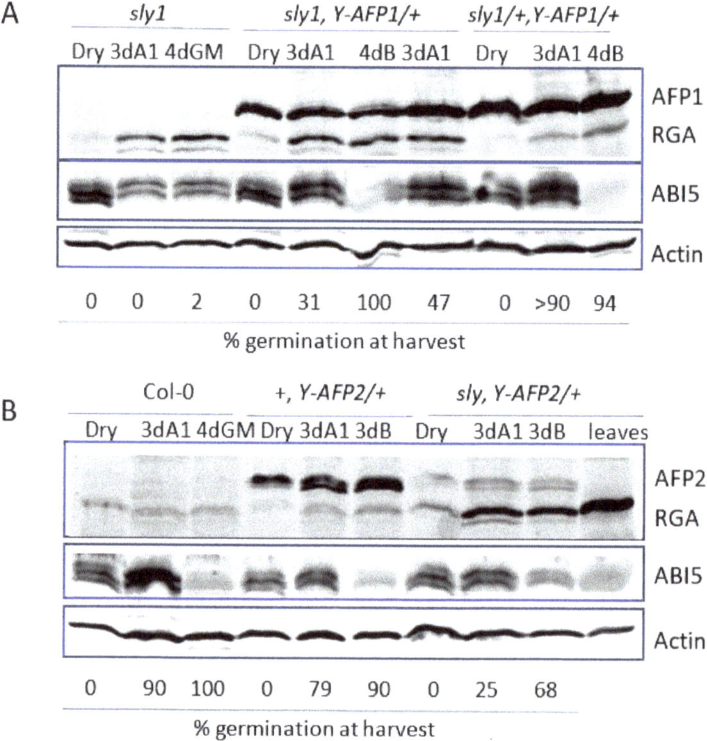

Figure 5. Immunoblot comparison of AFP, DELLA (RGA), and ABI5 protein accumulation in seeds and seedlings. Proteins were extracted from dry seeds or after 3d stratification, followed by 3d on GM + 1 µM ABA (3dA1) or 4d on GM, supplemented with BASTA (4dB) for the transgenic lines. Actin was used as a loading control. (**A**) Comparison of *sly1* background (lanes 1–3) with *YFP-AFP1* transgenic line #3 (lanes 4–6) or line #4 (lane 7) or the F2 from the backcross of *YFP-AFP1*#3 to Col-0 (lanes 8-10). (**B**) Comparison of wild-type (Col-0) (lanes 1–3), *YFP-AFP2* in wild type (lanes 4–6), or *sly1* backgrounds (lanes 7–10). Extract from rosette leaves is in lane 10.

YFP-AFP1 levels are relatively constant in seeds and seedlings, but are below detection in rosette leaves of older plants (Figures 5A and S4B,C). The YFP-AFP2 fusions accumulate as germination proceeds and show a slight shift toward higher mobility, likely reflecting changes in post-translational modifications, but are also not detected in older plants (Figure 5B). Levels of the YFP-AFP fusions also varied among independent transformants, but generally paralleled the degree of ABA resistance in a given background (Figure S4C,D). However, similar amounts of the fusion protein were far less effective in promoting germination in *sly1* mutants compared to the wild-type background. For example, the YFP-AFP2 fusion protein accumulated to at least three-fold higher levels in the wild-type background than the same construct in the *sly1* background (Figures 5B and S5) and permitted comparable germination on at least 50-fold higher ABA concentrations. Similarly, although the YFP-AFP1 fusion protein accumulated to similar levels in *sly1* and wild-type backgrounds, the transgene permitted a similar degree of germination on 1 μM ABA in the *sly* background as on 30 μM ABA in the wild-type background (Figure 2). Differences in YFP-AFP2 protein accumulation did not reflect different transcript levels (Figure S5), suggesting the possibility of post-transcriptional control.

The DELLA protein RGA1 is barely detectable in wild-type seeds and seedlings (Figure 5B). In contrast, RGA1 accumulates substantially post-imbibition in the *sly1* background, consistent with the roughly ten-fold increase in transcript levels reported for incubation in the presence of either water or ABA, and accumulates to even higher levels in rosette leaves (Figure 5 and Figure S4B). RGA1 levels are not substantially altered by the presence of either *YFP-AFP* transgene.

As previously reported, ABI5 levels increase post-imbibition when exposed to ABA and decrease during germination [29]. In the present study, they are mostly reflective of germination status, increasing during delayed germination of ABA-treated wild-type seeds, but showing only a slight, further delayed increase in *sly1* mutant seeds, which are not germinating at all (Figure 5 and Figure S6). In lines carrying the transgenes, germination proceeds more rapidly and by 3d post-stratification, seeds exposed to ABA are greening and ABI5 levels are not substantially higher than those in dry seeds. Similar to AFP2, ABI5 undergoes a shift in mobility in seeds incubated on media containing ABA, in this case, toward higher apparent molecular weight forms (Figure 5). ABI5 is fully degraded in all genotypes that have germinated on hormone-free media.

3. Discussion

ABA and GA act antagonistically throughout plant development. Plant responses to these conflicting signals are mediated through a combination of changing hormone levels and sensitivities. Levels of these hormones are controlled by environmental signals, e.g., light and temperature, and cross-regulation of hormone metabolism, while sensitivity reflects relative activities of a complex network of positive and negative regulators of response (reviewed in [2,36]). In the current work, we have focused on the interaction between two sets of negative regulators, each of which have been characterized as signaling hubs due to their extensive interactions with other signaling proteins: the DELLA repressors of GA response [37] and the AFP repressors of ABA response [32]. During seed maturation, *RGA2* and *RGL3* expression in seeds is promoted by ABI5 (Figure S2). In mature seeds *RGA1*, *RGA2*, and *RGL2* transcripts increase following imbibition with or without ABA (Figure S2) and RGL2 is subject to proteasomal degradation when GA is present, thereby permitting germination [25]. *AFP1* and *AFP2* transcripts also increase following imbibition, but are induced by ABA, high glucose and salt stress over several days, potentially acting in a feedback loop to gradually release seeds from ABA-inhibition of germination [30].

Physical interactions between subsets of the DELLA and AFP families were detected in yeast and BiFC assays in *N. benthamiana*. Although not all interactions were observed in both assays, it is not unusual to see discrepancies between their results [38]. The interactions in yeast were dependent on the GRAS domain, previously shown to be important for interactions with over 150 transcriptional regulators that can have either positive or negative effects on gene expression [39,40]. The stronger interactions observed with the full-length DELLAs might reflect contributions of additional domains not tested individually in these constructs or effects on the conformation of the GRAS domain.

Because both DELLAs and AFPs comprise families with partially redundant functions that can mask the effects of loss of function mutations, we chose to analyze interactions between the gain of function lines that over-accumulate the respective repressors. Although the use of the *sly1* background does not readily distinguish between DELLAs relevant for different developmental effects, previous studies have identified RGL2 as the major regulator of seed dormancy and germination [24], but whose function is enhanced by RGA1,

RGA2/GAI, and RGL1 [15]. RGL2 acts in an ABI5-dependent manner both indirectly by increasing *ABI5* expression [26,27] and directly through complex formation [20,28]. RGL2 and several ABI transcription factors all repress *AtGASA6*, thereby inhibiting the promotion of cell wall loosening needed for cell expansion in germination [41]. RGL2 also forms a complex with DNA BINDING1 ZINC FINGER6 (DOF6) that directly activates the expression of *GATA12*, another transcription factor implicated in maintaining primary seed dormancy [42]. The Circadian clock protein REVEILLE(RVE)1 also promotes dormancy by interactions with RGL2 that stabilize it, blocking its degradation by SLY1-mediated targeting to the COP9 signalosome [43,44]. However, in after-ripened *sly1* seeds, loss of RGL2 is not required to permit germination [45].

In the present study, overexpression of either *AFP1* or *AFP2* was sufficient to release dormancy and promote some degree of ABA-resistant germination in the *sly1* background despite no ability to degrade DELLA proteins. Furthermore, ABI5 was maintained at similar levels in transgenic seeds germinating on a low concentration of ABA, regardless of whether in the wild-type or *sly1* background. This indicates that AFP-induced germination does not require loss of either the DELLAs or ABI5, and further suggests that the AFPs are acting by inhibiting the function of ABI5 and/or the DELLAs, but does not distinguish between direct and indirect inhibition. Furthermore, YFP-AFP2 appears to be subject to post-transcriptional regulation resulting in reduced accumulation in the *sly1* background. Although the mechanism is not known in this case, the YFP-AFP2 fusion is subject to proteasomal degradation [34].

While the balance between ABA and GA signaling shifts toward GA response in controlling the transition from seed to seedling at maturity, ABA signaling predominates in maturing seeds. The maize VP1 transcription factor, the ortholog to Arabidopsis ABI3, has a dual role as a positive regulator of ABA response in maturing seeds and repressor of GA response post-imbibition [8]. DELLA proteins, although initially characterized as repressors of GA signaling, can also function as positive regulators of ABA synthesis via the promotion of XERICO expression [26,46]. Increased ABA levels further promote ABA response by induction of transcription factors, including ABI5 (reviewed in [7]). During seed maturation, when GA levels are low, RGL3 is highly expressed and interacts directly with ABI3 to promote seed storage protein accumulation; overexpression of RGL3 leads to slight increases in storage protein content [35]. The *sly1* mutant has a similarly mild increase in storage proteins. In contrast, overexpression of either AFP substantially reduced storage protein accumulation, consistent with recent proteomic studies of the AFP2 overexpression line [34]; the combination of excess AFP and DELLAs was again intermediate.

All of the DELLA proteins are expressed in flowers and siliques, and genetic evidence indicates that *RGA1*, *RGL1*, and *RGL2* all regulate floral development [25]. Some aspects of DELLA regulation promote fertility: *RGA1* and *RGA2/GAI* function is required for the production of viable pollen [47] and *RGL2* positively regulate ovule number and fertility in Arabidopsis [48]. However, DELLA proteins repress cell elongation in stamens, contributing to male sterility in GA-deficient mutants [49]. Whereas loss of function for *RGA1* or *RGA2/GAI* can suppress the defects in stem elongation, apical dominance, and flowering time in *sly1* mutants, this is not sufficient to restore fertility [18]. Although YFP-AFP2 overexpression slightly reduced branching in the *sly1* background, AFP overexpression generally failed to rescue the *sly1* mutant's growth defects.

In summary, although the mechanism is not yet clear, the *AFP* transgenes and the over-accumulated DELLAs exert additive antagonistic effects on seed development and germination. The minimal effects of the *YFP-AFP* transgenes on vegetative growth and fertility are consistent with the strong expression of RGA1 and failure to detect the YFP-AFP fusions at later stages in growth, e.g., rosette leaves. These results show a stage-specific antagonism between these two classes of signaling hubs.

4. Materials and Methods

4.1. Yeast Two-Hybrid Constructs and Assays

Full-length ORF cDNAs for RGA1 (AT2G01570), RGA2 (AT1G14920), and RGL3 (AT5G17490), (U13937, U14047, and U60167, respectively) [50] were obtained from the ABRC and an RGL2 (AT3G03450) cDNA was constructed by PCR of the coding sequence from genomic DNA, adding BglII sites to the ends using primers described in Table S1, then subcloned into the BamHI site of pUNI51. Subclones encompassing either the DELLA or GRAS domains of RGA2 and RGL2 were constructed by PCR using the full-length cDNAs as templates and primers as indicated (Table S1), and ligated into pUNI51. The RGA2 DELLA domain clone included codons 1–121, the RGA2 GRAS domain clone included codons 115–533, RGL2 DELLA domain clone encoded amino acids 1–145, and the RGL2 GRAS domain included codons 139–547. Fusions between the GAL4 activation domain (AD) and these DELLA clones were constructed by recombination with pACT2lox, catalyzed by GST-CRE as described in [51]. The GAL4 binding domain (BD) fusions to the AFPs and both AD- and BD-fusions for ABI5 have been described previously [32]. Interactions were tested by matings between pairs of haploid lines carrying BD- and AD-fusions, replica-plated onto selective media to test for activation of the HIS3 and ADE2 reporter genes, as described in [32].

4.2. Plant Materials and Transgenes

Arabidopsis plants were grown in pots in growth chambers under continuous light at 22 °C. *SLY1* loss of function lines was described in [16]; we used a line carrying the *sly1-2* allele that had been backcrossed three times into the Col-0 background. The *35S:YFP:AFP* fusions and split YFP fusions for ABI5, AFP1 and AFP2 were described in [32]. The *35S:YFP:AFP2* line used was #4A2. *Agrobacterium tumefaciens*-mediated direct transformation of *sly1-2* mutants was performed by the floral dip method [52], followed by the selection of BASTA-resistant seedlings. Essentially all germinating seeds were transgenic, reflecting suppression of the strong dormancy of *sly1* mutants. Progeny continued to segregate BASTA-resistance over multiple generations. A *35S:YFP:AFP1* transgene was backcrossed from the *sly1* background into Col-0 and F1s screened for BASTA resistance; within the F2 population, germination is limited to approximately 94% of seeds that have a wild-type *SLY1* allele and/or the transgene.

Split YFP fusions for RGA, RGA2, and RGL2 were constructed using the Gateway compatible pSITE-cEYFP-C1 (Acc# GU734652) vector and PCR products with attL ends added as described in [53], using primers described in Table S1, following manufacturer's instructions for LR Clonase reactions (Invitrogen). BiFC assays were conducted as described in [32].

4.3. Plant Growth Conditions

Germination assays testing ABA sensitivity of age-matched seeds were performed on minimal nutrient media supplemented with ABA at concentrations over the range from 0–50 µM, as described in [32]. Following sterilization and plating, plates were incubated 3d at 4 °C; this was sufficient to break residual dormancy for all except the hyperdormant *sly1* parental line. Accumulation of fusion proteins was assayed by immunoblots of seeds or seedlings harvested after 3–4d incubation on Germination Medium (GM: 0.5 × MS salts and vitamins, 1% sucrose) with or without 1 µM ABA or 8 µg/mL BASTA, solidified with 0.7% agar.

For analysis of adult plant development, seedlings were transferred to soil and photographed at weekly intervals. The number of shoots was scored for individual mature plants (n = 11−22), and the number of sterile pods on the main inflorescence and at least 4 lateral branches of each plant was counted.

4.4. Protein Analyses

Seeds or seedlings were ground directly in $1\times$ or $2\times$ Laemmli loading buffer, respectively, microfuged 10 min at 4 °C to pellet debris, then boiled 5 min prior to fractionation by SDS-PAGE (10% polyacrylamide). Proteins were transferred to nitrocellulose filters, as described in [32]. Filters were blocked with Casein blocking buffer (LI-COR Biosciences, Lincoln, NE, USA), then co-incubated with anti-GFP mAb (1:10,000, UBPBio, Aurora, CO, USA) and anti-RGA pAb (1:2000, AS11 1630, Agrisera, Vännäs, Sweden) primary antibodies, followed by anti-mouse and anti-rabbit secondary IRDye 800 conjugated IgGs, and visualized using the 800 channel of the Licor Odyssey Infrared Imaging System or the iBright FL1500 Imaging System (Invitrogen, ThermoFisher Scientific, Waltham, MA USA). Filters were subsequently probed with anti-ABI5pAb (1:10,000, Ab98831, AbCam, Cambridge, UK) and anti-actin mAb (A0480, Sigma, St. Louis, MO, USA), followed by anti-mouse secondary IRDye 800 conjugated IgGs (LI-COR Biosciences, Lincoln, NE, USA).

4.5. Transcript Analyses

RNA was extracted, as described in [32], then incubated with RQ1 DNase (Promega, Madison, WI, USA) and RNAsin for 15 min at room temperature. The reaction was stopped by the addition of EGTA (1.8 mM final), then RNA was purified over Zymo-Clean columns (Zymo Resesarch, Irvine, CA, USA) according to the manufacturer's instructions. Approximately 0.5 ug of RNA was used as a template for cDNA reactions using MMLV reverse transcriptase or GoScript (Promega, Madison, WI, USA) and a 10:1 mix of random hexamers and oligo dT as primers. cDNA concentrations were assayed by qRT-PCR using EvaGreen Master Mix (Midwest Scientific, Valley Park, MO, USA) or Forget-Me-Not EvaGreen Master Mix (Biotium, Fremont, CA, USA) in an iQ5 cycler (BioRad, Hercules, CA, USA) according to the manufacturer's instructions. Primers used for normalizing were selected for uniform expression in seeds and seedlings grown in a variety of conditions [54]. Reactions with both primer sets were quantified relative to a standard curve spanning the range of concentrations present in all samples, as described in [55].

Supplementary Materials: The following are available online at https://www.mdpi.com/article/10.3390/ijms23105537/s1.

Author Contributions: Conceptualization, R.R.F.; validation, T.J.L. and R.R.F.; formal analysis, R.R.F.; investigation, T.J.L. and R.R.F.; resources, R.R.F.; data curation, T.J.L. and R.R.F.; writing—original draft preparation, R.R.F.; writing—review and editing, T.J.L. and R.R.F.; visualization, R.R.F.; supervision, R.R.F.; project administration, T.J.L. and R.R.F.; funding acquisition, R.R.F. All authors read and agreed to the published version of the manuscript.

Funding: This work was supported by UCSB Academic Senate Grant to R.R.F., Faculty Research Assistance Program funds, National Science Foundation Grant# IOS1558011 to R.R.F.

Data Availability Statement: Transcriptome data from Nakabayashi et al. (2005), the Yamaguchi lab, and Schmid et al. (2005), is present on the Arabidopsis eFP browser at (bar.utoronto.ca) [22,23].

Acknowledgments: We thank Camille Steber for the *sly1* mutant lines and the Arabidopsis Biological Resource Center (ABRC) for cDNA clones. In addition, we thank Shirley Luo for assistance with the yeast two-hybrid fusion constructions and assays, and Eliana Tucker for assistance with construction of split YFP and His-fusion clones and scoring fertility in the transgenic lines.

Conflicts of Interest: The authors declare no conflict of interest.

References

1. Carrera-Castaño, G.; Calleja-Cabrera, J.; Pernas, M.; Gómez, L.; Oñate-Sánchez, L. An Updated Overview on the Regulation of Seed Germination. *Plants* **2020**, *9*, 703. [CrossRef] [PubMed]
2. Holdsworth, M.J.; Bentsink, L.; Soppe, W.J.J. Molecular networks regulating Arabidopsis seed maturation, after-ripening, dormancy and germination. *New Phytol.* **2008**, *179*, 33–54. [CrossRef] [PubMed]
3. Seo, M.; Nambara, E.; Choi, G.; Yamaguchi, S. Interaction of light and hormone signals in germinating seeds. *Plant Mol. Biol.* **2009**, *69*, 463–472. [CrossRef] [PubMed]

4. Penfield, S. Seed dormancy and germination. *Curr. Biol.* **2017**, *27*, R874–R878. [CrossRef]
5. Koornneef, M.; van der Veen, J. Induction and analysis of gibberellin sensitive mutants in *Arabidopsis thaliana* (L.) Heynh. *Theor. Appl. Genet.* **1980**, *58*, 257–263. [CrossRef]
6. Koornneef, M.; Jorna, M.; Brinkhorst-van der Swan, D.; Karssen, C. The isolation of abscisic acid (ABA)-deficient mutants by selection of induced revertants in non-germinating gibberellin sensitive lines of *Arabidopsis thaliana* (L.) Heynh. *Theor. Appl. Genet.* **1982**, *61*, 385–393. [CrossRef]
7. Finkelstein, R. Abscisic Acid Synthesis and Response. *Arab. Book* **2013**, *11*, e0166. [CrossRef]
8. Hoecker, U.; Vasil, I.K.; McCarty, D.R. Integrated control of seed maturation and germination programs by activator and repressor functions of Viviparous-1 of maize. *Genes Dev.* **1995**, *9*, 2459–2469. [CrossRef]
9. Wind, J.J.; Peviani, A.; Snel, B.; Hanson, J.; Smeekens, S.C. ABI4: Versatile activator and repressor. *Trends Plant Sci.* **2013**, *18*, 125–132. [CrossRef]
10. Eisner, N.; Maymon, T.; Sanchez, E.C.; Bar-Zvi, D.; Brodsky, S.; Finkelstein, R.; Bar-Zvi, D. Phosphorylation of Serine 114 of the transcription factor ABSCISIC ACID INSENSITIVE 4 is essential for activity. *Plant Sci.* **2021**, *305*, 110847. [CrossRef]
11. Collin, A.; Daszkowska-Golec, A.; Szarejko, I. Updates on the Role of ABSCISIC ACID INSENSITIVE 5 (ABI5) and ABSCISIC ACID-RESPONSIVE ELEMENT BINDING FACTORs (ABFs) in ABA Signaling in Different Developmental Stages in Plants. *Cells* **2021**, *10*, 1996. [CrossRef]
12. Koornneef, M.; Elgersma, A.; Hanhart, C.; Loenen-Martinet, E.V.; Rijn, L.V.; Zeevaart, J. A gibberellin insensitive mutant of Arabidopsis thaliana. *Physiol. Plant.* **1985**, *65*, 33–39. [CrossRef]
13. Ueguchi-Tanaka, M.; Ashikari, M.; Nakajima, M.; Itoh, H.; Katoh, E.; Kobayashi, M.; Chow, T.-Y.; Hsing, Y.; Kitano, H.; Yamaguchi, I.; et al. *GIBBERELLIN INSENSITIVE DWARF1* encodes a soluble receptor for gibberellin. *Nature* **2005**, *437*, 693–698. [CrossRef]
14. Jacobsen, S.E.; Olszewski, N.E. Mutations at the SPINDLY locus of Arabidopsis alter gibberellin signal transduction. *Plant Cell* **1993**, *5*, 887–896. [PubMed]
15. Cao, D.N.; Hussain, A.; Cheng, H.; Peng, J.R. Loss of function of four DELLA genes leads to light- and gibberellin-independent seed germination in Arabidopsis. *Planta* **2005**, *223*, 105–113. [CrossRef] [PubMed]
16. Steber, C.M.; Cooney, S.E.; McCourt, P. Isolation of the GA-response mutant sly1 as a suppressor of ABI1-1 in Arabidopsis thaliana. *Genetics* **1998**, *149*, 509–521. [CrossRef]
17. McGinnis, K.; Thomas, S.; Soule, J.; Strader, L.; Zale, J.; Sun, T.-P.; Steber, C. The Arabidopsis *SLEEPY1* gene encodes a putative F-box subunit of an SCF E3 ubiquitin ligase. *Plant Cell* **2003**, *15*, 1120–1130. [CrossRef]
18. Dill, A.; Thomas, S.G.; Hu, J.; Steber, C.M.; Sun, T.P. The Arabidopsis F-box protein SLEEPY1 targets gibberellin signaling repressors for gibberellin-induced degradation. *Plant Cell* **2004**, *16*, 1392–1405. [CrossRef]
19. Ariizumi, T.; Hauvermale, A.L.; Nelson, S.K.; Hanada, A.; Yamaguchi, S.; Steber, C.M. Lifting DELLA Repression of Arabidopsis Seed Germination by Nonproteolytic Gibberellin Signaling. *Plant Physiol.* **2013**, *162*, 2125–2139. [CrossRef]
20. Lim, S.; Park, J.; Lee, N.; Jeong, J.; Toh, S.; Watanabe, A.; Kim, J.; Kang, H.; Kim, D.H.; Kawakami, N.; et al. ABA-insensitive3, ABA-insensitive5, and DELLAs Interact to activate the expression of SOMNUS and other high-temperature-inducible genes in imbibed seeds in Arabidopsis. *Plant Cell* **2013**, *25*, 4863–4878. [CrossRef]
21. Vaistij, F.E.; Barros-Galvão, T.; Cole, A.F.; Gilday, A.D.; He, Z.; Li, Y.; Harvey, D.; Larson, T.R.; Graham, I.A. *MOTHER-OF-FT-AND-TFL1* represses seed germination under far-red light by modulating phytohormone responses in *Arabidopsis thaliana*. *Proc. Natl. Acad. Sci. USA* **2018**, *115*, 8442–8447. [CrossRef]
22. Winter, D.; Vinegar, B.; Nahal, H.; Ammar, R.; Wilson, G.; Provart, N. An "Electronic Fluorescent Pictograph" Browser for Exploring and Analyzing Large-Scale Biological Data Sets. *PLoS ONE* **2007**, *2*, e718. [CrossRef] [PubMed]
23. Bassel, G.W.; Fung, P.; Chow, T.-F.F.; Foong, J.A.; Provart, N.J.; Cutler, S.R. Elucidating the Germination Transcriptional Program Using Small Molecules. *Plant Physiol.* **2008**, *147*, 143–155. [CrossRef] [PubMed]
24. Lee, S.; Cheng, H.; King, K.E.; Wang, W.; He, Y.; Hussain, A.; Lo, J.; Harberd, N.P.; Peng, J. Gibberellin regulates Arabidopsis seed germination via RGL2, a GAI/RGA-like gene whose expression is up-regulated following imbibition. *Genes Dev.* **2002**, *16*, 646–658. [CrossRef]
25. Tyler, L.; Thomas, S.G.; Hu, J.; Dill, A.; Alonso, J.M.; Ecker, J.R.; Sun, T.P. DELLA proteins and gibberellin-regulated seed germination and floral development in Arabidopsis. *Plant Physiol.* **2004**, *135*, 1008–1019. [CrossRef] [PubMed]
26. Piskurewicz, U.; Jikumaru, Y.; Kinoshita, N.; Nambara, E.; Kamiya, Y.; Lopez-Molina, L. The Gibberellic Acid Signaling Repressor RGL2 Inhibits Arabidopsis Seed Germination by Stimulating Abscisic Acid Synthesis and ABI5 Activity. *Plant Cell* **2008**, *20*, 2729–2745. [CrossRef] [PubMed]
27. Liu, X.; Hu, P.; Huang, M.; Tang, Y.; Li, Y.; Li, L.; Hou, X. The NF-YC–RGL2 module integrates GA and ABA signalling to regulate seed germination in Arabidopsis. *Nat. Commun.* **2016**, *7*, 12768. [CrossRef]
28. Hu, Y.; Han, X.; Yang, M.; Zhang, M.; Pan, J.; Yu, D. The Transcription Factor INDUCER OF CBF EXPRESSION1 Interacts with ABSCISIC ACID INSENSITIVE5 and DELLA Proteins to Fine-Tune Abscisic Acid Signaling during Seed Germination in Arabidopsis. *Plant Cell* **2019**, *31*, 1520–1538. [CrossRef]
29. Lopez-Molina, L.; Mongrand, S.; Kinoshita, N.; Chua, N.-H. AFP is a novel negative regulator of ABA signaling that promotes ABI5 protein degradation. *Genes Dev.* **2003**, *17*, 410–418. [CrossRef]

30. Garcia, M.; Lynch, T.; Peeters, J.; Snowden, C.; Finkelstein, R. A small plant-specific protein family of ABI five binding proteins (AFPs) regulates stress response in germinating *Arabidopsis* seeds and seedlings. *Plant Mol. Biol.* **2008**, *67*, 643–658. [CrossRef]

31. Chang, G.; Yang, W.; Zhang, Q.; Huang, J.; Yang, Y.; Hu, X. ABI5-BINDING PROTEIN2 Coordinates CONSTANS to Delay Flowering by Recruiting the Transcriptional Corepressor TPR2. *Plant Physiol.* **2019**, *179*, 477–490. [CrossRef] [PubMed]

32. Lynch, T.J.; Erickson, B.J.; Miller, D.R.; Finkelstein, R.R. ABI5-binding proteins (AFPs) alter transcription of ABA-induced genes via a variety of interactions with chromatin modifiers. *Plant Mol. Biol.* **2017**, *93*, 403–418. [CrossRef] [PubMed]

33. Pauwels, L.; Barbero, G.F.; Geerinck, J.; Tilleman, S.; Grunewald, W.; Perez, A.C.; Chico, J.M.; Bossche, R.V.; Sewell, J.; Gil, E.; et al. NINJA connects the co-repressor TOPLESS to jasmonate signalling. *Nature* **2010**, *464*, 788–791. [CrossRef]

34. Lynch, T.; Nee, G.; Chu, A.; Kruger, T.; Finkemeier, I.; Finkelstein, R.R. ABI5 binding protein2 inhibits ABA responses during germination without ABA-INSENSITIVE5 degradation. *Plant Physiol.* **2022**, kiac096. [CrossRef]

35. Hu, Y.; Zhou, L.; Yang, Y.; Zhang, W.; Chen, Z.; Li, X.; Qian, Q.; Kong, F.; Li, Y.; Liu, X.; et al. The gibberellin signaling negative regulator RGA-LIKE3 promotes seed storage protein accumulation. *Plant Physiol.* **2021**, *185*, 1697–1707. [CrossRef] [PubMed]

36. Liu, X.; Hou, X. Antagonistic Regulation of ABA and GA in Metabolism and Signaling Pathways. *Front. Plant Sci.* **2018**, *9*, 251. [CrossRef]

37. Davière, J.-M.; Achard, P. A Pivotal Role of DELLAs in Regulating Multiple Hormone Signals. *Mol. Plant* **2016**, *9*, 10–20. [CrossRef]

38. Lumba, S.; Toh, S.; Handfield, L.-F.; Swan, M.; Liu, R.; Youn, J.-Y.; Cutler, S.R.; Subramaniam, R.; Provart, N.; Moses, A.; et al. A Mesoscale Abscisic Acid Hormone Interactome Reveals a Dynamic Signaling Landscape in Arabidopsis. *Dev. Cell* **2014**, *29*, 360–372. [CrossRef]

39. Marín-de la Rosa, N.; Sotillo, B.; Miskolczi, P.; Gibbs, D.J.; Vicente, J.; Carbonero, P.; Oñate-Sánchez, L.; Holdsworth, M.J.; Bhalerao, R.; Alabadí, D.; et al. Large-Scale Identification of Gibberellin-Related Transcription Factors Defines Group VII ETHYLENE RESPONSE FACTORS as Functional DELLA Partners. *Plant Physiol.* **2014**, *166*, 1022–1032. [CrossRef]

40. Hernández-García, J.; Briones-Moreno, A.; Dumas, R.; Blázquez, M.A. Origin of Gibberellin-Dependent Transcriptional Regulation by Molecular Exploitation of a Transactivation Domain in DELLA Proteins. *Mol. Biol. Evol.* **2019**, *36*, 908–918. [CrossRef]

41. Zhong, C.; Xu, H.; Ye, S.; Wang, S.; Li, L.; Zhang, S.; Wang, X. Gibberellic Acid-Stimulated Arabidopsis6 Serves as an Integrator of Gibberellin, Abscisic Acid, and Glucose Signaling during Seed Germination in Arabidopsis. *Plant Physiol.* **2015**, *169*, 2288–2303. [PubMed]

42. Ravindran, P.; Verma, V.; Stamm, P.; Kumar, P.P. A Novel RGL2-DOF6 Complex Contributes to Primary Seed Dormancy in *Arabidopsis thaliana* by Regulating a GATA Transcription Factor. *Mol. Plant* **2017**, *10*, 1307–1320. [CrossRef]

43. Yang, L.; Jiang, Z.; Liu, S.; Lin, R. Interplay between REVEILLE1 and RGA-LIKE2 regulates seed dormancy and germination in Arabidopsis. *New Phytol.* **2020**, *225*, 1593–1605. [CrossRef] [PubMed]

44. Jin, D.; Wu, M.; Li, B.; Bücker, B.; Keil, P.; Zhang, S.; Li, J.; Kang, D.; Liu, J.; Dong, J.; et al. The COP9 Signalosome regulates seed germination by facilitating protein degradation of RGL2 and ABI5. *PLoS Genet.* **2018**, *14*, e1007237. [CrossRef] [PubMed]

45. Ariizumi, T.; Steber, C.M. Seed germination of GA-insensitive *sleepy1* mutants does not require RGL2 protein disappearance in Arabidopsis. *Plant Cell* **2007**, *19*, 791–804. [CrossRef]

46. Zentella, R.; Zhang, Z.-L.; Park, M.; Thomas, S.G.; Endo, A.; Murase, K.; Fleet, C.M.; Jikumaru, Y.; Nambara, E.; Kamiya, Y.; et al. Global Analysis of DELLA Direct Targets in Early Gibberellin Signaling in Arabidopsis. *Plant Cell* **2007**, *19*, 3037–3057. [CrossRef]

47. Plackett, A.R.G.; Ferguson, A.C.; Powers, S.J.; Wanchoo-Kohli, A.; Phillips, A.L.; Wilson, Z.A.; Hedden, P.; Thomas, S.G. DELLA activity is required for successful pollen development in the Columbia ecotype of Arabidopsis. *New Phytol.* **2014**, *201*, 825–836. [CrossRef]

48. Gómez, M.D.; Fuster-Almunia, C.; Ocaña-Cuesta, J.; Alonso, J.M.; Pérez-Amador, M.A. RGL2 controls flower development, ovule number and fertility in Arabidopsis. *Plant Sci.* **2019**, *281*, 82–92. [CrossRef]

49. Cheng, H.; Qin, L.; Lee, S.; Fu, X.; Richards, D.E.; Cao, D.; Luo, D.; Harberd, N.P.; Peng, J. Gibberellin regulates Arabidopsis floral development via suppression of DELLA protein function. *Development* **2004**, *131*, 1055–1064. [CrossRef]

50. Yamada, K.; Lim, J.; Dale, J.M.; Chen, H.; Shinn, P.; Palm, C.J.; Southwick, A.M.; Wu, H.C.; Kim, C.; Nguyen, M.; et al. Empirical Analysis of the Transcriptional Activity in the Arabidopsis Genome. *Science* **2003**, *302*, 842–846. [CrossRef]

51. Liu, Q.; Li, M.Z.; Leibham, D.; Cortez, D.; Elledge, S.J. The univector plasmid-fusion system, a method for rapid construction of recombinant DNA without restriction enzymes. *Curr. Biol.* **1998**, *8*, 1300–1309. [CrossRef]

52. Clough, S.; Bent, A. Floral dip: A simplified method for *Agrobacterium*-mediated transformation of *Arab. Thaliana*. *Plant J.* **1998**, *16*, 735–743. [CrossRef] [PubMed]

53. Fu, C.; Wehr, D.R.; Edwards, J.; Hauge, B. Rapid one-step recombinational cloning. *Nucleic Acids Res.* **2008**, *36*, e54. [CrossRef] [PubMed]

54. Czechowski, T.; Stitt, M.; Altmann, T.; Udvardi, M.K.; Scheible, W.-R.D. Genome-Wide Identification and Testing of Superior Reference Genes for Transcript Normalization in Arabidopsis. *Plant Physiol.* **2005**, *139*, 5–17. [CrossRef]

55. Carr, A.C.; Moore, S.D. Robust Quantification of Polymerase Chain Reactions Using Global Fitting. *PLoS ONE* **2012**, *7*, e37640.

International Journal of
Molecular Sciences

Article

Arabidopsis G-Protein β Subunit AGB1 Negatively Regulates DNA Binding of MYB62, a Suppressor in the Gibberellin Pathway

Xin Qi [1,2,†], Wensi Tang [1,†], Weiwei Li [3], Zhang He [1], Weiya Xu [1], Zhijin Fan [2], Yongbin Zhou [1], Chunxiao Wang [1], Zhaoshi Xu [1], Jun Chen [1], Shiqin Gao [4], Youzhi Ma [1,*] and Ming Chen [1,*]

[1] Institute of Crop Sciences, Chinese Academy of Agricultural Sciences (CAAS)/National Key Facility for Crop Gene Resources and Genetic Improvement, Key Laboratory of Biology and Genetic Improvement of Triticeae Crops, Ministry of Agriculture, Beijing 100081, China; 13121253699@163.com (X.Q.); tang_wensi@yeah.net (W.T.); hz384298890@gmail.com (Z.H.); 15651912225@163.com (W.X.); zhouyongbin@caas.cn (Y.Z.); wcx51619@163.com (C.W.); xuzhaoshi@caas.cn (Z.X.); chenjun01@caas.cn (J.C.)
[2] State Key Laboratory of Elemento-Organic Chemistry, College of Chemistry, Nankai University, Tianjin 300071, China; fanzj@nankai.edu.cn
[3] Beijing Advanced Innovation Center for Food Nutrition and Human Health, Beijing Technology & Business University (BTBU), Beijing 100048, China; liweiwei.0304@163.com
[4] Beijing Engineering Research Center for Hybrid Wheat, Beijing Academy of Agriculture and Forestry Sciences, Beijing 100097, China; gshiq@126.com
* Correspondence: mayouzhi@caas.cn (Y.M.); chenming02@caas.cn (M.C.)
† Equally contributed.

Citation: Qi, X.; Tang, W.; Li, W.; He, Z.; Xu, W.; Fan, Z.; Zhou, Y.; Wang, C.; Xu, Z.; Chen, J.; et al. *Arabidopsis* G-Protein β Subunit AGB1 Negatively Regulates DNA Binding of MYB62, a Suppressor in the Gibberellin Pathway. *Int. J. Mol. Sci.* **2021**, 22, 8270. https://doi.org/10.3390/ijms22158270

Academic Editor: Karen Skriver

Received: 25 May 2021
Accepted: 27 July 2021
Published: 31 July 2021

Abstract: Plant G proteins are versatile components of transmembrane signaling transduction pathways. The deficient mutant of heterotrimeric G protein leads to defects in plant growth and development, suggesting that it regulates the GA pathway in *Arabidopsis*. However, the molecular mechanism of G protein regulation of the GA pathway is not understood in plants. In this study, two G protein β subunit (*AGB1*) mutants, *agb1-2* and *N692967*, were dwarfed after exogenous application of GA_3. AGB1 interacts with the DNA-binding domain MYB62, a GA pathway suppressor. Transgenic plants were obtained through overexpression of *MYB62* in two backgrounds including the wild-type (*MYB62/WT Col-0*) and *agb1* mutants (*MYB62/agb1*) in *Arabidopsis*. Genetic analysis showed that under GA_3 treatment, the height of the transgenic plants *MYB62/WT* and *MYB62/agb1* was lower than that of WT. The height of *MYB62/agb1* plants was closer to *MYB62/WT* plants and higher than that of mutants *agb1-2* and *N692967*, suggesting that *MYB62* is downstream of *AGB1* in the GA pathway. qRT-PCR and competitive DNA binding assays indicated that MYB62 can bind MYB elements in the promoter of *GA2ox7*, a GA degradation gene, to activate *GA2ox7* transcription. AGB1 affected binding of MYB62 on the promoter of *GA2ox7*, thereby negatively regulating th eactivity of MYB62.

Keywords: *Arabidopsis*; GA signaling; *AGB1*; *MYB62*; protein interaction

1. Introduction

The heterotrimeric G protein pathway is a conservative transmembrane signal transduction pathway found in both animals and plants [1,2]. In *Arabidopsis*, the heterotrimeric G protein is composed of an α subunit (*GPA1*), a β subunit (*AGB1*), and three γ subunits (*AGG1*, *AGG2*, and *AGG3*) [3]. During G protein signaling transduction, G-α binds to GDP and forms a coupling polymer with the G-β-γ dimer, leaving the G protein pathway in a resting state [4] (Anantharaman et al., 2011). When the G protein pathway is activated, the GTP-G-α monomer is dissociated from the G-β-γ dimer, while the GTP-G-α monomer and the G-β-γ dimer interact with a variety of downstream effectors to transmit signals for different cell and physiological functions [4] (Anantharaman et al., 2011). The heterotrimeric G proteins are involved in growth and developmental processes such as seed germination and seedling development [5] (Ullah et al., 2003), cell division and morphology [6] (Ullah et al., 2002), ion channel regulation [7] (Assmann and Yu, 2015), stomatal development [8] (Wang et al., 2011), and the response to environmental conditions such as phytohormones, sugar, ROS (reactive oxygen species), and light [9] (Li et al., 2012). Plants with mutated G protein complex components have altered morphology in their fruits, grain weight, roots, and leaves, and these mutants are sensitive to a variety of hormones, including IAA (auxin), GA (gibberellins), and BR (brassinosteroids) [2,5,10–12] (Ullah et al., 2003; Chen et al., 2004; Pandey et al., 2006; Chen, 2007; Urano et al., 2014).

The G-α subunit positively regulates the GA pathway by inhibiting the activity of SLR (slender rice), a negative regulator of GA signaling in rice [13] (Ueguchi-Tanka et al., 2000), and the *Arabidopsis* G-α mutant *gpa1* is less sensitive to GA [14] (Trusov et al., 2007). The G-β subunit is involved in a variety of signaling pathways in plants. In *Arabidopsis*, the interactions between AGB1 and ERECTA during silique development were the first evidence of the interaction between G proteins and receptor kinases [15] (Lease et al., 2001). We previously found that AGB1 is involved in regulating ABA and drought response by interacting with the protein kinase MPK6 and the transcription factor VIP1 [16] (Xu et al., 2015). We also found that AGB1 interacts with the transcription factor BBX21 to regulate photomorphogenesis in *Arabidopsis* [17] (Xu et al., 2017). In rice, suppression of the G-β subunit gene, *RGB1*, causes dwarfism and browning of the internodes and lamina joint regions [18] (Utsunomiya et al., 2011). In *Arabidopsis*, the G-β subunit gene mutant *agb1* has shorter mature plants than the wild-type (WT) [19] (Urano et al., 2016). These results suggest that the G-β subunit also regulates plant development through the GA pathway, but the specific mechanism is not clear.

GA functions directly in regulating plant growth and development as well as crop yield [20] (Singh et al., 2002). GA-related genes can be divided into two categories: either the pathways involved in GA synthesis or degradation, or the GA signaling transduction pathway. In plants, a variety of enzymes are involved in GA synthesis, while GGPP (geranylgeranyl pyrophosphate), a precursor of GA synthesis, is catalyzed by GGPS (geranylgeranyl pyrophosphate synthase) [21] (Lange et al., 2003). GGPP is further catalyzed by CPS (copalyl pyrophosphate synthase) and KS (ent-kaurene synthase), which forms kaurene [22] (Morrone et al., 2009). Additionally, KO (ent-kaurene oxidase) and KAO (ent-kaurenoic acid oxidase) have important roles in the GA synthesis intermediate GA_{12}-aldehyde [23,24] (Davidson et al., 2003; Sakamoto et al., 2004). GA_{12}-aldehyde is a branch point of GA, which hydroxylates to form GA_{12} and GA_{53} under the oxidation of *GA20ox1* (gibberellin 20-oxidase gene). GA_{12} and GA_{53} produce GA_9 and GA_{20}, while the final form of bioactive GA is catalyzed by *GA30ox1* [25] (Hedden et al., 2012).

In rice, the plant is dwarfed due to the mutation of *GA20ox*, a key gene for GA synthesis [26] (Ashikari et al., 2002). The deletion of *KAO* can lead to severe dwarfing in many plants, such as rice (*d35* mutant) [24] (Sakamoto et al., 2004), maize (*dwarf3*) [27] (Helliwell et al., 2001), pea (*na*) [23] (Davidson et al., 2003), and sunflower (*dwarf2*) [28] (Fambrini et al., 2011). The signal transduction of GA in plants is mediated by the receptor GID1, while DELLA is an inhibitor of the GA pathway via transcriptional regulation [29] (Ueguchi-Tanaka et al., 2005). GA promotes the formation of the GA–GID1–DELLA complex via conformational changes caused by GID1 binding, where the DELLA protein is then ubiquitinated and degraded by the 26S proteasome to open the GA pathway [30,31] (Murase et al., 2008; Shimada et al., 2008). In *Arabidopsis*, the transcription factor *MYB62* is involved in the GA pathway, and overexpression of *MYB62* results in a GA-deficient phenotype, which suggests that *MYB62* is a suppressor in the GA pathway [32] (Devaiah et al., 2009).

In this study, we found that plant heights in two G-β subunit mutants, *agb1-2* and *N692967*, were significantly lower than in the WT following GA_3 treatment, suggesting that the function of *AGB1* in the GA pathway is similar to that of *GPA1* in *Arabidopsis*. We found that AGB1 regulates the GA pathway by negatively regulating the DNA binding of MYB62, a GA pathway suppressor on the promoter of the GA degradation gene *GA2ox7*.

The G protein complex regulates the GA pathway through the *AGB1-MYB62-GA2ox7* pair in *Arabidopsis*.

2. Materials and Methods

2.1. Plant Materials and Growth Conditions

The *agb1* mutant *agb1-2* (CS6536) has been described by Ullah et al. (2003). In the other *agb1* mutant *N692967* (SALK_204268C), the T-DNA insert occurs in chr4 16,477,780 of the *Col-0* genome (Supplemental Figure S1). The expression of *AGB1* in mutants and *MYB62* in transgenic plants was identified by qRT-PCR using gene-specific primers (Table S1, Supplemental Figure S1). Because we could not obtain *MYB62* mutants, we overexpressed the *MYB62* gene in different genetic backgrounds to obtain transgenic plants, including *MYB62* transgenic plants in a WT background—*MYB62:GFP/WT-8* and *MYB62:GFP/WT-10*—and *MYB62* transgenic plants in an *agb1* background—*MYB62:GFP/agb1-2-1, MYB62:GFP/agb1-2-4*. The response of *Arabidopsis* plants to GA$_3$ treatment was observed at the seedling stage. Before planting *Arabidopsis* (*Col-0*) seeds on plates, the seeds were soaked with 10% sodium hypochlorite for 10 min and then washed 3 times with sterile distilled water. The sterilized *Arabidopsis* seeds were treated at 4 °C in the dark for 3 days and then germinated on 1/2 MS medium (0.8% agar) for 7 days. When *Arabidopsis* grew to the 4-leaf stage, seedlings were transplanted into 1/2 MS medium containing 1 μM GA$_3$, 10 μM GA$_3$, or 100 μM GA$_3$ and grown at 24 °C in the light for 16 h then at 20 °C in darkness for 8 h (Supplemental Figures S2–S5). The growth state was observed and recorded every day. Differences were observed when the seedings were grown under GA$_3$ treatment for 10 days.

2.2. Measurement of Gibberellin Content

Total GA content, including GA$_1$, GA$_2$, GA$_3$, GA$_4$, GA$_7$, and GA$_{20}$, was measured using an ELISA kit (Plant GA ELISA Kit, X-Y Biotechnology company) according to the manufacturer's instructions. Plant samples (0.1 g) were ground, then dissolved in 900 μL of PBS buffer, fully mixed and centrifuged at 12,000 rpm for 10 min at 4 °C. All standards and samples were added in duplicate to Micro-ELISA strip plate wells (Eppendorf, Germany). The volume for a standard curve sample and a measured sample was 50 μL, and nothing was added to the blank well. A total of 100 μL of HRP-conjugate reagent was added to each well, then the wells were covered with tinfoil and incubated for 60 min at 37 °C. After washing 5 times, 50 μL of chromogen Solution A and 50 μL of chromogen Solution B were added to each well. The plate was incubated for 15 min at 37 °C and 50 μL of a stop solution was added to each well. The optical density at 450 nm was read using a microtiter plate within 15 min. The standard curve was generated by plotting the average OD$_{450}$ obtained for each of the 6 standard concentrations on the vertical (*y*) axis versus the corresponding concentration on the horizontal (*x*) axis. The concentration of the sample was calculated according to the equation of the standard curve.

2.3. Construction of Plasmids and Transgenic Lines

The full length of *MYB62* (AT1G68320) was amplified using the primers 1302-MYB62-F and 1302-MYB62-R and inserted into the pCambia-1302 vector (Clontech, San Francisco, CA, USA) with GFP at the C-terminal end through the *BamHII* restriction enzyme cutting site. The constructs were transformed into wild-type (*Col-0*) and mutant *Arabidopsis agb1-2* at the flowering stage by *Agrobacterium tumefaciens* (GV3101)-mediated transformation [33] (Clough et al., 1998). The seeds of the T0 generation of transgenic *Arabidopsis* were sown on selective medium (MS medium with 40 mg/L hygromycin), and the seedings were transplanted into the soil in pots. The seeds of theT1 and T2 transgenic lines were further screened by hygromycin. More than 95% of the seeds of the T2 generation with hygromycin resistance were homozygous lines. The *MYB62:GFP/agb1-2* transgenic plants were identified using kanamycin and hygromycin for screening and identification. The phenotypes of homozygous lines were analyzed under GA$_3$ treatment in the T3 generation.

2.4. Extraction of RNA and Analysis of Gene Expression

Seedlings of *MYB62* transgenic lines and the mutants *agb1-2* and *N692967*, and the roots and leaves of WT *Arabidopsis* were used for gene expression analysis. The total plant RNA was extracted using the Trizol method (Zhuangmeng Total RNA Extraction Kit), and the RNA was reverse-transcribed into cDNA using the TransScript One-Step gDNA removal and cDNA Synthesis SuperMix kit (TransGen Biotech, Beijing, China). We performed qRT-PCR using cDNA as the template and the primers in Table S1 according to the Real Master Mix (SYBR Green) kit (TransGen Biotech, Beijing, China). Gene expression was calculated using the $2^{-\Delta\Delta CT}$ method [34] (Livak & Schmittgen, 2001). Relative quantitative results were calculated through normalization based on the control gene *ACT2* (AT3G18780).

2.5. Subcellular Localization Analysis

The subcellular localization of MYB62 protein was completed in WT *Col-0 Arabidopsis* protoplasts. Protoplasts were prepared from the wild-type seedling leaves of *Arabidopsis* before bolting, according to the methods used in a previous study [35] (Yoo, Cho, & Sheen, 2007). The full-length coding sequences of *AGB1* and *MYB62* were amplified using gene-specific primers (Table S1), and *AGB1* and *MYB62* were inserted into the vector 16318h-GFP to express the fused proteins GFP-AGB1 and GFP-MYB62, respectively. The vectors 16318-AGB1-GFP and 16318-MYB62-GFP were then separately transformed into protoplasts. A confocal laser scanning microscope (LSM700, Zeiss, Yena, Germany) was used to observe the experimental results.

2.6. Yeast Two-Hybrid Assay

To analyze the interaction between AGB1 and MYB62, *AGB1* and *MYB62* were, respectively, inserted into the pGADT7 vector and the pGBKT7 vector in the *EcoR1* and *BamH1* restriction enzyme cutting sites. We prepared the yeast cells and completed vector transformation according to the manufacturer's instructions (TaKaRa, Tokyo, Japan). The transformants were selected on a synthetic dextrose (SD) medium lacking leucine and tryptophan (SD/-Leu/-Trp). The yeast transformants from the SD (-Leu/-Trp) were then streaked onto a solid SD (-Leu/-Trp/-His/-Ade) medium, with or without 40 μg/mL X-α-gal, to observe and photograph their growth.

2.7. Bimolecular Fluorescence Complementation (BiFC) Assay

AGB1 and MYB62 were, respectively, fused to the N- and C-termini of the luciferase reporter gene LUC, while the constructed vector was transformed into a strain of *A. tumefaciens*, GV3101. The *A. tumefaciens* samples transformed with nLUC-AGB1 and MYB62-cLUC were selected and the OD value of *A. tumefaciens* was adjusted to 0.8 with the infection solution (10 mM MES, 150 μM AS, 10 mM MgCl$_2$ 6H$_2$O). The *A. tumefaciens* samples transformed using the control, *nLUC-AGB1*, and *MYB62-cLUC* were mixed as pairs of *nLUC* and *cLUC*, *nLUC-AGB1* and *cLUC*, *nLUC* and *MYB62-cLUC*, and *nLUC-AGB1* and *MYB62-cLUC*, then injected into *Nicotiana benthamiana* leaves. Before analyzing LUC activity, *N. benthamiana* was cultured in darkness for 72 h. We performed 3 biological replicates for each combination of nLUC- and cLUC-infected tobacco leaves as a control.

2.8. Pull-Down Assays

MYB62 was inserted into the pMAL-c2x (MBP-Tag) vector to express MBP-labeled fusion proteins (Takara, Japan), and *AGB1* was inserted into the pGEX4T-1 (GST-Tag) vector to express GST-labeled fusion proteins (Takara, Tokyo, Japan). The vectors pMAL-c2x-MYB62 and pGEX4T-1-AGB1 were then introduced into *Escherichia coli BL21* (DE3). The expression of GST and MBP fusion proteins was induced by isopropylthio-β-galactoside (IPTG) and expressed at 16 °C for a minimum of 16 h. The fusion proteins of GST-AGB1 and MBP-MYB62 were purified by glutathione-agarose 4B (GE Healthcare, Stockholm, Sweden) beads and MBP-agarose gel, according to the instructions of the manufacturer.

In total, 50 μL of each recombinant fusion protein was mixed with 1 mL of the binding buffer (40 mM HEPES, 10 mM KCl, 0.4 M sucrose, 3 mM MgCl$_2$ 6H$_2$O, 1 mM EDTA, 1 mM DTT) in the pull-down assay. After eluting the MYB62 protein from the MBP-agarose gel with a 10 μM maltose solution, the MYB62-MBP protein and the GST-AGB1 protein were incubated in a pull-down buffer for approximately 8 h at 4 °C. They were then centrifuged at 2000× g for 1 min and washed 5 times at 4 °C with a 1× PBS buffer (pH 7.4). The particles containing binding proteins were then boiled in a 1× PBS buffer, and the released proteins were separated with 10% SDS-PAGE. The antibodies MBP-Tag and GST Tag were analyzed using a Western blot analysis (Abcam, Cambridge, UK).

2.9. Transcriptional Activation Experiment in Yeast

The transcription activation experiment was performed according to the methods used in previous research [36] (Yamaji et al., 2009). In order to further explore whether *AGB1* affects the transcriptional activation of *MYB62*, we carried out transcriptional activation experiments in yeast cells (Figure 5B). *MYB62* was inserted into pBridge vector to construct pBridge-MYB62, and *MYB62* was fused with the binding domain of the GAL4 transcription factor (GAL4-BD), which can bind target sequences upstream of the reporter genes (histidine-deficient reporter gene) in yeast chromosomes. When *MYB62* was inserted into pBridge, the reporter gene could be activated depending on the transcriptional activation activity of *MYB62* (Figure 5B). The yeast transformed with the pBridge-*MYB62* vector could grow on the screening medium. When *AGB1* was inserted into another expression cassette of the pBridge-MYB62 vector to make pBridge-MYB62-AGB1, the effect of AGB1 on the transcriptional activation of *MYB62* could be detected (Figure 5B). These constructed bodies were then introduced into the yeast reporter strain AH109 (Yeast Protocols Handbook; TaKaRa, Japan). The transformed yeast cells were selected on the selective medium (SD/-Trp, SD/-Trp-Ade, SD/-Trp-His-Ade) and the growth status was recorded.

2.10. LUC Assay of MYB62 in Tobacco (N. benthamiana)

MYB62 was inserted into the pCambia-1302 vector as an effector vector, and the 2000 bp promoter sequences of the downstream genes, including *GA2ox7*, were fused into the pGreenII 0800-LUC vector as a reporter vector. The reporter gene and the effector gene were introduced into GV3101, a strain of *A. tumefaciens*, and then injected into tobacco leaves. The activity of *LUC* was observed after 72 h of growth. Each sample was injected into 8 tobacco leaves, with 3 biological replicates performed for each.

2.11. Electrophoretic Mobility Shift Assay (EMSA)

MYB62-MBP and AGB1-GST proteins were induced by IPTG in *E. coli BL21* (DE3). The fusion proteins of GST-AGB1 and MBP-MYB62 were purified by glutathione-agarose 4B (GE Healthcare, North Richland Hills, TX, USA) beads and MBP-agarose gel, according to the instructions of the manufacturer. The synthetic oligonucleotide probe was synthesized by the ShengGong Biotech Company. The LightShift chemiluminescence EMSA kit (Thermo Science, Waltham, MA, USA) was used for EMSA. The biotin-labeled probe was incubated for 30 min at room temperature in a binding buffer (2.5% glycerol, 50 mM KCl, 5 mM MgCl$_2$, and 10 mM EDTA) with or without MYB62 MBP or AGB1-GST fused protein. For unlabeled probe competition, an unlabeled probe was added to the reaction, and single GST and MBP tags were used as negative controls. The probe sequence is shown in Table S1.

2.12. Low Phosphate-Tolerant Phenotypic Assay of Plants

Sterilized *Arabidopsis* seeds were placed in a 1/2 MS medium and cultured in an incubator (24 °C/16 h, 20 °C/8 h). The seedlings were transplanted into a 1/2 MS medium (Caissonabs, Rexburg, ID, USA) and a phosphate-free medium (Caissonabs, USA); the media's formulations are shown in Tables S2 and S3, respectively. The seedlings were then cultured in an incubator (24 °C/16 h, 20 °C/8 h) for 7–10 days.

3. Results

3.1. AGB1 Mutants agb1-2 and N692967 Were Dwarfed Compared with the WT after Exogenous Application of GA₃

After treatment with 10 μM GA$_3$, we found that the plant height of *agb1-2* was lower than that of the WT (Supplementary Figure S2). In addition, we identified another homozygous *AGB1* mutant, *N692967* (SALK_204268C) (Supplementary Figure S1A), and carried out phenotypic experiments under te conditions of 1 μM GA$_3$, 10 μM GA$_3$, or 100 μM GA$_3$ treatment (Supplementary Figures S3–S5). Without GA$_3$ treatment, plant height and the rosette leaf of the mutants *agb1-2* and *N692967* were slightly smaller than in the WT (Figure 1A,B). When treated with different concentrations of GA$_3$ for 10 days, the plant height of the mutants *agb1-2* and *N692967* was significantly lower than that of the WT (Figure 1D,E and Figures S3–S5). Moreover, we found that following an increase in GA$_3$ concentration, the plant height of *AGB1* mutant partially recovered compared with the WT, indicating that *AGB1* may be involved in the GA synthesis or degradation pathway. Therefore, in order to analyze the downstream pathway of *AGB1*, we analyzed the endogenous GA content of the *AGB1* mutant. The results showed that without GA$_3$ treatment, there were no significant differences in GA content between the mutants *agb1-2* and *N692967*, and the WT (Figure 1C), but under GA$_3$ treatment, the GA contents of *agb1-2* and *N692967* were significantly lower than that of the WT (Figure 1F), consistent with the results of their height (Figure 1B,E).

Figure 1. Phenotypes of the wild-type (Col-0), *agb1-2*, and *N692967* under GA$_3$ treatment. (**A**) Plant height phenotypes of the wild-type (Col-0), *agb1-2*, and *N692967* under normal conditions. Scale bars, 1 cm. (**B,C**) Plant height and GA content of the wild-type (Col-0), *agb1-2*, and *N692967* under normal conditions. Data are the average of three independent experiments, and the error bars represent SE (n = 10). Significant differences were analyzed using Duncan's multiple range test (p < 0.05). (**D**) Plant height phenotypes of the wild-type (Col-0), *agb1-2*, and *N692967* treated with 10 μM GA$_3$. Scale bars, 1 cm. (**E,F**) Plant height and GA content of the wild-type (Col-0), *agb1-2*, and *N692967* under normal conditions. Data are the average of three independent experiments, and the error bar represents the SE (n = 10). Significant differences were analyzed using Duncan's multiple range test (p < 0.05).

3.2. AGB1 Interacts with the DNA-Binding Region of MYB62, a GA Pathway Suppressor

Overexpression of *MYB62* reduced the sensitivity of plants to GA treatment and reduced tolerance to low-phosphorus stress [32] (Devaiah, Madhuvanthi, Karthikeyan, & Raghothama, 2009). This phenotype, under GA_3 treatment and low-phosphorus stress, was similar to that of the *agb1-2* mutant in our study (Supplementary Figures S6–S8 and S11). Therefore, we tried to identify the interaction between AGB1 and MYB62 using a yeast two-hybrid assay. The yeast cells grew on selective media (SD/-Trp/-Leu/-His/-Ade) and selective media plus x-α-gels only when BD-AGB1 and AD-MYB62 fused proteins were co-expressed in yeast cells (Figure 2A). The pull-down experiment demonstrated that AGB1 and MYB62 interacted in vitro (Figure 2B). The firefly luciferase (LUC) complementary imaging (Lci) analysis demonstrated that when AGB1-nLUC and MYB62-cLUC were expressed in the leaves of *N. benthamiana*, strong LUC activity was observed, whereas there was no LUC activity in the negative control (including nLUC + cLUC, AGB1-nLUC + cLUC, and nLUC + MYB62-cLUC) (Figure 2C). The quantitative analysis results of LUC activity (Figure 2D) were consistent with those shown in Figure 2C. These results demonstrated that AGB1 can interact with MYB62 in plant cells.

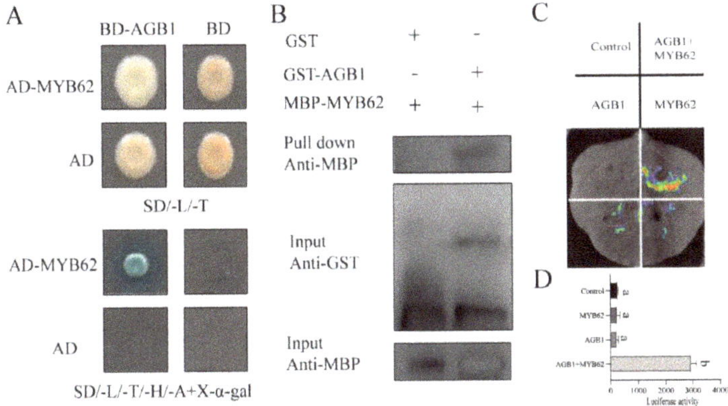

Figure 2. The interaction between AGB1 and MYB62. (**A**) Yeast two-hybrid interactions between full-length sequences of *AGB1* and *MYB62*. The transformed yeast cells were activated and cultured on SD/-Leu/-Trp and SD/-Leu/-Trp/-His/-Ade media. The yeast cells of the empty vectors AD and BD were used as negative controls. (**B**) Protein interaction of AGB1 and MYB62. In vitro GST pull-down assays showed that AGB1 interacted with MYB62 in vitro. (**C**) Interaction of AGB1 with MYB62. A luciferase (LUC) assay was performed to demonstrate that AGB1 and MYB62 can interact with *N. benthamiana* leaf cells. AGB1 and MYB62 were fused with nLUC and cLUC, respectively. nLUC-only and cLUC-only refer to the empty vectors used as negative controls. (**D**) Quantification of luminous intensity in C. Error bars represent the means $\pm$ SE ($n = 3$). Significant differences were analyzed using Duncan's multiple range test ($p < 0.05$).

To identify the interaction regions of MYB62, we inserted three truncated (M1–M3) segments as well as full-length *MYB62* into the AD vectors and inserted *AGB1* into the BD vector. These vectors were transformed into yeast cells to identify the interactions between different regions of *MYB62* and *AGB1*. Our results demonstrated that full-length *MYB62* interacted with AGB1, and the M1 and M2 truncated structures interacted slightly with AGB1, while M3 did not interact with AGB1 (Figure 3A), which suggests that the two DNA-binding regions of MYB62 were necessary for the interaction with AGB1.

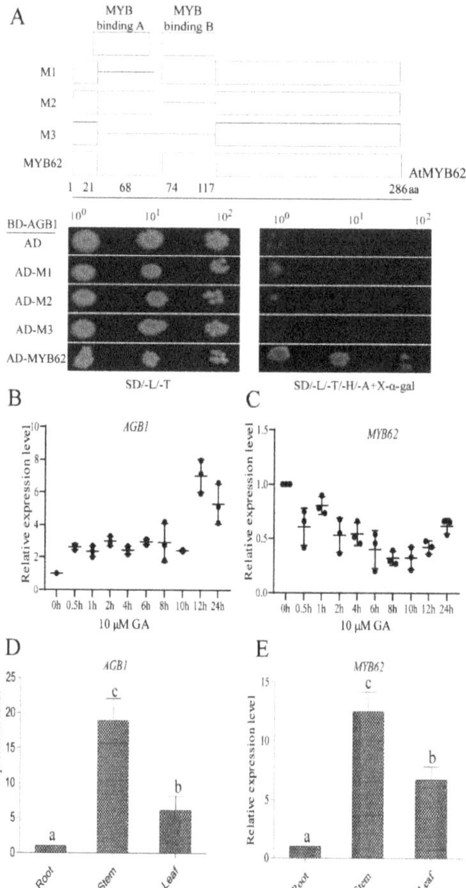

Figure 3. *AGB1* and *MYB62* expression analysis. (**A**) Yeast two-hybrid interactions between the full-length sequence of AGB1 and the piecewise sequence of MYB62. The transformed yeast cells were activated and cultured on SD/-Leu/-Trp and SD/-Leu/-Trp/-His/-Ade media. The yeast cells of the empty vectors AD and BD were used as negative controls. (**B**) Analysis of AGB1 expression in wild-type Col-0 leaves grown for 4 weeks after treatment with 10 μM GA3 at different time periods. The data are the average of three independent experiments, and the error bar represents the SE ($n = 3$). Significant differences were analyzed using Duncan's multiple range test ($p < 0.05$). Relative quantitative results were calculated by normalization using the control gene ACT2 (AT3G18780). (**C**) Analysis of MYB62 expression in wild-type Col-0 leaves, grown for 4 weeks after treatment with 10 μM GA3 at different time periods. The data are the average of three independent experiments, and the error bar represents the SE ($n = 3$). Significant differences were analyzed using Duncan's multiple range test ($p < 0.05$). (**D**) Analysis of AGB1 expression in the roots, stem, and leaves of wild-type Col-0 grown for 4 weeks under normal growth conditions. The data are the average of three independent experiments, and the error bar represents the SE ($n = 3$). a, b c indicate significant differences. Significant differences were analyzed using Duncan's multiple range test ($p < 0.05$). (**E**) Analysis of MYB62 expression in the roots, stem, and leaves of wild-type Col-0 grown for 4 weeks under normal growth conditions. The data are the average of three independent experiments, and the error bar represents the SE ($n = 3$). Significant differences were analyzed using Duncan's multiple range test ($p < 0.05$).

In addition, in seedlings of WT *Arabidopsis*, the expression of *AGB1* was induced under GA$_3$ treatment (Figure 3B), whereas the expression of *MYB62* was inhibited under GA$_3$ treatment (Figure 3C). The tissue-specific expression of *AGB1* and *MYB62* at the seedling stage was analyzed in *Arabidopsis* (*Col-0*) without GA$_3$ treatment. The results showed that *MYB62* was mainly expressed in *Arabidopsis* stems, which was similar to *AGB1* (Figure 3D,E). These results indicate that both *AGB1* and *MYB62* play an important role in stem development.

3.3. Genetic Analysis Indicated that MYB62 Was Downstream of AGB1 in the GA Pathway

In previous studies, researchers hoped to obtain homozygous mutants of *MYB62* by T-DNA insertion and antisense and RNAi-mediated *MYB62* silencing, but were unsuccessful and thus created the *MYB62* overexpression plant [32] (Devaiah, Madhuvanthi, Karthikeyan, & Raghothama, 2009). In order to study the genetic relationship between *AGB1* and *MYB62* further, we separately overexpressed *MYB62* in WT *Arabidopsis* (*MYB62:GFP/WT-8* and *MYB62:GFP/WT-10*) and *agb1-2* (*MYB62:GFP/agb1-2-1* and *MYB62:GFP/agb1-2-4*). Phenotypic analyses were carried out under the conditions of 1 μM GA$_3$, 10 μM GA$_3$, and 100 μM GA$_3$, and the plant height was recorded (Figure 4 and Figures S6–S8). Phenotypic analysis showed that without GA$_3$ treatment, the mutants *agb1-2* and *N692967*, and the transgenic plants *MYB62:GFP/WT-8*, *MYB62:GFP/WT-10*, *MYB62:GFP/agb1-2-1*, and *MYB62:GFP/agb1-2-4* were slightly smaller than the WT (Figure 4A). The GA content of *MYB62:GFP/WT-8*, *MYB62:GFP/WT-10*, *MYB62:GFP/agb1-2-1*, and *MYB62:GFP/agb1-2-4* was significantly lower than that of the WT and *AGB1* mutants (Figure 4C).

Under 10 μM GA$_3$ treatment, the height and the flowering time of the mutants *agb1-2* and *N692967* and the transgenic *Arabidopsis*, including *MYB62:GFP/WT-8*, *MYB62:GFP/WT-10*, *MYB62:GFP/agb1-2-1*, and *MYB62:GFP/agb1-2-4*, were significantly lower than in the WT (Figure 4D,E). The GA contents of the mutants *agb-1-2* and *N692967* and transgenic *Arabidopsis*, including *MYB62:GFP/WT-8*, *MYB62:GFP/WT-10*, *MYB62:GFP/agb1-2-1*, and *MYB62:GFP/agb1-2-4* were lower than that of the WT (Figure 4F), consistent with their height results (Figure 4D,E). In addition, the plant height of *MYB62:GFP/agb1-2* transgenic plants were more similar to that of *MYB62:GFP/WT*, and was higher than in the mutants *agb-1-2* and *N692967*, also consistent with their GA content results (Figure 4D–F). The height and GA content analysis was carried out under 1 μM GA$_3$ and 100 μM GA$_3$, and the results were consistent with those under 10 μM GA$_3$ (Supplementary Figures S9 and S10). The genetic analysis showed that *Arabidopsis MYB62* and *AGB1* belong to the same GA pathway, and *MYB62* is downstream of *AGB1* in this GA pathway.

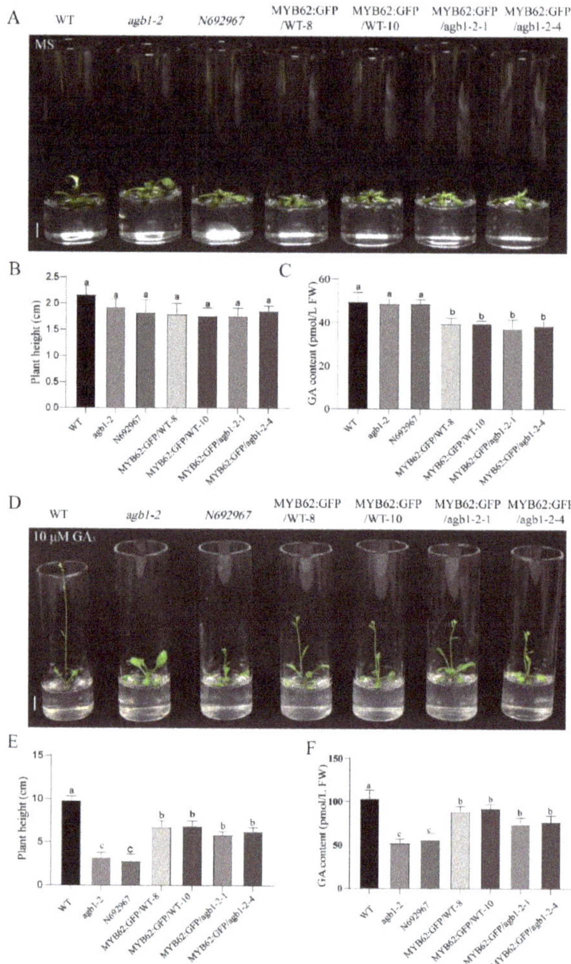

Figure 4. Phenotypic identification of the wild-type, *agb-1-2*, *N692967*, *MYB62*:GFP/WT-8, *MYB62*:GFP/WT-10, *MYB62*:GFP/*agb1-2-1*, and *MYB62*:GFP/*agb1-2-4* under normal conditions and 10 μM GA₃ treatment. (**A**) Plant height phenotype of the wild-type (Col-0), *agb-1-2*, *N692967*, *MYB62*:GFP/WT-8, *MYB62*:GFP/WT-10, *MYB62*:GFP/*agb1-2-1*, and *MYB62*:GFP/*agb1-2-4* under normal conditions. Scale bars, 1 cm. (**B,C**) Plant height and GA content of the wild-type (Col-0), *agb-1-2*, *N692967*, *MYB62*:GFP/WT-8, *MYB62*:GFP/WT-10, *MYB62*:GFP/*agb1-2-1*, and *MYB62*:GFP/*agb1-2-4* under normal conditions. The data are the average of three independent experiments, and the error bar represents the SE ($n = 10$). a, b indicate significant differences. Significant differences were analyzed using Duncan's multiple range test ($p < 0.05$). (**D**) Plant height phenotype of the wild-type (Col-0), *agb-1-2*, *N692967*, *MYB62*:GFP/WT-8, *MYB62*:GFP/WT-10, *MYB62*:GFP/*agb1-2-1*, and *MYB62*:GFP/*agb1-2-4* under 10 μM GA₃ treatment. Scale bars, 1 cm. (**E,F**) Plant height and GA content of the wild-type (Col-0), *agb-1-2*, *N692967*, *MYB62*:GFP/WT-8, *MYB62*:GFP/WT-10, *MYB62*:GFP/*agb1-2-1*, and *MYB62*:GFP/*agb1-2-4* under 10 μM GA₃ treatment. The data are the average of three independent experiments, and the error bar represents the SE ($n = 10$). a, b c indicate significant differences. Significant differences were analyzed using Duncan's multiple range test ($p < 0.05$).

3.4. The Interaction between AGB1 and MYB62 Did Not Affect the Transcriptional Activation of MYB62

The subcellular localization of MYB62 protein was completed in WT *Col-0 Arabidopsis* protoplasts. The results showed that green fluorescent protein (GFP) was expressed in the nucleus and cell membrane of the protoplasts transformed with the vector 16318-AGB1-GFP (Figure 5A), indicating that the AGB1 protein was located in the nucleus and cell membrane. GFP was expressed in the nucleus of the protoplasts transformed with the vector 16318-MYB62-GFP (Figure 5A), consistent with a previous report on the localization of MYB62 in plant cells [32] (Devaiah, Madhuvanthi, Karthikeyan, & Raghothama, 2009). These results indicated that AGB1 and MYB62 are co-located in the nucleus.

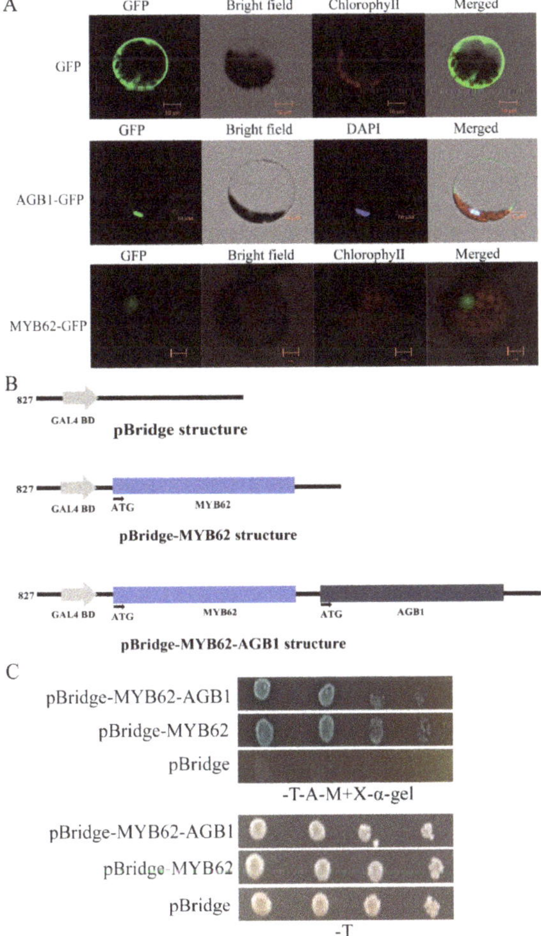

Figure 5. The interaction between AGB1 and MYB62 did not affect the transcriptional activation of MYB62; AGB1 and MYB62 are both located in the nucleus. (**A**) Analysis of the subcellular localization of AGB1 and MYB62 in the protoplasts of wild-type *Arabidopsis Col-0* leaves grown for 3 weeks. Scale bars, 10 μm. (**B**) The structure of the pBridge vector, pBridge-MYB62, and pBridge-MYB62-AGB1. (**C**) The vectors pBridge, pBridge-MYB62, and pBridgeMYB62-AGB1 were transferred into AH109 yeast cells, and the cells were diluted and transferred to selective media (SD/-Trp, SD/-Trp-Ade, SD/-Trp-His-Ade) to observe their growth.

In order to further explore whether *AGB1* affects the transcriptional activity of *MYB62*, we carried out transcriptional activation experiments in yeast cells (Figure 5B,C). We constructed a pBridge-MYB62 vector to detect the transcriptional activation activity of *MYB62* in yeast cells. In addition, a pBridge-MYB62-AGB1 vector was constructed by inserting *AGB1* into pBridge-MYB62, and the effect of *AGB1* on *MYB62* transcriptional activation was detected (Figure 5B). The results showed that the yeast transformed with the pBridge-MYB62 vector grew on the selective medium (Figure 5C). The growth of the yeast transformed with pBridge-MYB62-AGB1 was similar to that of pBridge-MYB62 (Figure 5C), indicating that *MYB62* had transcriptional activation activity, and *AGB1* had no effect on the transcriptional activation of *MYB62* in yeast cells.

3.5. MYB62 Can Bind the Promoter of GA Degradation Gene GA2ox7 to Enhance Its Expression, and AGB1 Negatively Regulates the DNA-Binding Activity of MYB62 on the Promoter of GA2ox7

AGB1 did not affect the transcriptional activation of *MYB62*. Therefore, we speculated that *AGB1* may affect the DNA binding activity of *MYB62*. Thus, we analyzed the expression of many genes related to GA synthesis and degradation in the WT, *agb1-2*, and *MYB62:GFP/WT-10* under GA_3 treatment. We found that the *GA2ox7* gene (At1g47990), which is related to the degradation process of GA, may be related to the regulation pathway of AGB1-MYB62. Gene expression analysis showed that the expression of *GA2ox7* in the WT was significantly lower than that in the mutants *agb1-2* and *N692967*, and transgenic plants, including *MYB62:GFP/WT-8*, *MYB62:GFP/WT-10*, *MYB62:GFP/agb1-2-1*, and *MYB62:GFP/agb1-2-4* (Figure 6A). This result was consistent with the GA content results (Figure 4F), suggesting that *GA2ox7* is regulated by AGB1-MYB62. Moreover, we found that the expression of *GA2ox7* in *MYB62:GFP/agb1-2-1* and *MYB62:GFP/agb1-2-4* was similar to that in *MYB62:GFP/WT-8* and *MYB62:GFP/WT-10*, but lower than in the mutants *agb1-2* and *N692967* (Figure 6A), also consistent with the GA content results (Figure 4F). In addition, EMSA analysis and LUC (luciferase analysis) showed that *MYB62* bound to the MYB element (TGGTTG) in the *GA2ox7* promoter and enhanced the expression of *GA2ox7* (Figure 6B–F). However, *AGB1* can negatively regulate the binding of *MYB62* to the *GA2ox7* promoter, thus negatively regulating the expression of the *GA2ox7* promoter in plants (Figure 6G–I). These results indicated that AGB1 interacted with the DNA-binding region of MYB62, further negatively regulating MYB62 binding on the downstream gene *GA2ox7*, and actively participating in the GA pathway.

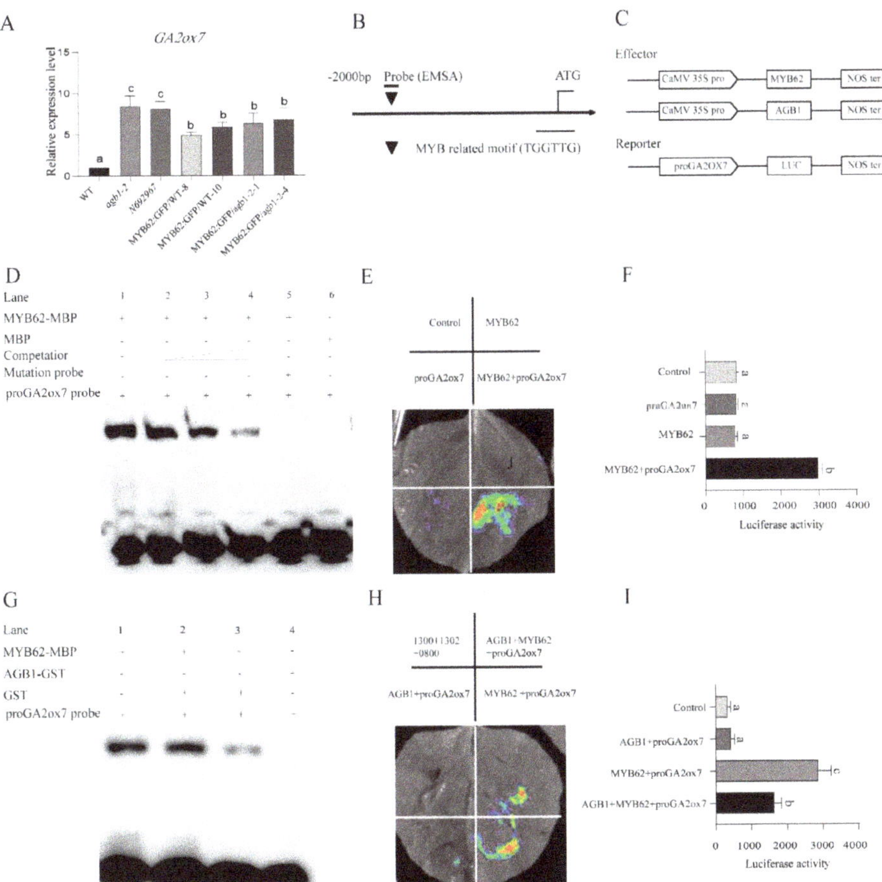

Figure 6. *AGB1* negatively regulates the expression of the downstream genes of *MYB62*. (**A**) The GA degradation-related gene, *GA2ox7*, in the wild-type (*Col-0*), *agb-1-2*, *N692967*, *MYB62*:GFP/WT-8, *MYB62*:GFP/WT-10, *MYB62*:GFP/*agb1-2-1*, and *MYB62*:GFP/*agb1-2-4* revealed by qRT-PCR analysis under GA$_3$ treatment. Error bars represent the means ± SE (n = 3). Significant differences were analyzed using Duncan's multiple range test (p < 0.05). Relative quantitative results were calculated by normalization using the control gene *ACT2* (AT3G18780). (**B**) Illustration of the MYB62 promoter region showing the presence of the MYB-binding site. (**C**) Schematic diagram of the effectors and the reporter structure of LUC (luciferase analysis) in E and H. (**D**) The EMSA (electrophoretic mobility shift assay) experiment showed that *MYB62* could bind to the promoter of *GA2ox7*. (**E**) The LUC experiment showed that *MYB62* could bind to the *GA2ox7* promoter and promote the expression of *GA2ox7*; 302 indicates the pCambia 1302 vector, and 0800 indicates the pGreenII 0800-LUC vector. Representative images of *N. benthamiana* leaves were taken 48 h after infiltration. All experiments were repeated three times with similar results. (**F**) Quantification of luminescence intensity in E. Error bars represent the means ± SE (n = 3). Significant differences were analyzed using Duncan's multiple range test (p < 0.05). (**G**) The EMSA experiment showed that *AGB1* negatively regulates the binding of *MYB62* to the e*GA2ox7* promoter. (**H**) The LUC experiment showed that *AGB1* could negatively regulate the binding of *MYB62* to the *GA2ox7* promoter and negative regulately the expression of *GA2ox7*; 1300 indicates the pCambia 1300 vector, 1302 indicates the pCambia 1302 vector, and 0800 indicates the pGreenII 0800-LUC vector. Representative images of *N. benthamiana* leaves were taken 48 h after infiltration. All experiments were repeated three times with similar results. (**I**) Quantification of luminescence intensity in H. Error bars represent the means ± SE (n = 3). Significant differences were analyzed using Duncan's multiple range test (p < 0.05).

4. Discussion

4.1. AGB1 Regulates the GA Pathway by Negatively Regulating MYB62 Activity

In this study, we demonstrated the molecular mechanism of how *AGB1* regulates the GA pathway (Figure 7). We found that *AGB1* was upregulated and *MYB62* was downregulated under GA$_3$ treatment (Figure 3B,C). Both *AGB1* and *MYB62* were mainly expressed in the stem of *Arabidopsis* (Figure 3D,E). Through phenotype analysis of the mutant *agb1* and *MYB62*-overexpressing plants under GA$_3$ treatment, *AGB1* was found to positively regulate the response to the GA pathway, and *MYB62* negatively regulated the response to the GA pathway (Figure 4). Genetic analysis showed that *AGB1* and *MYB62* were involved in the same GA pathway, and that *MYB62* was downstream (Figure 4). Biochemical experiments showed that AGB1 interacted with the DNA-binding region of MYB62, and that MYB62 could directly bind to the promoter of the GA degradation gene *GA2ox7* and promote its expression (Figures 3 and 6). AGB1 inhibited the binding of MYB62 on the *GA2ox7* promoter, thus negatively regulating the activity of MYB62 and positively regulating the GA pathway in *Arabidopsis* (Figure 6G–I).

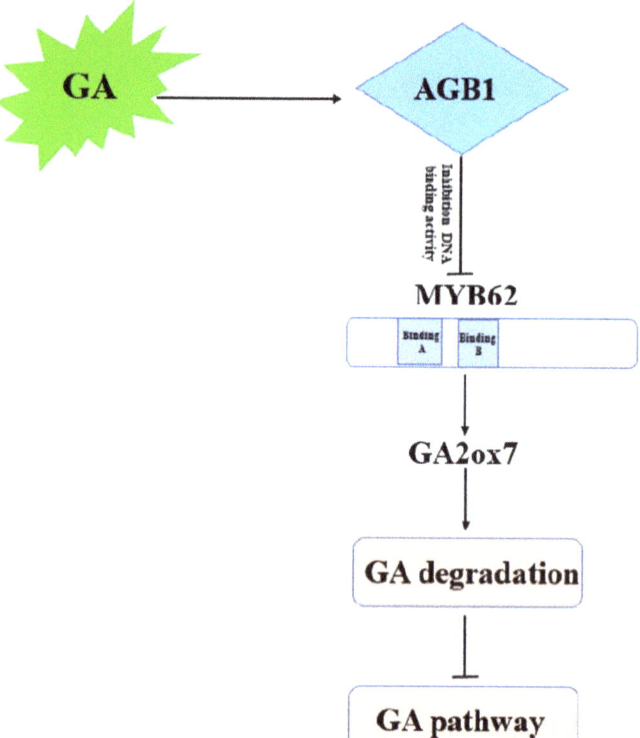

Figure 7. Model of how AGB1 regulates GA-related signaling pathways by controlling MYB62. This regulatory model shows the role of AGB1–MYB62 in regulation of the GA signaling pathway in *Arabidopsis thaliana*. AGB1 positively participates in the GA signaling pathway, interacts with the negative transcription factor MYB62, inhibits the binding of MYB62 to the promoter of GA metabolism related gene *GA2ox7*, and negative regulates *GA2ox7* expression.

We found that the interaction between AGB1 and MYB62 did not affect the transcriptional activation of *MYB62* (Figure 5C). AGB1 interacted with the DNA-binding region of MYB62 (Figure 3A) and negatively regulated the activity of MYB62 by affecting binding to the promoter region of the metabolic gene *GA2ox7* (Figure 6G–I), though neither affected the transcription activity of *MYB62*. Our previous research on *AGB1* in *Arabidopsis* determined that *AGB1* inhibits the transcriptional activation activity of *BBX21*, a positive regulatory factor, by binding to the transcriptional activation region of *BBX21*, thereby regulating the expression of the downstream genes [17] (Xu et al., 2017). Taken together, these results suggest that *AGB1* regulates the activity of downstream transcription factors by combining different regions of the downstream transcription factors. Other research has found that the G protein β subunit regulates plant hypocotyl elongation [17] (Xu et al., 2017), BR signal transduction [37] (Zhang et al., 2018), and plant nutrition regulation [38] (Wu et al., 2020) by interacting with different transcription factors. These results suggest that *AGB1* binding with different downstream transcription factors may be a general mechanism to regulate different signaling pathways in *Arabidopsis*.

In *Arabidopsis*, the G protein α subunit *GPA1* has an important role in regulating hypocotyl elongation, ABA inhibition of the stomatal opening, stomatal density, pollen tube development, and plant height, but the most common regulation mode is the GPA1–GCR1 coupling complex [39] (Chakraborty, Singh, Kaur, & Raghuram, 2015). In rice, the G-α subunit positively regulates the GA pathway by inhibiting the activity of a negative regulator, SLR (slender rice), in GA signaling [13] (Ueguchi-Tanaka et al., 2000). These studies suggested that different subunits of the G protein complex can positively regulate the GA pathway through different downstream genes in plants.

4.2. The AGB1–MYB62 Pair Is Involved in Regulating the Phosphate Starvation Response in Plants

MYB62 participates in the response to low-phosphorus stress by negatively regulating the synthesis of GA [32] (Devaiah, Madhuvanthi, Karthikeyan, & Raghothama, 2009). Therefore, in order to study whether AGB1–MYB62 also functions in phosphorus deficiency, we analyzed the phenotypes of the WT, the *agb1-2* mutant, *MYB62-GFP/WT-10*, and *MYB62-GFP/agb1-2-4* under phosphorus-free conditions (Supplementary Figure S11). The results showed that under normal conditions, the length of the main root and the number of lateral roots of *MYB62:GFP/WT-10* were lower than those of the WT (Supplementary Figure S11). This is consistent with the results of Devaiah et al. (2009). In this study, under normal conditions, there was no significant difference in main root length and lateral root number between *agb1-2* and the WT. Under phosphorus-free conditions, the main root length and lateral root number of *agb1-2* were significantly lower than those of the WT, while those of *MYB62-GFP/agb1-2-4* were significantly lower than those of the WT and consistent with *MYB62-GFP/WT-10* (Supplementary Figure S11). These results indicate that *AGB1* positively regulated the response to phosphorus starvation in plants by stimulating root growth. The AGB1–MYB62 pair is also involved in regulating the root growth process under phosphorus starvation. More evidence is needed to demonstrate how the AGB1–MYB62 pair coordinates the regulation of GA and low-phosphorus stress responses in plants.

Supplementary Materials: The following are available online at https://www.mdpi.com/article/10.3390/ijms22158270/s1. Figure S1 Verification of mutants and transgenic plants. Figure S2 Phenotypes of wild type (Col-0) and agb1-2 under 10 μM GA3. Figure S3 Phenotypes of wild type, agb1-2, and N692967 under 1 μM GA3 treatment at 1d, 7d, and 10d. Figure S4 Phenotypes of wild type, agb1-2, and N692967 under 10 μM GA3 treatment at 1d, 7d, and 10d. Figure S5 Phenotypes of wild type, agb1-2, and N692967 under 100 μM GA3 treatment at 1d, 7d, and 10d. Figure S6 Phenotypes of wild type, agb-1-2, N692967, MYB62:GFP/WT-8, MYB62:GFP/WT-10, MYB62:GFP/agb1-2-1, and MYB62:GFP/agb1-2-4 under 1 μM GA3 treatment at 1d, 7d, and 10d. Figure S7 Phenotypes of wild type, agb-1-2, N692967, MYB62:GFP/WT-8, MYB62:GFP/WT-10, MYB62:GFP/agb1-2-1, and MYB62:GFP/agb1-2-4 under 10 μM GA3 treatment at 1d, 7d, and 10d. Figure S8 Phenotypes of wild type, agb-1-2, N692967, MYB62:GFP/WT-8, MYB62:GFP/WT-10, MYB62:GFP/agb1-2-1, and MYB62:GFP/agb1-2-4 under 100 μM GA3 treatment at 1d, 7d, and 10d. Figure S9 Phenotypic identification of wild type, agb-1-2, N692967, MYB62:GFP/WT-8, MYB62:GFP/WT-10, MYB62:GFP/agb1-2-1 and MYB62:GFP/agb1-2-4 under 1 µM GA3 treatment at 10d. Figure S10 Phenotypic identification of wild type, agb-1-2, N692967, MYB62:GFP/WT-8, MYB62:GFP/WT-10, MYB62:GFP/agb1-2-1, and MYB62:GFP/agb1-2-4 under 100 µM GA3 treatment at 10d. Figure S11 Phenotypes of wild type, agb-1-2, N692967, MYB62:GFP/WT-8, MYB62:GFP/WT-10, MYB62:GFP/agb1-2-1, and MYB62:GFP/agb1-2-4 under normal conditions and phosphate-free conditions. Table S1 Primers used in this study. Table S2 Composition of MS powder. Table S3 Composition of MS powder phosphate-free.

Author Contributions: X.Q. and W.T. designed and performed experiments and wrote the manuscript; W.L. contributed to the implementation of the study; Z.H. and W.X. contributed valuable discussions; Z.F., Y.Z., C.W., Z.X., J.C., S.G. and Z.X. provided instructions for the experiments; M.C. coordinated the project, conceived and designed experiments, and edited the manuscript; Y.M. coordinated the project and edited the manuscript. All authors have read and agreed to the published version of the manuscript.

Funding: Key Projects of Genetic Modification (2018ZX08009-17B) and the National Key Project for Research on Transgenic Biology (2016ZX08002-002).

Acknowledgments: This work was supported by the Key Projects of Genetic Modification (2018ZX08009-17B), the National Key Project for Research on Transgenic Biology (2016ZX08002-002), and the Agricultural Science and Technology Innovation Program (ASTIP, Transgenic Technology and Application of Crops, and Development and Application of Molecular Markers in Crops).

Conflicts of Interest: The authors declare no conflict of interest.

References

1. Assmann, S.M. G proteins go green: A plant G protein signaling FAQ sheet. *Science* **2005**, *310*, 71–73. [CrossRef]
2. Chen, J.G. Heterotrimeric G-proteins in plant development. *Front. Biosci.* **2007**, *13*, 3321. [CrossRef]
3. Assmann, S.M. Heterotrimeric and unconventional GTP binding proteins in plant cell signaling. *Plant Cell* **2002**, *14* (Suppl. S1), S355–S373. [CrossRef]
4. Anantharaman, V.; Abhiman, S.; de Souza, R.F.; Aravind, L. Comparative genomics uncovers novel structural and functional features of the heterotrimeric GTPase signaling system. *Gene* **2011**, *475*, 63–78. [CrossRef]
5. Ullah, H.; Chen, J.G.; Wang, S.C.; Jones, A.M. Role of a Heterotrimeric G Protein in Regulation of Arabidopsis Seed Germination. *Plant Physiol.* **2004**, *129*, 897–907. [CrossRef]
6. Ullah, H.; Chen, J.G.; Temple, B.; Boyes, D.C.; Alonso, J.M.; Davis, K.R.; Jones, A.M. The b-subunit of the Arabidopsis G protein negatively regulates auxin-induced cell division and affects multiple developmental processes. *Plant Cell* **2003**, *15*, 393–409. [CrossRef] [PubMed]
7. Assmann, S.M.; Yu, Y.Q. The heterotrimeric g-protein beta subunit, agb1, plays multiple roles in the Arabidopsis salinity response. *Plant Cell Environ.* **2015**, *38*, 2143–2156. [CrossRef]
8. Wang, R.S.; Pandey, S.; Li, S.; Gookin, T.E.; Zhao, Z.X.; Albert, R.; Assmann, S.M. Common and unique elements of the ABA-regulated transcriptome of arabidopsis guard cells. *BMC Genom.* **2011**, *12*, 216. [CrossRef] [PubMed]
9. Li, S.J.; Liu, Y.J.; Zheng, L.Y.; Chen, L.L.; Li, N.; Corke, F.; Bevan, M.W. The plant-specific G protein γ subunit AGG3 influences organ size and shape in arabidopsis thaliana. *New Phytol.* **2012**, *194*, 690–703. [CrossRef]
10. Chen, J.G.; Pandey, S.; Huang, J.; Alonso, J.M.; Ecker, J.R.; Assmann, S.M.; Jones, A.M. GCR1 can act independently of heterotrimeric G-protein in response to brassinosteroids and gibberellins in Arabidopsis seed germination. *Plant Physiol.* **2004**, *135*, 907–915. [CrossRef] [PubMed]
11. Pandey, S.; Che, J.G.; Jones, A.M.; Assmann, S.M. G-protein complex mutants are hypersensitive to abscisic acid regulation of germination and postgermination development. *Plant Physiol.* **2006**, *141*, 243–256. [CrossRef]
12. Urano, D.; Jones, A.M. Heterotrimeric G protein–coupled signaling in plants. *Annu. Rev. Plant Biol.* **2014**, *65*, 365–384. [CrossRef]
13. Ueguchi-Tanaka, M.; Fujisaw, Y.; Kobayashi, M.; Ashikari, M.; Iwasaki, Y.; Kitano, H.; Matsuoka, M. Rice dwarf mutant d1, which is defective in the alpha subunit of the heterotrimeric g protein, affects gibberellin signal transduction. *Proc. Natl. Acad. Sci. USA* **2000**, *97*, 11638–11643. [CrossRef]
14. Trusov, Y.; Rookes, J.E.; Tilbrook, K.; Chakravorty, D.; Mason, M.G.; Anderson, D.; Botella, J.R. Heterotrimeric G protein γ subunits provide functional selectivity in Gβγ dimer signaling in arabidopsis. *Plant Cell* **2007**, *19*, 1235–1250. [CrossRef]
15. Lease, K.A.; Wen, J.Q.; Li, J.; Doke, J.T.; Liscum, E.; Walker, J.C. A mutant Arabidopsis heterotrimeric G-protein beta subunit affects leaf, flower, and fruit development. *Plant Cell* **2001**, *13*, 2631–2641. [CrossRef] [PubMed]
16. Xu, D.B.; Chen, M.; Ma, Y.N.; Xu, Z.S.; Li, L.C.; Chen, Y.F.; Ma, Y.Z. A G-Protein β Subunit, AGB1, negatively regulates the ABA response and drought tolerance by down-regulating AtMPK6-related pathway in Arabidopsis. *PLoS ONE* **2015**, *10*, e0116385. [CrossRef] [PubMed]
17. Xu, D.B.; Ma, Y.N.; Gao, S.Q.; Wang, X.T.; Feng, L.; Li, L.C.; Ma, Y.Z. The G-protein β subunit AGB1 promotes hypocotyl elongation through inhibiting transcription activation function of BBX21 in Arabidopsis. *Mol. Plant* **2017**, *10*, 1206–1223. [CrossRef]

18. Utsunomiya, Y.; Samejima, C.; Takayanagi, Y.; Izawa, Y.; Yoshida, T.; Sawada, Y.; Iwasaki, Y. Suppression of the rice heterotrimeric G protein beta-subunit, RGB1, causes dwarfism and browning of internodes and lamina joint regions. *Plant J.* **2011**, *67*, 907–916. [CrossRef] [PubMed]

19. Urano, D.; Miura, K.; Wu, Q.Y.; Iwasaki, Y.; Jackson, D.; Jones, A.M. Plant Morphology of Heterotrimeric G protein Mutants. *Plant Cell Physiol.* **2016**, *57*, 437–445. [CrossRef] [PubMed]

20. Singh, D.P.; Jermakow, A.M.; Swain, S.M. Gibberellins are required for seed development and pollen tube growth in Arabidopsis. *Plant Cell* **2002**, *14*, 3133–3147. [CrossRef]

21. Lange, B.M.; Ghassemian, M. Genome organization in Arabidopsis thaliana: A survey for genes involved in isoprenoid and chlorophyll metabolism. *Plant Mol. Biol.* **2003**, *51*, 925–948. [CrossRef]

22. Morrone, D.; Chambers, J.; Lowry, L.; Kim, G.; Anterola, A.; Bender, K.; Peters, R.J. Gibberellin biosynthesis in bacteria: Separate ent-copalyl diphosphate and ent-kaurene synthases in bradyrhizobium japonicum. *FEBS Lett.* **2009**, *583*, 475–480. [CrossRef]

23. Davidson, S.E.; Elliott, R.C.; Helliwell, C.A.; Poole, A.T.; Reid, J.B. The pea gene NA encodes ent-kaurenoic acid oxidase. *Plant Physiol.* **2003**, *131*, 335–344. [CrossRef]

24. Sakamoto, T.; Miyura, K.; Itoh, H.; Tatsumi, T.; Ueguchi-Tanaka, M.; Ishiyama, K.; Miyao, A. An overview of gibberellin metabolism enzyme genes and their related mutants in rice. *Plant Physiol.* **2004**, *134*, 1642–1653. [CrossRef]

25. Hedde, P.; Thomas, S.G. Gibberellin biosynthesis and its regulation. *Biochem. J.* **2012**, *444*, 11–25. [CrossRef]

26. Ashikari, M.; Sasaki, A.; Ueguchi-Tanaka, M.; Itoh, H.; Nishimura, A.; Swapan Datta, S.; Khush, S.G. Loss-of-function of a rice gibberellin biosynthetic gene, GA20 oxidase (GA20ox), led to the rice 'green revolution'. *Breed. Sci.* **2002**, *52*, 143–150. [CrossRef]

27. Helliwell, C.A.; Chandler, P.M.; Poole, A.; Dennis, E.S.; Peacock, W.J. The CYP88A cytochrome P450, ent-kaurenoic acid oxidase, catalyzes three steps of the gibberellin biosynthesis pathway. *Proc. Natl. Acad. Sci. USA* **2001**, *98*, 2065–2070. [CrossRef]

28. Fambrini, M.; Mariotti, L.; Parlanti, S.; Picciarelli, P.; Salvini, M.; Ceccarelli, N.; Pugliesi, C. The extreme dwarf phenotype of the GA-sensitive mutant of sunflower, dwarf2, is generated by a deletion in the ent-kaurenoic acid oxidase1 (HaKAO1) gene sequence. *Plant Mol. Biol.* **2011**, *75*, 431–450. [CrossRef] [PubMed]

29. Ueguchi-Tanaka, M.; Ashikari, M.; Nakajima, M.; Itoh, H.; Katoh, E.; Kobayashi, M.; Chow, T.Y.; Matsuoka, M. Gibberellin insensitive dwarf1 encodes a soluble receptor for gibberellin. *Nature* **2005**, *437*, 693–698. [CrossRef]

30. Murase, K.; Hirano, Y.; Sun, T.P.; Hakoshima, T. Gibberellin-induced DELLA recognition by the gibberellin receptor gid1. *Nature* **2008**, *456*, 459–463. [CrossRef] [PubMed]

31. Shimada, A.; Ueguchi-Tanaka, M.; Nakatsu, T.; Nakajima, M.; Naoe, Y.; Ohmiya, H.; Matsuoka, M. Structural basis for gibberellin recognition by its receptor DID1. *Nature* **2008**, *456*, 520–523. [CrossRef] [PubMed]

32. Devaiah, B.N.; Madhuvanthi, R.; Karthikeyan, A.S.; Raghothama, K.G. Phosphate starvation responses and Gibberellic Acid biosynthesis are regulated by the MYB62 transcription factor in Arabidopsis. *Mol. Plant* **2009**, *2*, 43–58. [CrossRef] [PubMed]

33. Clough, S.J.; Bent, A.F. Floral dip: A simplified method for Agrobacterium-mediated transformation of Arabidopsis thaliana. *Plant J* **2010**, *16*, 735–743. [CrossRef]

34. Livak, K.J.; Schmittgen, T.D. Analysis of relative gene expression data using real-time quantitative PCR and the $2^{-\Delta\Delta CT}$ method. *Methods* **2001**, *25*, 402–408. [CrossRef] [PubMed]

35. Yoo, D.S.; Cho, Y.H.; Sheen, J. Arabidopsi mesophyll protoplasts: A versatile cell system for transient gene expression analysis. *Nat. Protoc.* **2007**, *2*, 1565–1572. [CrossRef]

36. Yamaji, N.; Huang, C.F.; Nagao, S.; Yano, M.; Sato, Y.; Nagamura, Y.; Ma, J.F. A zinc finger transcription factor ART1 regulates multiple genes implicated in aluminum tolerance in rice. *Plant Cell* **2009**, *21*, 3339–3349. [CrossRef]

37. Zhang, T.; Xu, P.B.; Wang, W.X.; Wang, S.; Caruana, J.C.; Yang, H.Q.; Hongli Lian, H.L. Arabidopsis g-protein β subunit agb1 interacts with BES1 to regulate brassinosteroid signaling and cell elongation. *Front. Plant Sci.* **2018**, *8*, 2225. [CrossRef]

38. Wu, T.Y.; Krishnamoorthi, S.; Goh, H.; Leong, R.; Sanson, A.C.; Urano, D. Crosstalk between heterotrimeric G protein-coupled signaling pathways and WRKY transcription factors modulating plant responses to suboptimal micronutrient conditions. *J. Exp. Bot.* **2020**, *71*, 3227–3239. [CrossRef]

39. Chakraborty, N.; Singh, N.; Kaur, K.; Raghuram, N. G-protein signaling components GCR1 and GPA1 mediate responses to multiple abiotic stresses in arabidopsis. *Front. Plant Sci.* **2015**, *6*, 1000. [CrossRef]

International Journal of
Molecular Sciences

MDPI

Review

ABA Mediates Plant Development and Abiotic Stress via Alternative Splicing

Xue Yang [1,†], Zichang Jia [1,†], Qiong Pu [2,3], Yuan Tian [1], Fuyuan Zhu [2,*] and Yinggao Liu [1,*]

1   State Key Laboratory of Crop Biology, College of Life Science, Shandong Agricultural University, Tai'an 271018, China; xueyang202001@163.com (X.Y.); jiazc973@163.com (Z.J.); ty15610312143@163.com (Y.T.)
2   Co-Innovation Center for Sustainable Forestry in Southern China & Key Laboratory of National Forestry and Grassland Administration on Subtropical Forest Biodiversity Conservation, College of Biology and the Environment, Nanjing Forestry University, Nanjing 210037, China; qiongpu2022@163.com
3   College of Mechanical and Electronic Engineering, Shandong Agriculture and Engineering University, Jinan 250000, China
*   Correspondence: fyzhu@njfu.edu.cn (F.Z.); liuyg@sdau.edu.cn (Y.L.)
†   These authors contributed equally to this work.

Abstract: Alternative splicing (AS) exists in eukaryotes to increase the complexity and adaptability of systems under biophysiological conditions by increasing transcriptional and protein diversity. As a classic hormone, abscisic acid (ABA) can effectively control plant growth, improve stress resistance, and promote dormancy. At the transcriptional level, ABA helps plants respond to the outside world by regulating transcription factors through signal transduction pathways to regulate gene expression. However, at the post-transcriptional level, the mechanism by which ABA can regulate plant biological processes by mediating alternative splicing is not well understood. Therefore, this paper briefly introduces the mechanism of ABA-induced alternative splicing and the role of ABA mediating AS in plant response to the environment and its own growth.

Keywords: abscisic acid; alternative splicing; abiotic stress responses; plant development

Citation: Yang, X.; Jia, Z.; Pu, Q.; Tian, Y.; Zhu, F.; Liu, Y. ABA Mediates Plant Development and Abiotic Stress via Alternative Splicing. *Int. J. Mol. Sci.* **2022**, *23*, 3796. https://doi.org/10.3390/ijms23073796

Academic Editor: Karen Skriver

Received: 9 February 2022
Accepted: 27 March 2022
Published: 30 March 2022

Publisher's Note: MDPI stays neutral with regard to jurisdictional claims in published maps and institutional affiliations.

1. Introduction

Plant growth and development are significantly affected by environmental factors. Among them, the most important environmental factors are abiotic stresses such as drought and low temperature. Harsh environments affect plant physiology and distribution, especially crops [1]. Abiotic stress triggers a series of systemic responses in plants, and each stress response pathway is not independent. The responses generated by local stress are expressed in local and distal tissues, resulting in the systemic acquired domestication of plants [2]. When plants are subjected to environmental stimuli, such as drought, salt, and extreme temperature, ABA accumulates in different degrees. Therefore, ABA is considered the main plant hormone for plants to respond to external environmental stresses. The ABA signaling pathway is a central reaction pathway for environmental adversities, such as salt and drought in plants [3,4].

ABA is one of the most important plant hormones widely distributed in higher plants. It is named for its ability to promote leaf abscission. Under adversity stress, plants accumulate a large amount of ABA, thereby enhancing their resistance to abiotic stress [5]. Abscisic acid is also called a stress hormone. The main function of abscisic acid is to break seed dormancy, promote seed germination, and regulate stomatal closure, root development, and resistance to abiotic stress [6].

The genomic DNA of eukaryotes is a fragmented gene that contains a large number of introns that do not encode proteins. The precursor mRNA of the initial product of transcription contains this part of the long intron, and the intron itself does not have the ability to encode a protein, therefore, this part of the intron needs to be removed. Precursor mRNA becomes mature mRNA through splicing [7].

Alternative splicing (AS) is the process of generating different mature mRNAs from the same mRNA precursor by selecting different combinations of splicing sites. AS produces various transcripts and diverse proteins in eukaryotic cells [8]. AS involves many biological processes in plants, especially adverse abiotic environmental stress. In adverse environments, such as drought and cold, certain genes in plants will undergo alternative splicing to respond to the living environment [9]. AS is a significant mediator of gene expression and increases proteome diversity.

The process of AS is mainly catalyzed by a spliceosome. A spliceosome mainly contains ribonucleoproteins. The composition of U1, U2, U4, U5, and U6 is the main type of spliceosome complex at present. The spliceosome assembles at one intron and recognizes splice sites to complete alternative splicing [10]. There are eight manners of AS, and the common four are: intron retention (retained introns, RI), variable acceptor sites (alternative $3'$ splicing sites, A3), variable donor sites (alternative $5'$ splicing sites, A5), exon skipping (skipped exons, SE). The remaining four include alternative splicing of the transcription start site (alternative first exon, AF), alternative splicing of the transcription termination region (alternative last exon, AL) alternative exon (AE), and mutually exclusive exon (MX) [11,12]. The regulation of AS is accomplished by the recognition of specific elements in the precursor mRNA by alternative splicing regulatory proteins, namely trans-acting factors. Splicing factors include serine/arginine-rich (SR) proteins and heterogeneous nuclear ribonucleoprotein (hnRNP) family proteins, RNA helicases, kinases, and many other factors [13]. Cis-regulatory RNA sequences can be further divided into enhancers that promote splicing at specific splice sites and splicing silencers that reduce splicing. Alternative splicing is a mechanism for regulating gene expression at the mRNA level [14].

Recently, sequencing technology has been widely used in plant genome sequencing and transcriptome sequencing. It has been found that alternative splicing has a profound impact on plant growth and abiotic stress [11]. AS is very widespread in eukaryotes. It has been demonstrated that in mammals, over 95% of genes undergo intron-containing transcription. Research now shows that in plants such as *Arabidopsis*, rice, and maize, 60% of genes that contain introns are alternatively spliced [13,15,16]. More transcriptome data and RNA-seq data have shown that the AS of genes can be induced by environmental conditions [17–19]. In the research, the AS pattern of 9200 genes changed in response to heat stress in tomatoes, while 1000 genes had differential splicing under cold-stressed conditions [20–22]. The gene expression induced by abiotic stress was more prone to AS. Moreover, abiotic stresses imposed by high temperatures induced different splice sites [23,24].

The promoters of ABA-responsive genes contain cis-acting elements ABA-responsive element (ABRE) and a coupling element (CE). ABRE-binding factors (AREB/ABF) involved in ABA-dependent stress responses. As the AREB/ABF subfamily, AREB2 exists in five different isoforms by the use of splice acceptor sites in exons 4. Isoform1 and isoform3 represent the major and minor forms, respectively. Isoform3 is expressed in root differentiation and elongation [9,25,26]. Sixteen transcripts are produced by 13 PYL ABA receptor genes in maize B73, of which two genes undergo alternative splicing. Nineteen transcripts generate from 14 A-type *PP2C* genes in maize B73, four of which undergo alternative splicing. There are thirty-two transcripts from the 12 *SnRK2* genes in maize B73, 10 of which undergo alternative splicing [27]. Additionally, HYPERSENSITIVE TO ABA1 (*HAB1*) encodes PP2Cs. The PP2Cs negatively regulate abiotic stress via mediating ABA signaling. The splicing of *HAB1* can be controlled by the splicing factor, RBM25, by RNA-seq [28]. *HAB1* produces four transcripts via AS [28]. The first two transcripts are involved in seed germination by mediating ABA signaling [28]. The first transcript, *HAB1.1*, inhibited is the OST1 protein kinase by interacting with the OST1 protein kinase. The second transcript, *HAB1.2*, which retains the third intron, formed is a truncated protein. *HAB1.2* positively regulates ABA signaling by interacting with OST1 [9]. Pladienolide B (PB) is involved in the splicing and ABA response. PB affects the splicing of PP2C mRNAs, which reduce the activity of PP2Cs [29]. However, PB mediates SnRK2.6 activity by binding amino acid residue to Leu-46 [30].

Alternative splicing is widely involved in rapidly regulating plant developmental and environmental changes [31]. When the external environment changes, such as high temperatures, drought, and other harsh conditions, plants can improve their ability to resist stress through alternative splicing of their own transcriptomes. Furthermore, ABA is an important phytohormone in response to various stress signals. Research shows that both the ABA signaling pathway and AS mediate plant growth and abiotic stress [30,32,33]. However, how ABA regulates plant development and stress tolerance through AS is still unclear. In this review, we discuss that ABA mediates plant abiotic stress tolerance and development, particularly by alternative splicing events.

2. ABA Mediates Plant Development and Abiotic Stress
2.1. ABA in Responses to Abiotic Stress

Abiotic stress seriously endangers crop production around the world. Abiotic stresses usually lead to the rapid accumulation of ABA in plants, which can cause a large number of cellular and molecular responses [34]. The main functions of ABA are: (1) to make seeds dormant and delay seed germination. (2) to induce stomatal closure. Its most important function is to regulate the water potential balance and osmotic pressure balance of cells. Environmental stresses activate the expression of the ABA biosynthesis gene. The 9-cis-epoxy compound that catalyzes this reaction, carotene dioxygenase (*NCED*), encodes the ABA synthesis enzyme in higher plants. Hai et al. observed that the overexpression of *NCED3* in plants significantly increased the ABA content and improved salt tolerance [35]. Research shows that *NCED3* responds to water deficits by regulating ABA synthesis [36]. ABA levels in plants are regulated by both synthesis and decomposition. *CYP707A* plays a dominant role in abscisic acid catabolism, and it was demonstrated in *Arabidopsis* that a rapid increase in *CYP707A* mRNA levels is closely related to a sharp decrease in abscisic acid levels [37]. Therefore, both promoting the synthesis of ABA and inhibiting the decomposition of ABA can increase the level of ABA in plants.

The regulatory network of abscisic acid signaling is very complex, and its key regulatory mechanism lies in the interaction of its core components. The core members of the ABA signal are usually three components, which are the PYR/PYL/RCAR receptors (PYLs), protein phosphatase 2c (PP2C), and sucrose nonfermenting 1-related protein kinase 2s (SnRK2s) [38]. When plants encounter abiotic stress, ABA inhibits PP2Cs by binding to the PYR/PYL, and PP2C phosphatases bind to dephosphorylates in SnRK2s protein kinases, and then the SnRK2s phosphorylate several transcription factors and membrane proteins involved in the ion channels [39]. The AREB/ABF transcription factors (TFs) mediate the expression of stress-responsive genes to increase plant tolerance [40,41]. AREB/ABFs and ABSCISIC ACID-INSENSITIVE5 (ABI5) are the main transcription factors that regulate ABA-mediated plant growth and adaptation to the external environment.

SnRK2s phosphorylate AREB/ABFs to activate the expression of ABA-responsive genes. In *Arabidopsis*, the AREB/ABFs family contains three members: AREB1/ABF2, AREB2/ABF4, and ABF3/DPBD5 [42]. The phosphorylated AREB regulates the expression of ABA-responsive genes by binding to ABRE cis-elements. The AREB/ABFs can regulate the expression of *atRD29B*, *atAIL1*, *atEM1*, *atEM6*, *atRAB18*, and other LEA-like genes in response to cold and dehydration stress [43]. The AREB/ABFs can regulate the expression of *atAHG1*, *atAHG3*, *atHAI1*, *atHAI2*, *atHAI3*, and other PP2C genes in response to drought stress. The AREB/ABFs can regulate expression of *atADF5*, *atTPPE*, and *atTPPI* genes to mediate stomatal in response to abiotic stress, and regulate the expression of *atHsfA6A*, *atHsfA6B*, and *atMYB102* genes in response to osmotic stress [44]. In addition, PP2Cs genes mainly contain *ABI1*, *ABI2*, *HAB1* and *HAB2*. The SnRK2 kinases mainly include SnRK2.2, SnRK2.3, and SnRK2.6 [45–47]. SnRK2s phosphorylate the membrane proteins of ion channels. The ion channels, such as the potassium channel KAT1 and the slow anion channel SLAC1, regulate guard cells [48–50]. Guard cells improve plant resistance to abiotic stress mainly by regulating guard cell ion channels [51].

ABA functional mutants show marked insensitivity to all aspects of ABA response, especially resistance to stressful environments. The expression of the ABA receptor *msPYL* gene is increased under low-temperature stress [52]. *PP2C* genes are involved in salt stress and fruit development [46,53]. The *CBF* gene is highly expressed in grape tissues under low-temperature stress and ABA. ABA and low temperature have synergistic effects on the activation of these transcription factors and simultaneously affect the increase of shoot cold tolerance and antioxidants and dehydrin gene expression. Alkaline stress can cause the accumulation of reactive oxygen species, which is the main cause of root damage and seedling wilting. Salt stress leads to the accumulation of ABA in plant roots, which is transmitted upward through the xylem, regulates stomatal closure, and reduces plant transpiration and water loss.

Stomatal closure is considered one of the ABA responses to abiotic stress [54]. SMALL AUXIN UP-REGULATED RNA32 (SAUR32) encodes the *AtSAUR32* gene, which is induced by ABA. The AtSAUR32-overexpressed line increases the tolerance of drought stress by stomatal closure [55]. Studies have shown that the transcriptome sequencing of ABA- and salt-stressed tobacco had similar differential gene expressions in many ways [56]. The *msPYL* gene of the ABA receptor may be a response to cold stress by the bioinformatics methods in alfalfa (*Medicago sativa* L.) [52].

The accumulation of ABA in plants has a circadian rhythm and peaks in the afternoon to evening. The results of chromatin immunoprecipitation sequencing (ChIP-seq) showed that the early morning component of the circadian clock LHY could bind to the promoter regions of genes related to ABA synthesis and signaling pathways, regulating their transcription [57]. EEL (ENHANCED EM LEVEL) is a transcription factor that responds to ABA during plant dehydration, and GI binds to the promoter region of *NCED3* by forming a complex with EEL, promoting the expression of NCED3 [58]. The expression of *NCED3* enhances drought tolerance via the regulation of abscisic acid synthesis. The accumulation of ABA is lower in the *gi-1* mutant than that of Col-0. The degree of stomatal closure in the *gi-1* mutant is weakened during the dehydration process, resulting in an increase in the rate of leaf water loss and a decrease in plant survival [58].

In addition, Epigenetic regulation helps plants resist abiotic stresses. The ABA-dependent gene may confer drought stress tolerance due to CG and CHG hypomethylation changes in maize [27]. DNA methylation of many stress-responsive genes regulates ABA-mediated abiotic stress signaling.

2.2. ABA Mediates Plant Development

The role of ABA in the growth and development of plants is also multi-faceted. ABA inhibits seed germination by promoting seed drying and dormancy [59] and is involved in floral transition, lateral root formation, and seedling growth [60].

Studies have shown that at the seed maturity stage, the ABA content was higher in up-regulated ABA anabolic genes and at the germination stage, the ABA content was low-level in the up-relegated ABA catabolism gene expression [61,62]. Studies have also shown that the *ABI5* overexpression is hypersensitive to ABA treatment while the *abi5* mutant is insensitive to ABA at the germination stage [63]. The regulation of seeds by ABA is mainly achieved through the *ABI5* transcription factor, which prevents germination and post-germination growth under adverse conditions. In the ABA signaling pathway, SnRK2 kinase phosphorylates *ABI5*, and the phosphorylated ABI5 regulates the expression of downstream genes [44]. *ABI5* mediates seed germination by regulating the downstream genes, *atPGIP1* and *atPGIP2*. ABI5 mediates the stability of seed germination by regulating the downstream genes, *atPHO1* and *atCAT1* [41]. *ABI5* mediates plant resistance to abiotic stress by regulating LEA protein genes, such as *atEM1* and *atEM6*. *ABI5* mediates the vegetative growth of plants by regulating *atSGR1*, *atNYC1*, and *atSAG29* [64].

Flowering is another important biological process for plants. Plant flowering is usually regulated by internal signals and the external environment. The flowering time is affected by the synthesis and signal transduction of ABA, in which ABA usually inhibits flowering. The regulation of the flowering transition by ABA is mainly achieved by the ABI5 transcription factor. In the ABA signaling pathway, SnRK2.6 kinase phosphorylates ABI5, and the phosphorylated ABI5 binds to the ABRE cis-acting element in the *atFLC* promoter region to regulate the expression of *atFLC* and delay the flowering transition [65]. In the ABA signaling pathway, ABI5 binds to the *atSOC1* promoter region and induces the expression of *atSOC1* to promote flowering [41]. Studies have shown that ABSCISIC ACID-INSENSITIVE4 (*ABI4*)-overexpression delayed the flowering time while the early flowering phenotype of the *abi4* mutant [66]. This suggests that *ABI4* negatively regulates plant flowering time. A schematic diagram of the transcriptional regulation of ABA-mediated plant abiotic stress and development is summarized in Figure 1.

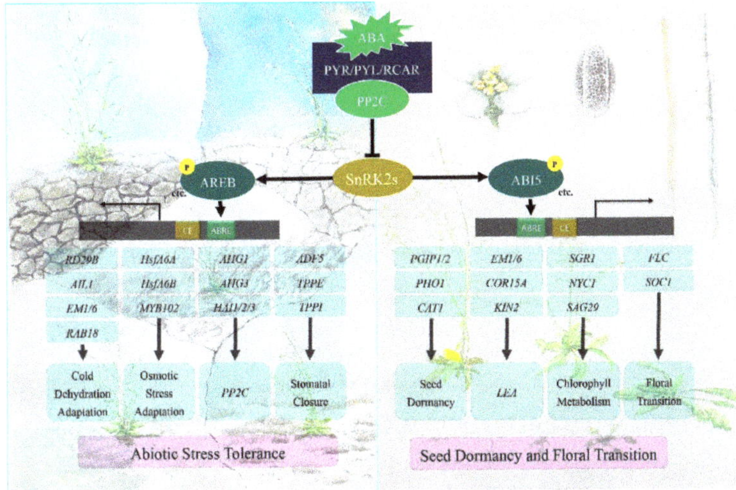

Figure 1. Schematic diagram of ABA-mediated transcriptional regulation of plant. The PYL-PP2C-SnRK2s-TFs complex regulate the ABA-mediated ABRE-dependent gene expression. Two distinct family TFs (ABI5 and AREB/ABF) regulate the ABA-mediated gene expression in response to abiotic stress and development in *Arabidopsis* by interacting with cis-elements in the upstream promoter regions of target genes. See text for details.

3. ABA Mediates Plant Development and Abiotic Stress via Alternative Splicing

Numerous transcriptome results show that ABA can significantly affect numerous and key transcription factors through the transcriptional regulatory pathway of signal transduction by regulating the expression of a large number of genes [58]. Studies have shown that ABA-induced differentially expressed genes have little intersection with differentially alternatively spliced genes, suggesting that the mechanism by which ABA regulates alternative splicing to help plants adapt to the environment is an independent mechanism [29,67]. At present, the mechanism of ABA regulation of AS has not been clearly elucidated.

Widespread use of RNA-seq data provides a favorable resource for analyzing AS. In the ABA treatment group and the ABA control group at the seed germination stage of *Populus hopeiensis*, the proportion of the RI differential variable splicing events was still the highest, accounting for 43.16%, the A3 event type accounted for 24.01%, A5 event type accounted for 13.98%, SE event type accounted for 8.51%, AF event type accounted for 7.90%, AL event type accounted for 2.43%, and MX type events did not have differential variable splicing events [68]. By analyzing the transcriptome sequencing results of ABA-treated *Nicotiana tabacum*, the SE event type was the highest proportion, followed by RI and A3, also occupying a relatively high proportion; MEX was the lowest proportion [58]. Through the results of the differential alternative splicing analysis of ABA-treated *Arabidopsis* sequencing, we found that the proportion of the AL differential alternative splicing events in the ABA-treated group was the highest compared with the control group, accounting for about 35.4%. Followed by the AF event types, the proportion of variable splicing was 34.9%, the proportion of AE events was 14%, the proportion of RI events was 10%, and the proportion of other variable splicing was 10%. The type events, SE and MX, accounted for very little, both less than 10% [67].

From the above, ABA treatment was shown to induce dramatic changes in AS in many dicot species. ABA treatment not only increased the total number of genes with AS but also increased the percentage of genes with AS at the transcriptional level. Different types of ABA-induced differential alternative splicing events occur in different plants, but the most predominant type of splicing event is RI. We speculate that RI-type alternative splicing events play a certain biological role in Hu plants affecting environmental complexity and adaptability, while MX-type events may not play a major regulatory role in this regard. Differential splicing types (DAS) induced by abscisic acid in different plants are shown in Table 1.

Table 1. Differential splicing type (DAS) statistics of different plants induced by abscisic acid.

Plants	Major DAS Type	References
Arabidopsis	AL > AF > AE > RI > SE > MX	[67]
Nicotiana tabacum	SE > RI > A3 > A5 > MX	[56]
Populus hopeiensis	RI > A3 > A5 > SE > MX	[68]

ABA-induced alternative splicing is mostly derived from transcriptome sequencing results. The functional significance of most AS events regulated by ABA in plants has not been demonstrated. The ABA-induced alternative splicing of genes in the ABA signaling pathway in *Arabidopsis* is summarized in Table 2.

Table 2. Summary of the alternative splicing of genes in the ABA-induced abscisic acid (ABA) signaling pathway in *Arabidopsis thaliana* [67].

Locus	Gene	Transcript Type
AT1G17550	*HAB2*	*HAB2-iso1,2,3*
AT1G45249	*ABRE1*	*ABRE1-iso1,2,3*
AT4G34000	*ABF3*	*ABF1-iso1,2,3*
AT5G25610	*RD22*	*RD22-iso1,2*
AT4G27410	*RD26*	*RD26-iso1,2,3,4*
AT1G20620	*CAT3*	*CAT3-iso1,2,3,4*
AT5G62470	*MYB96*	*MYB96-iso1,2*
AT1G20630	*CAT1*	*CAT1-iso1,2*
AT4G46270	*GBF3*	*GBF3-iso1,2,3,4,5,6*
AT4G19230	*CYP707A1*	*CYP707A1-iso1,2*
AT4G26080	*ABI1*	*ABI1-iso1,2*
AT5G57050	*ABI2*	*ABI2-iso1,2,3*

ABA-induced AS occurs primarily in regulatory genes, such as transcription factors, protein kinases, and splicing factors. Splicing factors have long been shown to be involved in spliceosome assembly and AS site selection; in particular, the U1snRNA and U2AF are major players in 5′ and 3′ splice site recognition, respectively [69,70]. ABA-induced differential alternative splicing is involved in ABA responses by increasing the number of unconventional splicing sites. Changes in increased splice sites under ABA treatment may be due to the recruitment of different splicing factor isoforms in the splicing complex.

SR genes are an important part of the splicing complex, which are involved in splicing mainly by co-recognizing the 5′ cleavage site with U1snRNA and U1-70K. The splicing mode of SR genes can be affected by various abiotic stresses and hormones. Among them, ABA can induce the splicing of most SR genes, which is summarized in Table 3. Among the several affected SR genes, the genes that can receive ABA that significantly affect the splicing pattern are *SR34*, *SR34b*, *RS31a*, and *SCL33* [67,71]. However, the functional significance of the alternative splicing products of SR genes is currently unclear. The current study shows that both ABA and abiotic stress can express the SR gene, thereby affecting the splicing of its own and other genes [72].

Table 3. Summary of ABA-induced alternative splicing events of *Arabidopsis* splicing factors [67,71].

Locus	Splicing Factors
AT1G02840	SR34
AT4G02430	SR34b
AT1G09140	SR30
AT3G61860	RS31
AT2G46610	RS31a
AT4G25500	RS40
AT5G52040	RS41
AT2G24590	RSZ22A
AT3G53500	RSZ32
AT2G37340	RSZ33
AT5G64200	SC35
AT1G55310	SCL33
AT3G55460	SCL30
AT3G13570	SCL30A
AT1G16610	SR45
AT3G50670	U1-70K
AT1G27650	U2AF35A
AT1G60900	U2AF65B
AT2G43810	LSM6B

Among them, SR45, as a spliceosome protein, interacts with the U1-70K protein to play a splicing function. ABA regulates the splicing of the key genes of ABA signaling, such as *HAB1*, by mediating the expression of SR45, thereby regulating salt stress and drought stress; at the same time, ABA can further mediate the splicing of SRP30 and SR34 by regulating the expression of SR45, thereby affecting a series of gene splicing, regulating flowering and root development in *Arabidopsis* [73,74]. Studies have shown that ABA may mediate gene splicing by regulating the coupling of SR proteins to protein kinases and transcription factors, thereby regulating plants [75]. The nuclear distribution of *Arabidopsis* SR proteins is affected by abiotic stress. Therefore, we speculated that the effect of ABA on splicing might act through the distribution of SR proteins [76]. However, the mechanism by which ABA mediates alternative splicing through SR genes remains to be elucidated.

Alternative splicing factors typically responded to ABA treatment. *U2AF35A* and *U2AF65A/B* are important subunits that help recruit U2 small nuclear ribonucleoproteins (snRNP). ABA not only regulated and induced the transcription of U2AF but also affected its splicing. Alternatively, *U2AF* regulated the splicing of key flowering genes. The splicing factor *AtU2AF65B* mediated ABA-mediated flowering by regulating the AS of *ABI5* and *FLC* [70]. *ABI5* regulated ABA-mediated seed germination and flowering transition. *ABI5* regulated floral transition by activating the expression of *FLC* [70]. The splicing efficiencies of *FLC* and *ABI5* were regulated by ABA in the *atU2AF65B* mutants, which indicates that *atU2AF65B* could regulate ABA-mediated flowering [70,77]. *AtU2AF65B* regulated flowering by mediating the splicing of *MAF1* (AT1G77080, flowering regulator) and *MAF2* (AT5G65050, flowering regulator) [78]. Additionally, U2AF65B interacted with U2AF35 to mediate the splicing of *HAB1* (AT1G72770, ABA signal regulator) and DWA3 (AT1G61210, ABA signal regulator), affecting salt and drought stress [79].

LUC7 is a subunit of U1 snRNP in plants that respond to both ABA and abiotic stresses. LUC7 interacts with U1-70K to participate in the splicing function. Studies have shown that ABA can regulate cold and salt stress by affecting the expression of the splicing factor, LUC7, and then regulating the alternative splicing of the stress genes *ACA4* (AT2G41560, salt stress regulator) and *K9L2.5* (AT5G44290, cold stress regulator). ABA can mediate the development of plant stems and leaves by regulating the splicing of genes *LUH* (shoot and leaf development regulator) and *TRM3* (shoot and leaf development regulator) by regulating the expression of LUC7 [80].

The Sm-like protein (LSm) is the core component of U6 RNPs and is involved in pre-mRNA splicing [9]. LSm5 is encoded and is super sensitive to ABA and the drought 1 gene (SAD1). *sad1* mutants are sensitive to the ABA inhibition of germination and root growth, and SAD1 depletion showed canonical 5′ and 3′SS recognition. SAD1-OE plants exhibited more salt tolerance via strengthening the AS of salt-induced genes and the recognition accuracy [81]. LSM5 interacts with LSM6/7 to participate in splicing. LSM5 regulates the salt stress resistance of *Arabidopsis* by regulating the splicing of stress-responsive genes, *SnRK2.1/2.2* (AT5G08590, osmotic stress regulator), *SOS2* (AT5G35410, salt stress regulator), and *DREB2A* (AT5G05410, drought stress regulator). In addition, LSM5 regulates *Arabidopsis* root development by regulating the splicing of *AUX1* (AT2G38120, root development regulator) and *EIN2* (AT5G03280, root development regulator) genes.

STABILIZED1 (*STA1*) encodes a nuclear protein and interacts with U5-snRNP. STA1 responds to ABA and abiotic stress by modulating splicing. STA1 is important for the heat shock transcription factor (*HSFA3*) and *COR15A* expression [82,83]. U5 snRNP normally interacts with U4/6 snRNP to mediate splicing. U5 snRNP responds to external temperature stress by regulating the splicing of *COR15A* (AT2G42540, cold stress regulator) and *HSFA3* (AT5G03720, heat stress regulator). However, it is currently unclear how U5 snRNP regulates splicing to mediate root development.

FERONIA (FER), a receptor-like kinase, is a receptor for the rapid alkalinization factor 1 (RALF1) peptide. FER and RALF1 form a complex. The complex interacts with and phosphorylates RNA-binding proteins, which triggers AS [84]. Among them, the glycine-rich RNA binding protein 7 (GRP7) is a splicing factor and regulates seed germination, seedling growth, and flowering [85]. In our study, GRP7 phosphorylation enhanced its interaction with U1-70K, modulating the AS in response to external stimuli and ABA signaling [84]. Both GRP7 and FER are involved in ABA signaling. GRP7 regulated drought resistance in plants by mediating the splicing of *ABF1* (AT1G49720, ABA signal regulator). In addition, GRP7 regulated the root development of plants by mediating the splicing of *CUB9* (AT3G07360, root development regulator).

RBM25 is a splicing factor that responds to salt, drought, and osmotic stress. RBM25 affects plant aridity by regulating ABA-mediated stomatal movement. RBM25 normally interacts with LSM6/7 for splicing. RBM25 responds to osmotic stress by affecting *HAB1* splicing. RBM25 responds to plant stem and leaf development by affecting the splicing of *PAC* (AT2G48120, leaf development regulator) and *PSBQA-2* (AT4G05180, photosynthesis regulator). In summary, a simplified mechanism diagram of ABA mediating abiotic stress and plant development by affecting splicing is shown in Figure 2.

ABA-induced alternative splicing is mostly derived from results of the transcriptome sequencing. The functional significance of most AS events regulated by ABA in plants has not been demonstrated. The following studies have shown that ABA-induced splicing of key genes plays an important role in plant adaptation to the environment. Sanyal et al. showed that ABA treatment significantly enhanced the induction of five splice variants of *CIPK3*. CIPK3 is a protein kinase, which is a serine-threonine protein whose expression increases in response to abscisic acid, cold, drought, high salt, and trauma conditions. *CIPK3* regulates stress and development by participating in Ca^{2+} signaling and ABA pathways. Among them, *CIPK3.1* and *CIPK3.4* splice variants were the two most dominant transcripts under the ABA treatment, while *CIPK3.1* showed an interaction with ABR1 in response to ABA, osmotic stress, and drought stress [79,86]. The study by Zhan et al. showed that ABA has an effect on the splicing pattern of *Arabidopsis* wild-type, which can significantly induce the alternative splicing of 27 genes. Among them, the *AtSPL2* gene is involved in shoot maturation at the reproductive stage. *AtATERDJ3B* is involved in protein folding, and *AtNUP50* is involved in intracellular transport [29]. ABA mediates plant development and abiotic stress responses by affecting splicing of genes, as shown in Table 4.

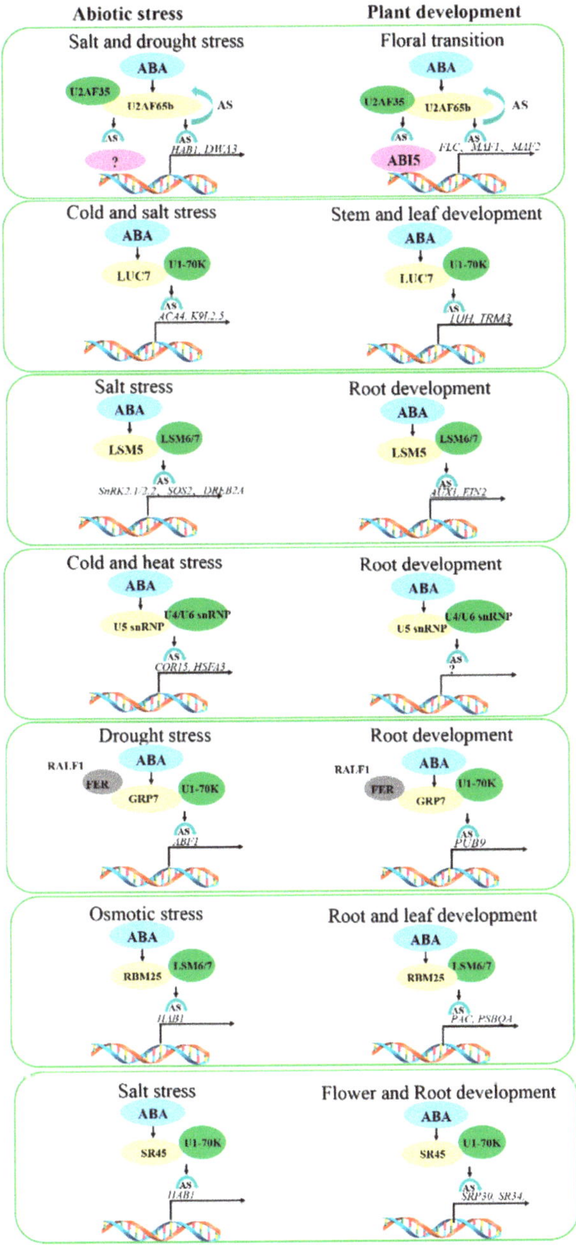

Figure 2. ABA mediates abiotic stress and plant development by affecting the expression or splicing of the same splicing factor, thereby affecting the splicing of different genes. Orange ovals represent splicing factors acted on by ABA, green ovals represent interacting splicing factors, purple ovals represent transcription factors, white ovals represent alkalizing factors, and gray ovals represent kinases.

Table 4. Summary of the fundamental studies on AS in ABA-mediated plant development and abiotic stress responses.

Gene ID	Gene Name	AS Mediated Development and Abiotic Stress	References
AT2G26980	*AtCIPK3*	ABA, osmotic stress, and drought stress	[79,86]
At5g43270	*AtSPL2*	shoot maturation	[29]
At3g62600	*AtATERDJ3B*	protein folding	[29]
At3g15970	*AtNUP50*	intracellular transport	[29]

AS plays an important regulatory role in plant cells, and AS can regulate the transcriptional level by introducing premature termination codons (PTC). ABA-induced alternative splicing is often accompanied by a shift of the reading frame, resulting in premature codon termination. Most aberrantly spliced RNAs are degraded through the NMD (nonsense-mediated decay/nonsense-mediated degradation) pathway. The NMD pathway can prevent abnormally spliced RNAs from being translated into abnormal proteins, thereby inhibiting abnormal biological effects in plants. AS produces truncated non-functional proteins through RI or PTC to adjust the amounts of functional proteins in plants in time so that plants have stronger adaptability.

AS is an important mechanism for altering the transcriptome in response to the environment and improving reproduction. Dysregulation of alternative splicing is a hallmark feature of plant adaptation. Studies have shown that there is little intersection between ABA-induced differentially expressed genes and key differentially spliced genes [67]. The stress-induced transcriptome results of wheat and rice are consistent with the results [16,87]. This suggests that AS is an independent mechanism by which ABA regulates plant growth and environmental responses. The ABA-induced alternative splicing mechanism will be a potential target for improving the ability of plants to adapt to the environment. The study of the ABA-induced variable shearing mechanism will offer an important mechanism for crops to adapt to the environment and improve crop quality through the variable shearing mechanism in the future, and, at the same time, make a strong guarantee for future food security.

4. Conclusions

ABA is an important hormone in response to plant growth and abiotic stress. There are different mechanisms by which ABA affects the environment. First, from the transcriptional level, ABA regulates the up-regulation or down-regulation of a large number of genes by regulating kinases and a large number of transcription factors in the signaling pathway, thereby increasing the abundance of proteins and improving the adaptability of plants. Secondly, from the post-transcriptional level, ABA affects the splicing of its own and other genes by regulating the expression of splicing factors, such as SR genes and U2AF, to produce different splicing isoforms, thereby increasing the abundance of proteins and improving plant adaptation.

Author Contributions: Conceptualization, Y.L.; writing original draft preparation, X.Y., Z.J., Y.T. and F.Z.; writing review and editing, Q.P., F.Z. and Y.L.; funding, Y.L. All authors have read and agreed to the published version of the manuscript.

Funding: This work was supported by the Jiangsu Agricultural Science and Technology Innovation Fund (CX (21) 2023), Major project of natural science research in colleges of Jiangsu Province (20KJA220001), Program for Scientific Research Innovation team of Young scholar in colleges and Universities of Shandong Province(2019KJE011). Hong Kong Research Grant Council (AoE/M-05/12, AoE/M-403/16, GRF 12100318, 12103219, 12103220).

Institutional Review Board Statement: Not applicable.

Informed Consent Statement: Not applicable.

Data Availability Statement: Not applicable.

Conflicts of Interest: The authors declare no conflict of interest.

References

1. Hada, A.; Jaabir, M.; Singh, N.; Changwal, C.; Kumar, A. Functional genomics approaches for combating the effect of abiotic stresses. In *Stress Tolerance in Horticultural Crop*; Woodhead Publishing: Sawston, UK, 2021; Volume 1, pp. 1–19.
2. Chi, D.N.; Chen, J.; Clark, D.; Perez, H.; Huo, H.A. Effects of Maternal Environment on Seed Germination and Seedling Vigor of Petunia × hybrida under Different Abiotic Stresses. *Plants* **2021**, *10*, 581.
3. Zhang, A.; Yang, X.; Lu, J.; Song, F.; Zhao, B. *OsIAA20*, an Aux/IAA protein, mediates abiotic stress tolerance in rice through an ABA pathway. *Plant Sci.* **2021**, *308*, 110903. [CrossRef] [PubMed]
4. Ullah, F.; Xu, Q.; Zhao, Y.; Zhou, D.X. Histone deacetylase HDA710 controls salt tolerance by regulating ABA signaling in rice. *J. Integr. Plant Biol.* **2021**, *63*, 451–467. [CrossRef] [PubMed]
5. Min, M.K.; Kim, R.; Hong, W.J.; Jung, K.H.; Kim, B.G. *OsPP2C09* Is a Bifunctional Regulator in Both ABA-Dependent and Independent Abiotic Stress Signaling Pathways Citation. *Int. J. Mol. Sci.* **2021**, *22*, 393. [CrossRef]
6. Shi, X.; Tian, Q.; Deng, P.; Zhang, W.; Jing, W. The rice aldehyde oxidase *OsAO3* gene regulates plant growth, grain yield, and drought tolerance by participating in ABA biosynthesis. *Biochem. Biophys. Res. Commun.* **2021**, *548*, 189–195. [CrossRef]
7. Black, D.L. Mechanisms of alternative pre-messenger RNA splicing. *Annu. Rev. Biochem.* **2003**, *72*, 291–336. [CrossRef]
8. Lee, G.H.; Montez, J.M.; Carroll, K.M.; Darvishzadeh, J.G.; Jun, L.J.F.; Proenca, R. Abnormal splicing of the leptin receptor in diabetic mice. *Nature* **1996**, *379*, 632–635. [CrossRef]
9. Punzo, P.; Grillo, S.; Batelli, G. Alternative splicing in plant abiotic stress responses. *Biochem. Soc. Trans.* **2020**, *48*, 2117–2126. [CrossRef]
10. Nilsen, T.W.; Graveley, B.R. Expansion of the eukaryotic proteome by alternative splicing. *Nature* **2010**, *463*, 457–463. [CrossRef]
11. Stamm, S.; Ben-Ari, S.; Rafalska, I.; Tang, Y.; Zhang, Z.; Toiber, D.; Thanaraj, T.A.; Soreq, H. Function of alternative splicing. *Gene* **2005**, *344*, 1–20. [CrossRef]
12. Keren, H.; Lev-Maor, G.; Ast, G. Alternative splicing and evolution: Diversification, exon definition and function. *Nat. Rev. Genet.* **2010**, *11*, 345–355. [CrossRef] [PubMed]
13. Zhang, H.; Li, G.; Fu, C.; Duan, S.; Guo, X. Genome-wide identification, transcriptome analysis and alternative splicing events of Hsf family genes in maize. *Sci. Rep.* **2020**, *10*, 8073. [CrossRef] [PubMed]
14. Li, J.; Wang, Y.; Rao, X.; Wang, Y.; Liu, Y. Roles of alternative splicing in modulating transcriptional regulation. *BMC Syst. Biol.* **2017**, *11*, 89. [CrossRef] [PubMed]
15. Romanowski, A.; Schlaen, R.G.; Perez-Santangelo, S.; Mancini, E.; Yanovsky, M.J. Global transcriptome analysis reveals circadian control of splicing events in *Arabidopsis thaliana*. *Plant J.* **2020**, *103*, 889–902. [CrossRef]
16. Ganie, S.A.; Reddy, A. Stress-Induced Changes in Alternative Splicing Landscape in Rice: Functional Significance of Splice Isoforms in Stress Tolerance. *Biology* **2021**, *10*, 309. [CrossRef]
17. Wang, T.Y.; Liu, Q.; Ren, Y.; Alam, S.K.; Yang, R. A pan-cancer transcriptome analysis of exitron splicing identifies novel cancer driver genes and neoepitopes. *Mol. Cell* **2021**, *81*, 2246–2260. [CrossRef]
18. Wang, Q.; Xu, Y.; Zhang, M.; Zhu, F.; Sun, M.; Lian, X.; Zhao, G.; Duan, D. Transcriptome and metabolome analysis of stress tolerance to aluminium in Vitis quinquangularis. *Planta* **2021**, *254*, 105. [CrossRef]
19. Holliday, M.; Singer, E.S.; Ross, S.B.; Lim, S.; Bagnall, R.D. Transcriptome Sequencing of Patients with Hypertrophic Cardiomyopathy Reveals Novel Splice-altering Variants in MYBPC3. *Circ. Genom. Precis. Med.* **2021**, *14*, e003202. [CrossRef]
20. Noam, L.; Noam, A.; Dena, L.; Robert, F.; Shin-Han, S. Genome-Wide Survey of Cold Stress Regulated Alternative Splicing in *Arabidopsis thaliana* with Tiling Microarray. *PLoS ONE* **2013**, *8*, e66511. [CrossRef]
21. Mario, K.; Hu, Y.; Anida, M.; Sotirios, F.; Enrico, S.; Stefan, S. Alternative splicing in tomato pollen in response to heat stress. *DNA Res.* **2017**, *24*, 205–217. [CrossRef]
22. Seo, P.J.; Park, M.J.; Park, C.M. Alternative splicing of transcription factors in plant responses to low temperature stress: Mechanisms and functions. *Planta* **2013**, *237*, 1415–1424. [CrossRef] [PubMed]
23. Takechi, H.; Hosokawa, N.; Hirayoshi, K.; Nagata, K. Alternative 5′ splice site selection induced by heat shock. *Mol. Cell. Biol.* **1994**, *14*, 567–575. [CrossRef] [PubMed]
24. Barashkov, N.A.; Dzhemileva, L.U.; Fedorova, S.A.; Teryutin, F.M.; Posukh, O.L.; Fedotova, E.E.; Lobov, S.L.; Khusnutdinova, E.K. Autosomal recessive deafness 1A (DFNB1A) in Yakut population isolate in Eastern Siberia: Extensive accumulation of the splice site mutation IVS1+1G>A in GJB2 gene as a result of founder effect. *J. Hum. Genet.* **2011**, *56*, 631–639. [CrossRef]
25. Li, S.; Yamada, M.; Han, X.; Ohler, U.; Benfey, P. High-Resolution Expression Map of the *Arabidopsis* Root Reveals Alternative Splicing and lincRNA Regulation. *Dev. Cell* **2016**, *39*, 508–522. [CrossRef] [PubMed]
26. Swarup, R.; Crespi, M.; Bennett, M.J. One Gene, Many Proteins: Mapping Cell-Specific Alternative Splicing in Plants. *Dev. Cell* **2016**, *39*, 383–385. [CrossRef]
27. Sallam, N.; Moussa, M. DNA methylation changes stimulated by drought stress in ABA-deficient maize mutant vp10. *Plant Physiol. Biochem.* **2021**, *160*, 218–224. [CrossRef]
28. Zhan, X.; Qian, B.; Cao, F.; Wu, W.; Yang, L.; Guan, Q.; Gu, X.; Wang, P.; Okusolubo, T.A.; Dunn, S.L.; et al. An *Arabidopsis* PWI and RRM motif-containing protein is critical for pre-mRNA splicing and ABA responses. *Nat. Commun.* **2015**, *6*, 8139. [CrossRef]

29. Ling, Y.; Alshareef, S.; Butt, H.; Lozano-Juste, J.; Mahfouz, M.M. Pre-mRNA splicing repression triggers abiotic stress signaling in plants. *Plant J.* **2016**, *89*, 291–309. [CrossRef]

30. Punkkinen, M.; Mahfouz, M.M.; Fujii, H. Chemical activation of *Arabidopsis* SnRK2.6 by pladienolide B. *Plant Signal. Behav.* **2021**, *16*, 1885165. [CrossRef]

31. Xue, X.; Jiao, F.; Xu, H.; Jiao, Q.; Wang, M. The role of RNA-binding protein, microRNA and alternative splicing in seed germination: A field need to be discovered. *BMC Plant Biol.* **2021**, *21*, 194. [CrossRef]

32. Laloum, T.; Martín, G.; Duque, P. Alternative Splicing Control of Abiotic Stress Responses. *Trends Plant Sci.* **2017**, *23*, 140–150. [CrossRef] [PubMed]

33. Cruz, T.; Carvalho, R.F.; Richardson, D.N.; Duque, P. Abscisic Acid (ABA) Regulation of *Arabidopsis* SR Protein Gene Expression. *Int. J. Mol. Sci.* **2014**, *15*, 17541–17564. [CrossRef] [PubMed]

34. Lee, S.C.; Sheng, L. ABA signal transduction at the crossroad of biotic and abiotic stress responses. *Plant Cell Environ.* **2012**, *35*, 53–60. [CrossRef] [PubMed]

35. Hai, A.T.; Lee, S.; Cao, S.T.; Lee, W.J.; Lee, H. Overexpression of the *HDA15* Gene Confers Resistance to Salt Stress by the Induction of NCED3, an ABA Biosynthesis Enzyme. *Front. Plant Sci.* **2021**, *12*, 631.

36. Toups, H.S.; Cochetel, N.; Galdamez, K.; Deluc, L.; Cramer, G.R. *Abscisic Acid Metabolism in Leaves and Roots of Four Vitis Species in Response to Water Deficit*; Research Square: Durham, NC, USA, 2021.

37. Hwang, S.G.; Chen, H.C.; Huang, W.Y.; Chu, Y.C.; Shii, C.T.; Cheng, W.H. Ectopic expression of rice *OsNCED3* in *Arabidopsis* increases ABA level and alters leaf morphology. *Plant Sci.* **2010**, *178*, 12–22. [CrossRef]

38. Chuong, N.N.; Xuan, L.; Nghia, D.; Dai, T.; Thao, N.P. Protein Phosphatase Type 2C Functions in Phytohormone-Dependent Pathways and in Plant Responses to Abiotic Stresses. *Curr. Protein Pept. Sci.* **2021**, *22*, 430–440. [CrossRef]

39. Collin, A.; Daszkowska-Golec, A.; Szarejko, I. Updates on the Role of ABSCISIC ACID INSENSITIVE 5 (ABI5) and ABSCISIC ACID-RESPONSIVE ELEMENT BINDING FACTORs (ABFs) in ABA Signaling in Different Developmental Stages in Plants. *Cells* **2021**, *10*, 1996. [CrossRef]

40. Nakashima, K.; Yamaguchi-Shinozaki, K.; Shinozaki, K. The transcriptional regulatory network in the drought response and its crosstalk in abiotic stress responses including drought, cold, and heat. *Front. Plant Sci.* **2014**, *5*, 170. [CrossRef]

41. Soma, F.; Takahashi, F.; Yamaguchi-Shinozaki, K.; Shinozaki, K. Cellular Phosphorylation Signaling and Gene Expression in Drought Stress Responses: ABA-Dependent and ABA-Independent Regulatory Systems. *Plants* **2021**, *10*, 756. [CrossRef]

42. Nakashima, K.; Yamaguchi-Shinozaki, K. ABA signaling in stress-response and seed development. *Plant Cell Rep.* **2013**, *32*, 959–970. [CrossRef]

43. Yoshida, T.; Fujita, Y.; Maruyama, K.; Mogami, J.; Todaka, D.; Shinozaki, K.; Yamaguchi-Shinozaki, K. Four *Arabidopsis* AREB/ABF transcription factors function predominantly in gene expression downstream of SnRK2 kinases in abscisic acid signalling in response to osmotic stress. *Plant Cell Environ.* **2015**, *38*, 35–49. [CrossRef] [PubMed]

44. Yoshida, T.; Fujita, Y.; Sayama, H.; Kidokoro, S.; Maruyama, K.; Mizoi, J.; Shinozaki, K.; Yamaguchi-Shinozaki, K. AREB1, AREB2, and ABF3 are master transcription factors that cooperatively regulate ABRE-dependent ABA signaling involved in drought stress tolerance and require ABA for full activation. *Plant J.* **2010**, *61*, 672–685. [CrossRef] [PubMed]

45. Lin, Z.; Li, Y.; Wang, Y.; Liu, X.; Wang, P. Initiation and amplification of SnRK2 activation in abscisic acid signaling. *Nat. Commun.* **2021**, *12*, 2456. [CrossRef] [PubMed]

46. Wang, J.; Xu, Y.; Zhang, W.; Zheng, Y.; Yuan, B.; Li, Q.; Leng, P. Tomato SlPP2C5 Is Involved in the Regulation of Fruit Development and Ripening. *Plant Cell Physiol.* **2021**, *11*, 1760–1769. [CrossRef] [PubMed]

47. Liu, C.; Hu, J.; Fan, W.; Zhu, P.; Cao, B.; Zheng, S.; Xia, Z.; Zhu, Y.; Zhao, A. Heterotrimeric G-protein γ subunits regulate ABA signaling in response to drought through interacting with PP2Cs and SnRK2s in mulberry (*Morus alba* L.). *Plant Physiol. Biochem.* **2021**, *161*, 210–221. [CrossRef]

48. Geiger, D.; Scherzer, S.; Mumm, P.; Marten, I.; Ache, P. Guard cell anion channel SLAC1 is regulated by CDPK protein kinases with distinct Ca^{2+} affinities. *Proc. Natl. Acad. Sci. USA* **2010**, *107*, 8023–8028. [CrossRef]

49. Chan, C.; Panzeri, D.; Okuma, E.; Tldsepp, K.; Zimmerli, L. STRESS INDUCED FACTOR 2 Regulates *Arabidopsis* Stomatal Immunity through Phosphorylation of the Anion Channel SLAC1. *Plant Cell* **2020**, *32*, 2216–2236. [CrossRef]

50. Sato, A.; Sato, Y.; Fukao, Y.; Fujiwara, M.; Uozumi, N. Threonine at position 306 of the KAT1 potassium channel is essential for channel activity and is a target site for ABA-activated SnRK2/OST1/SnRK2.6 protein kinase. *Biochem. J.* **2009**, *424*, 439–448. [CrossRef]

51. Kuromori, T.; Seo, M.; Shinozaki, K. ABA Transport and Plant Water Stress Responses. *Trends Plant Sci.* **2018**, *23*, 513–522. [CrossRef]

52. Nian, L.; Zhang, X.; Yi, X.; Liu, X.; Yang, Y.; Li, X.; Haider, F.U.; Zhu, X. Genome-wide identification of ABA receptor PYL/RCAR gene family and their response to cold stress in *Medicago sativa* L. *Physiol. Mol. Biol. Plants* **2021**, *27*, 1979–1995. [CrossRef]

53. Ibrahim, A.K.; Xu, Y.; He, Q.; Niyitanga, S.; Zhang, L. *Comparative Transcriptomes and Genome-Wide Identification Reveal Salt Stress-Responsive PP2C in Jute (Corchorus capsularis)*; Research Square: Durham, NC, USA, 2021.

54. Chen, K.; Li, G.J.; Bressan, R.A.; Song, C.P.; Zhu, J.K.; Zhao, Y. Abscisic acid dynamics, signaling, and functions in plants. *J. Integr. Plant Biol.* **2020**, *62*, 27–56. [CrossRef] [PubMed]

55. Kurihara, Y.; Yokohama, R.; Reed, J.J.; Liu, H.Y.; Lu, G. Citation: The *Arabidopsis* SMALL AUXIN UP RNA32 Protein Regulates ABA-Mediated Responses to Drought Stress. *Front. Plant Sci.* **2021**, *12*, 259.

56. Wu, H.; Li, H.; Zhang, W.; Tang, H.; Yang, L. Transcriptional regulation and functional analysis of Nicotiana tabacum under salt and ABA stress—ScienceDirect. *Biochem. Biophys. Res. Commun.* **2021**, *570*, 110–116. [CrossRef] [PubMed]

57. Adams, S.; Grundy, J.; Veflingstad, S.R.; Dyer, N.P.; Hannah, M.A.; Ott, S.; Carré, I.A. Circadian control of abscisic acid biosynthesis and signalling pathways revealed by genome-wide analysis of LHY binding targets. *New Phytol.* **2018**, *220*, 893–907. [CrossRef]

58. Baek, D.; Kim, W.Y.; Cha, J.Y.; Park, H.J.; Yun, D.J. The GIGANTEA-ENHANCED EM LEVEL complex enhances drought tolerance via regulation of abscisic acid synthesis. *Plant Physiol.* **2020**, *184*, 443–458. [CrossRef]

59. Wang, Z.; Ren, Z.; Cheng, C.; Wang, T.; Ji, H.; Zhao, Y.; Deng, Z.; Zhi, L.; Lu, J.; Wu, X.; et al. Counteraction of ABA-Mediated Inhibition of Seed Germination and Seedling Establishment by ABA Signaling Terminator in *Arabidopsis*. *Mol. Plant* **2020**, *13*, 1284–1297. [CrossRef]

60. Hong, J.H.; Seah, S.W.; Xu, J. The root of ABA action in environmental stress response. *Plant Cell Rep.* **2013**, *32*, 971–983. [CrossRef]

61. Shu, K.; Qian, C.; Wu, Y.; Liu, R.; Zhang, H.; Wang, S.; Tang, S.; Yang, W.; Xie, Q. ABSCISIC ACID-INSENSITIVE 4 negatively regulates flowering through directly promoting *Arabidopsis* FLOWERING LOCUS C transcription. *J. Exp. Bot.* **2016**, *67*, 195–205. [CrossRef]

62. Lee, Y.; Do, V.G.; Kim, S.; Kweon, H.; Mcghie, T.K. Cold stress triggers premature fruit abscission through ABA-dependent signal transduction in early developing apple. *PLoS ONE* **2021**, *16*, e0249975. [CrossRef]

63. Lopez-Molina, L.; Mongrand, S.; Chua, N.H. A postgermination developmental arrest checkpoint is mediated by abscisic acid and requires the ABI5 transcription factor in *Arabidopsis*. *Proc. Natl. Acad. Sci. USA* **2001**, *98*, 4782–4787. [CrossRef]

64. Skubacz, A.; Daszkowska-Golec, A.; Szarejko, I. The Role and Regulation of ABI5 (ABA-Insensitive 5) in Plant Development, Abiotic Stress Responses and Phytohormone Crosstalk. *Front. Plant Sci.* **2016**, *7*, 1884. [CrossRef] [PubMed]

65. Wang, Y.; Li, L.; Ye, T.; Lu, Y.; Chen, X.; Wu, Y. The inhibitory effect of ABA on floral transition is mediated by *ABI5* in *Arabidopsis*. *J. Exp. Bot.* **2013**, *64*, 675–684. [CrossRef] [PubMed]

66. Günter, T.E.; Heinz, S. The golden decade of molecular floral development (1990–1999): A cheerful obituary. *Dev. Genet.* **2015**, *25*, 181–193.

67. Zhu, F.Y.; Chen, M.X.; Ye, N.H.; Shi, L.; Ma, K.L.; Yang, J.F.; Cao, Y.Y.; Zhang, Y.; Yoshida, T.; Fernie, A.R. Proteogenomic analysis reveals alternative splicing and translation as part of the abscisic acid response in *Arabidopsis* seedlings. *Plant J.* **2017**, *91*, 518–533. [CrossRef]

68. Chen, Z.; Ji, L.; Wang, J.; Jin, J.; Yang, X.; Rao, P.; Gao, K.; Liao, W.; Ye, M.; An, X. Dynamic changes in the transcriptome of Populus hopeiensis in response to abscisic acid. *Sci. Rep.* **2017**, *7*, 42708. [CrossRef]

69. Kondo, Y.; Oubridge, C.; Van Roon, A.M.; Nagai, K. Crystal structure of human U1 snRNP, a small nuclear ribonucleoprotein particle, reveals the mechanism of 5′ splice site recognition. *eLife* **2015**, *4*, e04986. [CrossRef]

70. Wang, Y.Y.; Xiong, F.; Ren, Q.P.; Wang, X.L. Regulation of flowering transition by alternative splicing: The role of the U2 auxiliary factor. *J. Exp. Bot.* **2019**, *71*, 751–758. [CrossRef]

71. Palusa, S.G.; Ali, G.S.; Reddy, A.S. Alternative splicing of pre-mRNAs of *Arabidopsis* serine/arginine-rich proteins: Regulation by hormones and stresses. *Plant J.* **2007**, *49*, 1091–1107. [CrossRef]

72. Filichkin, S.A.; Priest, H.D.; Givan, S.A.; Shen, R.; Bryant, D.W.; Fox, S.E.; Wong, W.-K.; Mockler, T.C. Genome-wide mapping of alternative splicing in *Arabidopsis thaliana*. *Genome Res.* **2010**, *20*, 45–58. [CrossRef]

73. Albaqami, M.; Laluk, K.; Reddy, A. The *Arabidopsis* splicing regulator SR45 confers salt tolerance in a splice isoform-dependent manner. *Plant Mol. Biol.* **2019**, *100*, 379–390. [CrossRef]

74. Zhang, X.N.; Mount, S.M. Two Alternatively Spliced Isoforms of the *Arabidopsis* SR45 Protein Have Distinct Roles during Normal Plant Development. *Plant Physiol.* **2009**, *150*, 1450–1458. [CrossRef] [PubMed]

75. Xing, D.; Wang, Y.; Hamilton, M.; Ben-Hur, A.; Reddy, A. Transcriptome-Wide Identification of RNA Targets of *Arabidopsis* SERINE/ARGININE-RICH45 Uncovers the Unexpected Roles of This RNA Binding Protein in RNA Processing. *Plant Cell* **2015**, *27*, 3294–3308. [CrossRef] [PubMed]

76. Tillemans, V.; Dispa, L.; Remacle, C.; Collinge, M.; Motte, P. Functional distribution and dynamics of *Arabidopsis* SR splicing factors in living plant cells. *Plant J.* **2005**, *41*, 567–582. [CrossRef] [PubMed]

77. Xiong, F.; Ren, J.J.; Yu, Q.; Wang, Y.Y.; Lu, C.C.; Kong, L.J.; Otegui, M.S.; Wang, X.L. *AtU2AF65b* functions in abscisic acid mediated flowering via regulating the precursor messenger RNA splicing of *ABI5* and *FLC* in *Arabidopsis*. *New Phytol.* **2019**, *223*, 277–292. [CrossRef] [PubMed]

78. Airoldi, C.A.; McKay, M.; Davies, B. MAF2 Is Regulated by Temperature-Dependent Splicing and Represses Flowering at Low Temperatures in Parallel with *FLM*. *PLoS ONE* **2015**, *10*, e0126516. [CrossRef]

79. Sanyal, S.K.; Kanwar, P.; Samtani, H.; Kaur, K.; Jha, S.K.; Pandey, G.K. Alternative Splicing of *CIPK3* Results in Distinct Target Selection to Propagate ABA Signaling in *Arabidopsis*. *Front. Plant Sci.* **2017**, *8*, 1924. [CrossRef]

80. Marcella, D.; Willing, E.M.; Szabo, E.X.; Francisco-Mangilet, A.G.; Droste-Borel, I.; Macek, B.; Schneeberger, K.; Laubinger, S. The U1 snRNP subunit LUC7 modulates plant development and stress responses via regulation of alternative splicing. *Plant Cell* **2018**, *30*, 2838–2854. [CrossRef]

81. Peng, C.; Zhang, S.; Feng, D.; Ali, S.; Xiong, L. Dynamic regulation of genome-wide pre-mRNA splicing and stress tolerance by the Sm-like protein LSm5 in *Arabidopsis*. *Genome Biol.* **2014**, *15*, R1. [CrossRef]

82. Lee, B.H.; Kapoor, A.; Zhu, J.; Zhu, J.K. STABILIZED1, a Stress-Upregulated Nuclear Protein, Is Required for Pre-mRNA Splicing, mRNA Turnover, and Stress Tolerance in *Arabidopsis*. *Plant Cell* **2006**, *18*, 1736–1749. [CrossRef]

83. Kim, G.D.; Cho, Y.H.; Lee, B.H.; Yoo, S.D. STABILIZED1 modulates pre-mRNA splicing for thermotolerance. *Plant Physiol.* **2017**, *173*, 2370–2382. [CrossRef]

84. Wang, L.; Yang, T.; Wang, B.; Lin, Q.; Yu, F. RALF1-FERONIA complex affects splicing dynamics to modulate stress responses and growth in plants. *Sci. Adv.* **2020**, *6*, eaaz1622. [CrossRef] [PubMed]

85. Jin, S.K.; Jung, H.J.; Lee, H.J.; Kim, K.A.; Kang, H. Glycine-rich RNA-binding protein 7 affects abiotic stress responses by regulating stomata opening and closing in *Arabidopsis thaliana*. *Plant J.* **2010**, *55*, 455–466. [CrossRef]

86. Sanyal, S.K.; Kanwar, P.; Yadav, A.K.; Sharma, C.; Kumar, A.; Pandey, G.K. *Arabidopsis* CBL interacting protein kinase 3 interacts with ABR1, an APETALA2 domain transcription factor, to regulate ABA responses. *Plant Sci.* **2017**, *254*, 48–59. [CrossRef] [PubMed]

87. Harb, A.; Simpson, C.; Guo, W.; Govindan, G.; Kakani, V.G.; Sunkar, R. The Effect of Drought on Transcriptome and Hormonal Profiles in Barley Genotypes With Contrasting Drought Tolerance. *Front. Plant Sci.* **2020**, *11*, 618491. [CrossRef]

International Journal of
Molecular Sciences

Review

Mechanisms of Abscisic Acid-Mediated Drought Stress Responses in Plants

Mehtab Muhammad Aslam [1,2,†], Muhammad Waseem [3,4,†], Bello Hassan Jakada [5], Eyalira Jacob Okal [6], Zuliang Lei [1], Hafiz Sohaib Ahmad Saqib [7], Wei Yuan [2,*], Weifeng Xu [1,2] and Qian Zhang [1,*]

[1] Joint International Research Laboratory of Water and Nutrient in Crop and College of Resources and Environment, Fujian Agriculture and Forestry University, Fuzhou 350002, China; mehtabmuhammadaslam@yahoo.com (M.M.A.); 1200525014@fafu.edu.cn (Z.L.); wfxu@fafu.edu.cn (W.X.)
[2] College of Agriculture, Yangzhou University, Yangzhou 225009, China
[3] Department of Botany, University of Narowal, Narowal 51600, Pakistan; m.waseem.botanist@gmail.com
[4] College of Horticulture, Hainan University, Haikou 570100, China
[5] Key Laboratory of Genetics, Breeding and Multiple Utilization of Crops, College of Life Science, Fujian Agriculture and Forestry University, Ministry of Education, Fuzhou 350002, China; bellojakada@gmail.com
[6] Center for Mountain Futures, Kunming Institute of Botany, Chinese Academy of Sciences, Kunming 650201, China; eyalirajac@gmail.com
[7] Guangdong Provincial Key Laboratory of Marine Biology, College of Science, Shantou University, Shantou 515063, China; sohaibsaqib@gmail.com
* Correspondence: yuanwei@fafu.edu.cn (W.Y.); qian_z@fafu.edu.cn (Q.Z.)
† These authors contributed equally to this work.

Abstract: Drought is one of the major constraints to rain-fed agricultural production, especially under climate change conditions. Plants evolved an array of adaptive strategies that perceive stress stimuli and respond to these stress signals through specific mechanisms. Abscisic acid (ABA) is a premier signal for plants to respond to drought and plays a critical role in plant growth and development. ABA triggers a variety of physiological processes such as stomatal closure, root system modulation, organizing soil microbial communities, activation of transcriptional and post-transcriptional gene expression, and metabolic alterations. Thus, understanding the mechanisms of ABA-mediated drought responses in plants is critical for ensuring crop yield and global food security. In this review, we highlighted how plants adjust ABA perception, transcriptional levels of ABA- and drought-related genes, and regulation of metabolic pathways to alter drought stress responses at both cellular and the whole plant level. Understanding the synergetic role of drought and ABA will strengthen our knowledge to develop stress-resilient crops through integrated advanced biotechnology approaches. This review will elaborate on ABA-mediated drought responses at genetic, biochemical, and molecular levels in plants, which is critical for advancement in stress biology research.

Keywords: ABA; drought; metabolites; signaling; crop breeding

Citation: Muhammad Aslam, M.; Waseem, M.; Jakada, B.H.; Okal, E.J.; Lei, Z.; Saqib, H.S.A.; Yuan, W.; Xu, W.; Zhang, Q. Mechanisms of Abscisic Acid-Mediated Drought Stress Responses in Plants. *Int. J. Mol. Sci.* **2022**, *23*, 1084. https://doi.org/10.3390/10.3390/ijms23031084

Academic Editor: Ricardo Aroca

Received: 1 December 2021
Accepted: 13 January 2022
Published: 19 January 2022

Publisher's Note: MDPI stays neutral with regard to jurisdictional claims in published maps and institutional affiliations.

1. Introduction

Drought stress reduces soil water content which restricts water uptake by the plant root thereby limiting plant growth and productivity [1]. Plants have evolved a wide range of morpho-physiological, metabolic, and molecular mechanisms to resist long- or short-term responses to drought stress [2]. Phytohormones are important plant growth regulators and mediators of environmental stresses such as drought which adversely influence crop yield and pose threats to global food security [2]. To cope with drought stress, potent and novel approaches should be introduced, and phytohormone engineering could be a method of choice for sustainable crop production and breeding programs. In the last decade, the interest to understand the spatiotemporal changes of ABA to modulate plant responses is growing [3]. Abscisic acid (ABA) is critical for plant development and can redesign various physiological and biochemical signal transduction cascades in plants to cope with environmental stresses particularly drought [4,5]. Additionally, ABA plays a critical role in biomolecules biosynthesis, senescence, seed germination, stomatal closure, and root architecture modification [6,7].

ABA is classified as an isoprenoid molecule, synthesized from carotenoids (C40) derivative of isopentenyl diphosphate (IPP) through the methylerythritol phosphate (MEP) pathway in plastids [8]. The synthesis of ABA undergoes a series of steps, and each step is catalyzed by a specific enzyme. The conversion of zeaxanthin to all trans-violaxanthin is the first step in ABA biosynthesis occurring in the plastid. This cyclic hydroxylation of epoxycarotenoids to all-xanthin is catalyzed by *zeaxanthin epoxidase* (ZEP) through an intermediate antheraxanthin. In the next step, cis-isomerization of all trans-violaxanthin to violaxanthin or cisneoxanthin through an unknown enzymatic reaction. After that, 9-cis-epoxycarotenoid dioxygenase (NCED) enzymes split the cis-isomers of violaxanthin and neoxanthin to generate a C15 intermediate product called xanthoxin, finally exported to cytosol. In the cytosol, xanthoxin is converted into ABA through two enzymatic reactions. Next, xanthoxin is first converted to an abscisic aldehyde catalyzed by short-chain alcohol dehydrogenase/reductase (SDR). Finally, the oxidation of abscisic aldehyde to ABA by *aldehyde oxidase* (AAO) (Figure 1a) [9].

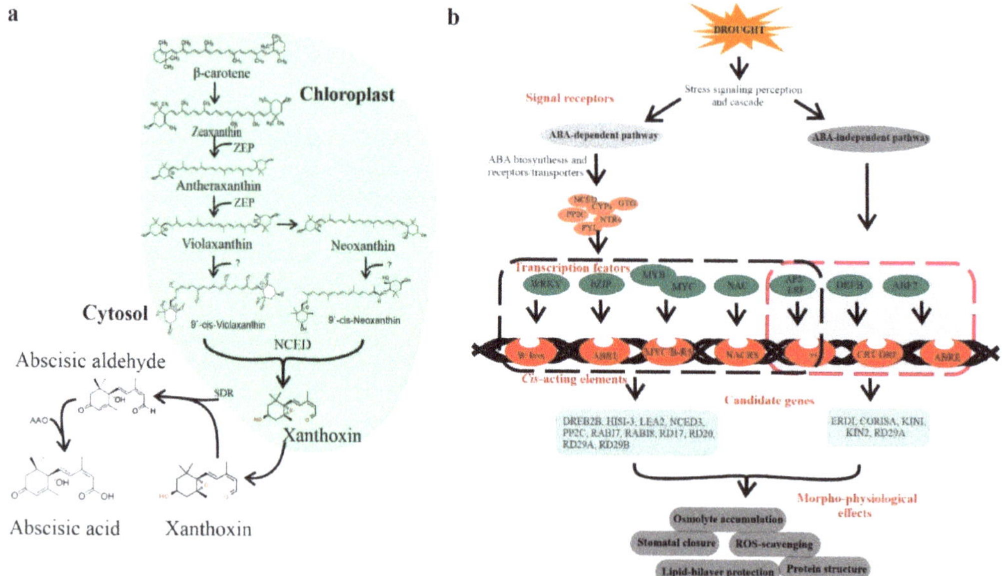

Figure 1. ABA biosynthesis and ABA-mediated drought-responsive pathways in plants. (**a**) Scheme of ABA biosynthesis. The precursors of ABA, β-carotene undergoes a series of oxidative reactions in the plastids and each step is catalyzed by specific enzyme such as ZEP (zeaxanthin epoxidase) or NCED (9-*cis*-epoxycarotenoid dioxygenase). The derived xanthoxin is exported to the cytosol and converted into ABA through an oxidation reaction mediated by AAO (aldehyde oxidase) and SDR (alcohol dehydrogenase/reductase), (**b**) ABA-dependent and -independent signaling pathways in the plant, which consists of several core components including ABA receptors and regulators. The ABA-dependent and -independent pathways are indicated by black and red arrows, respectively. Transcription factors (TFs) include bZIPs, MYB/MYC2, NAC (RD26), and WRKY bind to their corresponding *cis*-acting elements W-box, ABRE, MYB, MYC, DREB2, AREB/ABF, and NACRs.

Plants show a significant increase in ABA levels under drought stress, changes in expression of genes, and induction of ABA biosynthesis enzymes corresponding to mRNA level lead to enhanced ABA accumulation [10]. The transcript abundances of several ABA biosynthesis genes, such as *ZEP/ABA1, AAO3, 9-cis-epoxycarotenoid dioxygenase (NCED3),* and *molybdenum cofactor sulfurase (MCSU/LOS5/ABA3)*, has been upregulated through an ABA-dependent or ABA-independent pathway [11] assisted by binding factors such as ABF, MYC MYB, NAC, ERF, bZIP, and DREB/CBF transcription factors (TFs) (Figure 1b) [12].

ABA is a prime mediator of drought [13] and plays an important role in regulating plant growth, development, and responses to several environmental stresses [14]. Under drought conditions, ABA-mediated stomatal conductance prevents transpiration water loss [10]. Zhang et al. [15] found that multidrug and toxic compound extrusion (MATE) transporter family, *detoxification efflux carrier (AtDTX50)*, participate in ABA transport. ABA receptors such as PYRABACTIN RESISTANCE (PYR), or regulatory component of ABA receptor (RCAR) enhanced ABA responses and confer drought tolerance in Arabidopsis [16]. Similarly, ABA responsive-element binding protein (ABP9) a member of the bZIP family (Figure 1) improves photosynthetic capacity under drought [17]. Histone acetylation has been reported to be critical in ABA-mediated gene regulation to acclimatize plants to drought [18]. It has been shown that mitogen-activated protein kinase (MAPK) signaling cascade plays a critical role in ABA-mediated drought regulation at transcription and proteome level in various plant species including rice, maize, and Arabidopsis [19,20]. Altogether, ABA is a pivotal hormone governing plant responses to drought through complex molecular signaling mechanisms. Therefore, exploration of ABA regulators could assist in developing drought-tolerant crops through breeding programs.

The roles of ABA have been extensively studied through the implementation of molecular and genetics approaches, which enable the development of drought-resilient crops. This review is an attempt to underline the role of ABA in drought adaption, avoidance, in different biochemical and physiological responses. Moreover, we also highlighted a concise overview of the molecular mechanism of ABA actions and ABA crosstalk with other hormones in regulating drought stress responses.

2. ABA: A Key Player under Drought

Abscisic acid is of prime importance due to its stress-related responses and its involvement in various plant growth processes, making it possible to adapt to drought conditions. Upon drought stress, ABA-mediated stomatal closure reduces water loss by decreasing transpiration rate. Moreover, ABA progressively increases hydraulic conductivity and stimulates root cell elongation, enabling plants recovery from water-limited conditions [21]. Recent advancements in plant genomics accelerated the identification and functional characterization of ABA-dependent candidate genes responsive to drought. For instance, Zhang et al. [15] found that the MATE transporter gene, *AtDTX50*, is involved in ABA efflux, while mutants of *dtx50* show enhanced tolerance to drought with reduced stomatal conductance relative to WT plants. It is widely acknowledged that ABA binds to pyrabactin-resistance 1/pyrabactin resistance like/regulatory component of aba receptor (PYR/PYL/RCAR) receptors, the initial step of the core ABA signaling pathway, concerning previously characterized protein phosphatases 2C (PP2Cs) and sucrose nonfermenting related kinases 2 (SnRK2s) (Figure 1b) [22–24]. The PYR/PYL/RCAR) proteins are reported to be involved in improving drought tolerance in many species such as *Arabidopsis*, tomato, and rice [25–28].

ABA has also been reported to regulate calcium-dependent protein kinases (CPK) signaling by inducing *CPK6* expression under drought stress. CPKs interact and phosphorylates some core ABA-related TFs, ABFs/AREBs (ABA-responsive element-binding factors) enhancing their transcriptional activities [29]. Similarly, transgenic plants overexpressing ZEP confers tolerance to stresses such as drought [30]. Overexpression of *OsbZIP72* showed increased expression of ABA-responsive gene *LEAs (late embryogenesis abundant genes)* and improved drought resistance in rice, which may be useful for the engineering of drought-resilient crops [31]. Arabidopsis plants overexpressing *ABCG25* showed reduced water loss under drought by limiting evapotranspiration. Likewise, mutants of *AtABCG40* exhibited more sensitivity to drought [32], indicating the prime importance of ABA-related genes in regulating ABA responses to drought conditions.

The ABA hormone has mainly been associated with the regulation of water deficiency in plants. A plethora of studies have shown the critical roles of ABA in regulating genes expression, proteins, and enzymatic activities involved in plant cell dehydration tolerance [33,34]. For instance, the ABA levels were exponentially elevated in *Arabidopsis*, wheat, rice, tomato, soybean, maize, and sesame under drought [35,36]. Similarly, Wang et al. [37] and Baek et al. [38] demonstrated how multiple genes regulate ABA-mediated drought responses in *Arabidopsis*, *Vigna. radiata*, and *V. angularis*. These findings suggest that ABA-mediated drought tolerance is required for plants to fully respond to drought stress.

3. ABA-Mediated Drought Responses through Physio-Biochemical Alteration

Plants have evolved distinct adaptive mechanisms to survive and minimize the adverse effect of drought stress [39]. Reactive oxygen species (ROS) serve as a signal molecule that regulates plant responses to stresses. Upon drought, plants synthesize an array of secondary metabolites (SMs) assisting plant survival [40]. ABA is able to synchronize a wide range of functions in plants, facilitating to overcome drought stress [4]. Therefore, to tackle water limitations, dynamic and novel strategies should be formulated and engineered including ROS and SMs as an adaptive strategy to maintain plant growth and productivity.

3.1. ROS Scavenging System

Drought and ABA have an intricate relationship [41] that triggers various downstream responses to plant assisting adaption to drought in an ABA-dependent manner [14]. Drought may alter the metabolic and cellular redox status of plants that influence the cellular susceptibility to ABA accumulation [42] suggested the link between metabolic status and ABA signaling [43]. ABA is an indicator of soil water deficit and endogenous ABA concentration rapidly increases to initiate stomatal closure in the plant [44]. Previous studies have also been demonstrated that drought escape induced by water stress depends on ABA. For instance, ABA could improve the plant ability to scavenge ROS by activating antioxidant enzymes [45] such as SOD (superoxide dismutase), POD (peroxidase), CAT (catalase), APX (ascorbate peroxidase), and GR (glutathione reductase) in wheat seedlings under drought, thus regulating the osmotic adjustment, reducing oxidative damage, and improving the conductivity of roots by inducing aquaporin gene expression [45–47]. Kwak et al. [48] showed that ABA activates H_2O_2 biosynthesis in stomata guard cells via a membrane-bound NADPH oxidase causing stomata closure by activating plasma membrane Ca^{2+} channels [49].

3.2. Primary Metabolism

Land plants synthesize diverse primary metabolites (PMs) having higher medicinal and nutritional value which are essential for survival [50]. In general, PMs function in protein–disulfide linkage, redox regulation, methylation reactions, including DNA methylation, mRNA capping, synthesis of phosphatidylcholine, and synthesis of polyamines [51]. Primary metabolites and their associated metabolic genes are considered pivotal factors that contribute to drought tolerance via the involvement of different metabolic pathways [52]. To date, in planta, an estimate of 200,000 metabolites are reported [53]. Among those carbohydrates, nucleosides/nucleotides, and sulfur-containing metabolites were mainly induced by ABA [54]. The major pathways responsible for PMs are glycolysis, the TCA cycle, pentose phosphate pathway, shikimate pathway, aliphatic, and aromatic amino acids which produce secondary metabolites (SMs). Abscisic acid is tightly associated with changes in water availability to fine-tune plant growth [55–60] acting as a signaling molecule for plants to adjust their metabolism and growth in response to drought stress [55].

A. thaliana and *Camelina sativa* ABA-inducible WSD1 (Wax synthase/acyl-CoA:diacylglycerol acyltransferase) enhanced drought tolerance through leaf and stem wax loading and epicuticular wax accumulation [61]. Canola crop is sensitive to drought, which leads to severe yield losses. However, understanding the genetic basis of ABA-mediated drought tolerance will pave the way to engineering crops with improved drought resistance. Recently, the application of Omics approaches identified various ABA-induced and suppressed proteins involved in metabolism, photosynthesis, protein synthesis, membrane transport processes, protein folding/transport and degradation, and stress/defense responsiveness [62]. This finding suggests that various ABA-induced and suppressed metabolites were used as indicators in improving our knowledge of ABA signaling to drought tolerance.

3.3. Secondary Metabolites

Plants are surrounded by a complex set of environmental stresses and respond equally to them. Plant metabolites are sensitive to changing environments such as drought [63]. The metabolic profiles of plants have been analyzed to predict their role under drought [64,65]. Plants have adapted two distinct strategies including osmotic adjustment [66] and accumulation of specialized secondary metabolites [67] to mitigate drought responses [68]. It has been shown that metabolites such as phenolic compounds, proline, glycine-betaine, soluble sugars, and other compatible solutes accumulated by plants during stress responses. These metabolites maintain water potential, cell turgor maintenance, osmotic adjustment, survival, stabilize proteins and membrane lipid bilayer structures under drought assisting to retain normal physiological processes (Figure 2) [66]. On the other hand, secondary metabolites act as scavengers of free radicals to mitigate oxidative stress in plants under drought stress.

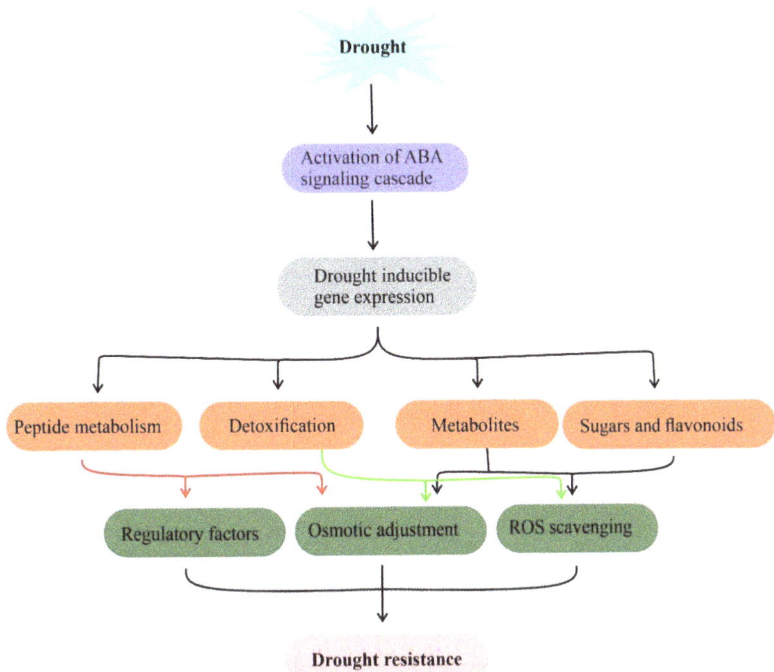

Figure 2. Metabolites and their functions in drought stress tolerance. Drought-induced accumulation of compatible solutes such as sugars, flavonoids, and amino acids for osmotic adjustment, free radical (ROS) scavenging to mitigate drought stress in plants. Genes involved in this metabolite biosynthesis against drought stress are useful in the metabolic engineering of drought resistance.

Metabolic profiling revealed ABA-inducible metabolic networks in response to drought which encourages the accumulation of dehydration-inducible branched-chain amino acids, and key dehydration-inducible genes such as *lysine ketoglutarate reductase/saccharopine dehydrogenase (AtLKR/SDH), branch-chain aminotransferase (AtBCAT2), arginine decarboxylase,* and *delta 1-pyrroline-5-carboxylase (P5CS)* [69]. For instance, the accumulation of most amino acids such as tryptophan, glutamine, alanine, proline, aspartate, leucine, isoleucine, ornithine, valine, citric acid cycle precursors including cis-aconitate, succinate, and 2-oxoglutarate; flavonoids such as cyanidin and quercetin; and lipids such as acylated sterylglycosides and glycosyl inositol phosphoceramides were increased under drought in Arabidopsis [70–72] and a few crop plants, such as maize, barley, and rice [73–75]. In maize, *ZmPIS*, a phosphatidylinositol synthase, efficiently improved drought tolerance by altering membrane lipid composition and ABA biosynthesis [76]. Overexpression of *ABF3* in *Glycine max* significantly altered various primary and secondary metabolites such as glycerophospholipids, glycolipids, fatty acyls, prenol-lipids, and their derivatives [77].

4. Alteration in Root System under Drought

4.1. Root System Architecture

The plant root system anchors the plant into the soil and acquires water and minerals from the surrounding for the plant to develop and differentiate [78]. The primitive role of ABA is to modify root architecture, root growth pattern [79], and limit root growth [80], directly correlated with environmental perturbations [78], and in regulating lateral root emergence [81]. Genetic studies on ABA deficient mutants (*aba2-1* and *aba3-1*) exhibit an increased rate of lateral root emergence [82]. This was further evidenced by Gou et al. [83] who found that endogenous ABA biosynthesis requires inhibition of lateral roots in peanuts. Interestingly, ABA-insensitive (*abi4*) shows an enhanced number of lateral roots [84]. Therefore, instead of ABA levels, ABA signaling is also involved in lateral root formation.

Contrastingly, water stress regulates root cellular integrity and elongation by monitoring respiratory burst oxidase homolog (RBOH) gene expression via ROS secretions [85]. Although, molecular studies on *Medicago truncatula* mutants specified a significant role of ABA in maintaining root meristem [80]. Studies on ABA-deficient mutants have revealed that ABA is important to maintain root growth under drought [86]. Recently, Zhang et al. [87] concluded that ABA-mediated root growth of tomatoes under soil drying may involve auxin-dependent processes. ABA deficient mutants *vp5* and *vp14* develop stunt primary root in maize [88], but primary root elongation is resorted by exogenous IAA in tomato mutant notabilis (*not*) under drought [87]. However, ABA affects root architecture either positively or negatively based on the genotypic background and environmental conditions, offering a nuanced way to fine-tune the root system to compete with drought stress.

4.2. Root Secretions

Root exudates are key drivers that play important role in responses to environmental perturbation [89] and serve as a nutritional or chemoattractant source with 11% to 40% of photosynthetically derived carbon [90,91]. Root exudates are enriched in organic acids, sugars, fatty acids, amino acids, and secondary metabolites [92], but the type and rate of exudation vary from species to species such as wild oat, grassland species, *Brassica napus*, and *A. thaliana* [93–97]. Plant exuded compounds assist in establishing diverse plant–microbial relationships [98] thereby enhancing plant ability to cope with environmental stresses (Figure 3) [99]. ABA acts as a mediator of drought via enhanced osmolytes biosynthesis including proline, organic acids, and protective proteins [100]. Yang et al. [101] demonstrated that ABA concentration was positively correlated with enhanced sucrose phosphate synthase activity and carbon remobilization. As expected, an increased endogenous root ABA concentration may enhance root exudates under water stress [102]. Garriga et al. [103] described that plant metabolites, more specifically ABA was significantly enhanced in root exudation upon drought stress [103]. These findings suggest the possible role of ABA in carbon remobilization and root exudation in mitigating plant stress tolerance (Figure 3).

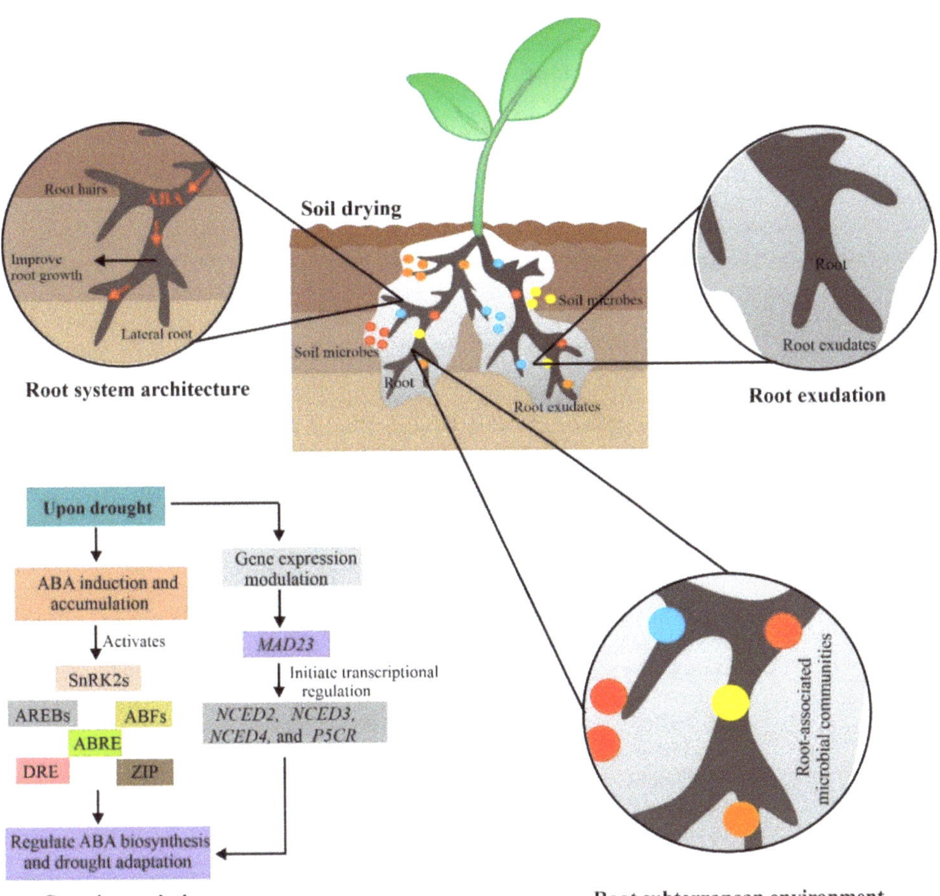

Figure 3. Role of ABA in mitigating drought tolerance at root system architecture that includes root exudates, microbial communities at the root-soil interface, and genetic and molecular regulation of various ABA-responsive genes and proteins. Endogenous ABA modulates the root system architecture by promoting root growth, soil microbial communities, and root exudation in response to soil drying. ABA accumulation upon drought led to the activation of TFs and modulates expression of genes responsible for improving ABA-mediated drought tolerance.

Root exudates can adopt specific drought-resistant bacteria. Plant growth-promoting rhizobacteria (PGPR) encouraged the growth of wild-type plants and showed an opposite response in ABA-deficient mutant tomato [104]. Some bacterial species such as *Bacillus* [105] and *B. megaterium* [104] establish interaction with plant roots by endogenous ABA [105] promoting plant growth [106,107]. Henry et al. [108] reported that in maize organic exudation (including fumaric, malonic, succinic, and oxalic acids) increases under drought, which attracts *B. subtilis*, an important beneficial bacterium that enhances drought resistance of *Phleum pretense* via osmolytes secretion [109]. These findings suggest that endogenous ABA content is essential for promoting growth.

4.3. Root Subterranean Environment

The subterranean environment not only represents the dark region in the soil but also denotes variation in numerous abiotic stress such as temperature, water, and minerals availability. One of the major challenges in ecology is to identify and predict how different species respond to environmental perturbations [110]. Abscisic acid mediates root growth, rhizobacteria abundance, and influences root-hormonal status. For instance, Dodd et al. [111] reported that rhizobacteria can produce, metabolize, and utilize phytohormones/phytohormonal precursors as a nutrient source. The hitherto study showed that rice seed inoculated with P6W and P1Y bacterial strains, ABA concentration was reduced to 14% by P6W and 22% by P1Y in the shoot, while root ABA concentration remain intact [112]. In the same study, tomato ABA-deficient mutants flacca (*flc*) notabilis (*not*) inoculated with P6W strains inhibited primary root extension and significantly increased biomass of WT plants [112].

Recently, Gowtham et al. [113] found that CRDT-EB-1 (*B. marisflavi*) extracted from the rhizosphere of mustard seedlings remarkably increased drought resistance by secreting ABA analog/xanthoxin. Additionally, a study by Porcel et al. [104] described that ABA-deficient mutants *flacca* and *sitiens* inoculated with *B. megaterium* showed restricted growth. This study concludes that ABA concentration is essential for *B. megaterium* growth, which promotes the growth of wild-type plants. Therefore, it may suggest that ABA-metabolizing rhizobacteria can modulate root-phytohormonal status and stimulate plant growth.

4.4. Biphasic Root Growth Responses

ABA is reported to either promote or inhibit root growth and is an inhibitor of plant shoot and root growth under well-watered conditions depending on its concentration. Mild soil drying stimulates root growth but inhibits root growth when it becomes more severe [44]. Similarly, ABA acts as an inhibitor of plant growth under water deficit. The biphasic effects of dry soil on root growth were mild, while water deficit stimulated root growth but severe water deficit inhibited root growth. Furthermore, complex biphasic effects of exogenous ABA on root growth under well-watered conditions were demonstrated by Li et al. [44]; low concentrations of ABA stimulated root growth while high concentrations inhibited root growth.

ABA also inhibits root growth in Arabidopsis by promoting ethylene biosynthesis and auxin influx [114]. In addition, ethylene insensitive mutants *etr1-1*, *ein2-1*, and *ein3-1*, auxin influx (*aux1-7*, *aux1-T*), and auxin-insensitive mutants (*iaa7/axr2-1*) unable to respond to the inhibitory effect of elevated ABA [44]. Furthermore, the stimulatory effect of low ABA concentrations was blocked by auxin efflux inhibitors (*pin2/eir1-1*) but less pronounced in auxin efflux mutant (*iaa7/axr2-1*). Taken together, ABA may play a pivotal role in biphasic root growth in plants.

5. Molecular Mechanism of ABA-Mediated Drought Regulation

ABA regulates different physiological and molecular responses under drought. Plants perceive stress stimulus in their roots and leaves and transmit the signal to their shoots to synthesize ABA. Protein kinases are positive regulators of ABA signaling, metabolism, and transport in response to drought [115]. For instance, an ABA-responsive G protein receptor (GCR2) caused insensitivity in germination and influenced the expression of ABA inducible genes [116]. Moreover, ABA has been reported to interact with the flowering-time control protein (FCA) and the Mg-chelatase H subunit [117]. Sensors sense drought stimuli on the membrane, and these signals are then passed down through multiple signal transduction pathways, resulting in the expression of drought-responsive genes. Recently, several classes of ABA biosynthesis *NCED1*, *NCED3*, *ABA2*, and *AAO3* and transporter proteins AtABCG25 and AtABCG40 localizing in the plasma membrane indicated that plants possess a complex system to sense and respond to fluctuating environmental conditions [118]. Nitrate transporter 1/peptide transporter protein (NPF) named AIT1 initially, mediates ABA uptake into cells, suggesting that this protein could also have a role in intercellular ABA transport [119].

To gain an insight into the genetic regulation of the root system under drought stress, several studies have been investigated. Recently, Zhang et al. [120] reported transcriptomic analysis at three time scales (1 h, 3 h, and 7 h) under drought in *Pearl millet*. It was observed that a total of 2004, 1538, and 605 genes were differentially expressed at 1 h, 3 h, and 7 h, respectively, while 12 genes were upregulated at all the time scales. Some of these highly expressed genes were related to the MAPK signaling pathway, metabolic processes, and plant hormonal signaling such as the ABA signal transduction pathway. This may provide a genetic basis to understand drought resistance mechanisms in other plants. Similarly, Li et al. [121] showed that *OsMAD23* serves as a positive regulator for altering ABA biosynthesis to enhance drought tolerance in rice. *OsMAD23* encourages endogenous ABA accumulation through its biosynthesis and proline accumulation by activating several ABA and proline biosynthesis genes including *OsNCED2/3/4* and *OsP5CR*. These findings suggest a new way to enhance drought/salt resistance in rice.

The expression of several TFs is controlled by ABA such as AREBs, ABFs, DRE, ABRE, and several others (Figure 3). Indeed, transactivation analysis revealed that the ABA-responsive TF factor *OsBZIP72* activates the expression of *OsSWEET13* and *OsSWEET15* by binding with their promoters mediating sucrose transportation and distribution in rice under drought and other stresses [122]. It has also been found that Arabidopsis *ZmbZIP33* remarkably improved chlorophyll content and root length as a drought-adapted strategy in maize [123]. A few popular studies demonstrating the role of ABA-mediated drought tolerance are listed in Table 1.

Table 1. Some important genes or transporters for improved drought tolerance through modulating ABA signaling pathway in plants.

Plant Species	Gene/Transporter	Function	References
Agrostis grass	*VuNCED1*	Increased 3–4-fold plant biomass	[124]
	AtABCG17-18	Stomatal conductance, and increased water use efficiency	[125]
	AtPDR12/ABCG40	ABA uptake transporter	[32]
	AIT1/NRT1.2	ABA importer important for stomatal aperture	[126]
Arabidopsis thaliana	*AtBBD1*	Increased expression of ABA and drought-responsive genes	[127]
	AtSAUR32	Highly induced by abscisic acid and drought treatment	[128]
	AtbHLH68	Lateral root elongation in response to drought	[129]
	AtHDA9 and *ABI4*	Regulate *CYP707A1* and *CYP707A2* expression under drought	[130]
	AtDTX/MATE	Facilitate ABA efflux and tolerance to drought	[15]
Brachypodium distachyon	*BdABCG25*	Regulate intercellular ABA transport	[131]
Brassica napus	*BnFTA*	Improved under drought conditions	[132]
Glycine max	*GmCIPK2*	ABA signaling and drought tolerance	[133]
Nicotiana tabacum	*SgNCED1*	Enhanced ABA accumulation increased drought tolerance	[134]
	Oshox22	Increased ABA content, and enhanced drought tolerance	[135]
Oryza sativa	*OsbZIP46CA1*	Improved drought resistance	[136]
	OsPM1	ABA influx carrier is important in drought responses	[137]
Petunia	*LeNCED1*	Elevated levels of ABA and proline, increases drought resistance	[138]
Setaria italica	*SiARDP* target of *SiAREB*	ABA-dependent signal pathways	[139]
Solanum lycopersicum	*SlGRAS42s*	ABA signaling	[140]
	PYR/PYL/RCAR	Role in seed germination and basal ABA signaling	[141]
Triticum aestivum	*HVA1*	Improved growth characteristics under water deficit	[142]
Vigna unguiculata	*VuABCG25*	Involved in ABA signaling pathway under water stress	[143]
Vitis vinifera	*VviNCED1, VviNCED2*	ABA synthesis in response to plant water status	[144–146]
Xanthoceras sorbifolium	*XsWRKY20*	Regulate drought tolerance by ABA signaling pathway	[147]
Zea mays	*ZmXerico1-2*	Improved water use efficiency, yield under drought stress	[148]

ABA Dependent Translational and Posttranslational Modification

Post-transcriptional regulation such as alternative splicing and RNA-mediated silencing, and post-translational regulation such as protein activity, subcellular localization, and protein half-life significantly contribute to the fine-tuning of ABA-dependent plant response drought stress.

Post-transcriptional regulation of an ABA-responsive basic leucine zipper (bZIP) TF ABI5 has been studied extensively in response to various stresses, particularly drought. For instance, ABI5 represses/activates the expression of various ABA-dependent receptors and kinases and stimulates plant adaptation to drought stress [149,150]. Several members of the bZIP family (ABF1-4) were shown to be highly redundant to ensure ABA-mediated adaptation to drought in various species such as in *Arabidopsis*, rice, cotton, carrot, and barley [151–155]. ABI5 interaction is activated through phosphorylation by kinases SRK2D/SnRK2.2, SRK2E/SnRK2.6/OST1, SRK2I/SnRK2.3, SOS2-like protein kinase 5/CIPK11, glycogen synthase kinase 3-like kinase BRASSINOSTEROID INSENSITIVE 2 (BIN2) and calcineurin B-like interacting protein kinase 26 (CIPK26), which suppress or activate ABI5 post-translation under ABA treatment. Therefore, ABI5 protein regulation could serve as a model to study the activity and stability of critical components engaged in drought tolerance affecting posttranslational modification. Various key components in drought and ABA signaling under post-translational modifications result in modulation of drought responses indicating their role in stress adaptation (Table 2).

Table 2. Post-translational modifications and their predicted role in plants responsive to drought and ABA signaling in different plant species.

Plant Species	Protein	Target	Role in Plants	References
Arabidopsis	PUB22/23	RPN12a	Drought tolerance and ABA signaling	[156]
Arabidopsis	PUB19	nd	Drought tolerance	[157]
Arabidopsis	AIRP1	nd	ABA-dependent drought tolerance	[158]
Arabidopsis	Rma1	PIP2;1	Drought tolerance	[159,160]
Arabidopsis	SDIR1	SDIRIP1	Drought and salinity tolerance, ABA signaling	[161]
Arabidopsis	XERICO	nd	Drought stress tolerance, ABA biosynthesis	[162]
Arabidopsis	DOR	nd	Drought stress tolerance	[163]
Rice	RING-1	nd	Drought tolerance and ABA response	[164]
Maize	ZF1	nd	Drought tolerance and ABA signaling	[165]

6. Hormone Crosstalk

In plants, hormones secreted from predominantly vascular cells and guard cells and transported to distant target sites through the plant body circulatory system, suggesting that most plant hormones are mobile. When plants are exposed to environmental stress the ABA signaling cascade is rapidly activated, which in return activates ABA-responsive TFs and induces the expression of ABA-responsive genes [166]. ABA interacts with other hormones including auxin, gibberellins (GA), cytokinin (CK), ethylene (ET), salicylic acid (SA), and jasmonic acid (JA) to help the plant to withstand abiotic stresses such as drought. Phytohormonal engineering is a major environmental stress mediator, particularly ABA, a prime regulator of environmental stress tolerance. ABA is produced in shoot and roots but enhanced accumulation is stimulated by a decrease in cellular turgor [167]. However, ABA accumulation is directly correlated with plant/tissue water status, but the underlying molecular mechanism of ABA accumulation and drought sensation is still unclear.

Stress response in plants is regulated by integrating ABA and auxin signals [168]. Some studies revealed that ABA and auxin distribution within the primary roots and lateral roots are independent of each other because they show different localization patterns. Furthermore, auxin biosynthesis was inhibited by ABA, but ABA accumulation modulates auxin transport in the root tip of Arabidopsis during drought [169,170]. Some authors argue that auxin and ET are required for ABA response in the root [171]. Furthermore, auxin/ET/ABA crosstalk was examined, and it was found that auxin and ET likely operate in a linear pathway to affect ABA-responsive inhibition of root elongation, and probably act independently to affect ABA-responsive inhibition of seed germination [171]. Exogenous application of JA encourages foliar ABA accumulation while JA deficiency suppresses ABA levels [172,173], suggesting extensive crosstalk between JA–ABA pathways. Puértolas et al. [174] demonstrated that ABA concentration increase with declining plant water status while JA accumulates during early stages of stress, however, the molecular and physiological involvement of JA in concert with ABA is still unclear.

Abscisic acid enables plants to cope with several abiotic stresses by interacting with important ABA-related TFs and by coordinating with other plant hormones. For instance, ABA-related DREB (dehydration-responsive element binding), nced (9-cis-epoxycarotenoid dioxygenases), and MYB (myeloblastosis) TFs are known to regulate intracellular pathways involved in CK homeostasis [175]. AtMYB2 knockout led to enhanced expression of IPT1/4/5/6/8 (adenosine phosphate-isopentenyl transferases), suggesting the role of AtMYB2 in CK synthesis [175]. Similarly, overexpression of MsDREB6.2 (dehydration-responsive gene) resulted in increased *MdCKX4a* expression, thereby reducing endogenous CK and enhancing drought tolerance of transgenic apple plants [176].

In *Arabidopsis*, the SnRK2 protein interacts with CK signaling-ARR5 (type-A RR5) and regulates ABA-mediated drought tolerance [177]. In addition, ET crosstalk with ABA in an antagonistic manner via regulating root and shoot growth under drought [178]. Upon abiotic stress, along with ROS such as drought, inositol phosphate is produced, resulting in enhanced endogenous ABA levels [179]. Crosstalk between ABA and SA assists plant-water budget, osmotic adjustment, stomatal conductance, distribution of photoassimilates, and leaf senescence [178]. The expression of ERF1 (ethylene response factor1) is rapidly induced by ET and JA individually or synergistically [180]. Overexpression of ERF1 encourages drought tolerance and increases the accumulation of ABA levels and Pro, which restrict water loss and significantly contribute to stress resistance [181]. ERF1 activates the expression of LEA4-5 (late-embryogenesis abundant protein4-5) thereby resulting in drought tolerance (Figure 4) [182].

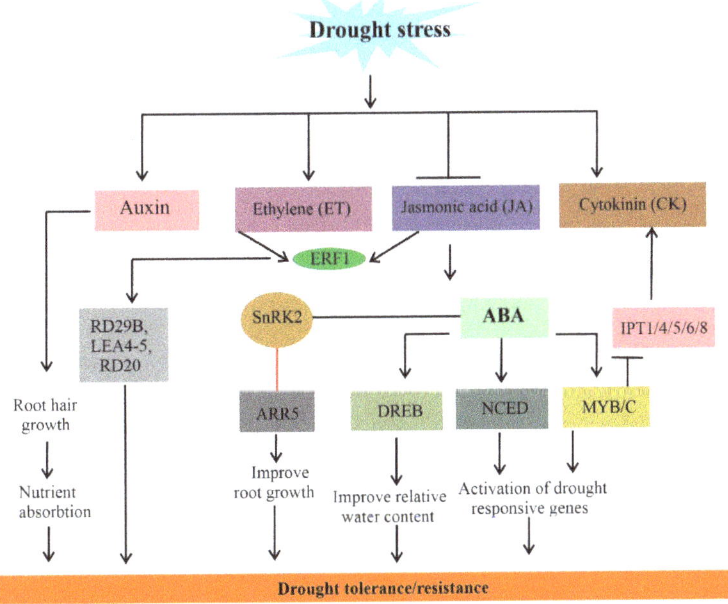

Figure 4. Proposed model of auxin, ethylene (ET), jasmonic acid (JA), cytokinin (CK), and ABA cross talk under drought stress. ERF1 (ETHYLENE RESPONSE FACTOR1) modulates expression of

potential genes involved in drought tolerance. ABA-responsive component SnRK2, directly phosphorylates type-A RR5 (ARR5), ABA activates several transcription factors (TFs) which could result in enhanced drought tolerance.

7. Other Related Mechanisms

Several studies on drought have shown that the ABA hormone has a significant role in regulating the biochemical mechanisms that enable drought-prone plants to rapidly grow, flower, and produce mature seeds just before the onset of drought. Hwang et al. [183] confirmed that ABA plays an important role in regulating flowering during accelerated floral transition in *A. thaliana*. The study further showed that ABA-binding factors, ABF3 and ABF4 in *A. thaliana* enhance flowering by inducing SOC1 transcription, thus enabling the plant to complete its lifecycle under drought stress [183].

Besides ABA hormone responses, plants possess other ABA-independent mechanisms through which they mitigate drought effects [184]. The role of nutritional stress in drought stress alleviation remains a key area of focus. Drought normally results in depleted nutrients which lead to poor plant growth and development, and finally low crop yields. Studies have shown that plants associated with high nutrient use efficiency can ameliorate stress effects and cope with drought [185,186]. Micro and macronutrients including phosphorus [187], calcium [188], potassium [189], nitrogen [190], silicon [191], magnesium, and zinc [192] were reported to help in alleviating the adverse effects of drought. Although several studies have shown how different micro and macronutrients in plants are affected by drought, more research is needed to decipher the interactive mechanisms of these molecules with plants.

ABA-deficient mutant rice, however, induced low ABA and delayed flowering. The study reported some of the light receptors, circadian components, and genes related to flowering including $OsTOC_1$, Ghd_7, and $PhyB$ involved in drought stress in an ABA-dependent manner. Several studies [183,193,194] have also reported ABA to modulate drought escape by activating the expression of florigen genes FT and TSP, and the floral integrator SOC_1 (ABF$_3$ and ABF$_4$) in *Arabidopsis*. Understanding the connection between ABA and drought escape is of importance in practical circumstances whereby crops are under irrigation, in such a scenario, the ABA responses could be targeted to ensure maximum crop production. The knowledge tapped from studies on drought escape mechanisms in plants needs to be utilized in developing drought-resistant crops that can easily withstand the dry seasons thus enhancing food production for the increasing world population (Figure 5).

Drought—at the early stage of plant development—could trigger early flowering and reduce tiller numbers by inducing the accumulation of ABA [195] by activating ABA signaling components that control water status and stomatal closure, promoting plant escape or adaptation to drought stress. Similarly, trehalose treatment upregulates ABA signaling-related gene expression via activating the ABA signaling pathway, regulating stomatal aperture, and preventing transpirational water loss [196]. It was reported that drought has feedback effects on the circadian clock by simultaneously regulating many flowering-related genes such as *OsTOC1, Ghd7, OsGI, OsELF3, OsPRR37, OsMADS50,* and *PhyB,* which promotes early flowering [195,197] in an ABA-dependent manner [195].

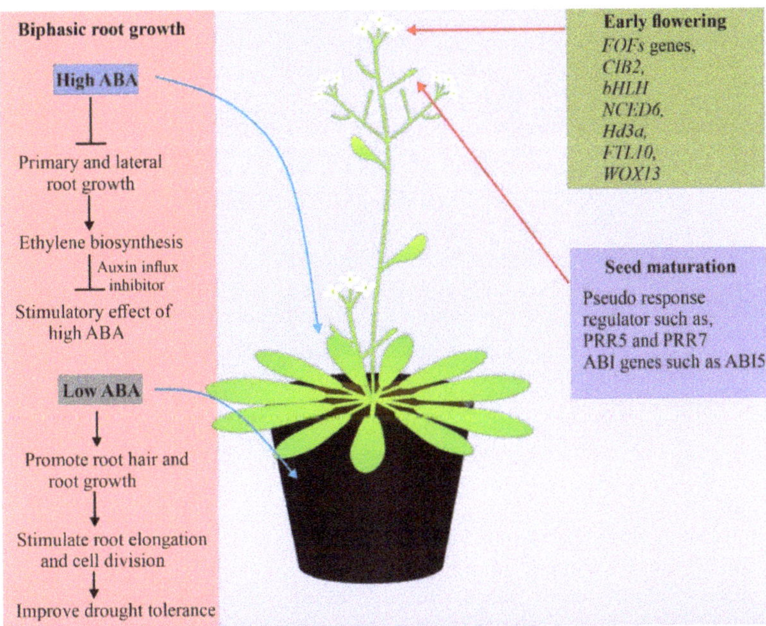

Figure 5. The molecular basis of ABA-mediated plant response to drought. A high concentration of ABA imposed a negative effect on auxin influx, while a low concentration of ABA promotes root growth resulted in enhanced drought tolerance.

8. Conclusions and Perspective

Environmental stresses, particularly drought, impair plant growth and productivity threatening global food security. Therefore, the development of drought-resistant crops is an important implication to sustainable crop production. Plants have evolved a series of adaptive strategies at cellular and molecular levels to cope with environmental cues including drought. The underlying genetic and molecular mechanisms of drought tolerance and escape are still an emerging topic in plant biology. Drought escape is the adaptive mechanism through which plants undergo rapid development to complete their life cycle before the onset of serious water deficits [198]. ABA is a key hormone and critical regulator of different stresses, including drought and salinity in various plant species. The overall progress of research on ABA-mediated drought responses has revealed its key role in growth and development during stress conditions. However, how plants perceive and transmit drought stress to different cells to initiate ABA accumulation for drought resistance remains unclear.

ABA receptors are present in almost every tissue and expressed specifically in each tissue allowing the plant to sense environmental signals and transmit them accurately to the target tissues. ABA regulates many aspects of plant growth, development, regulation of stomatal closure, channel activities in guard cells, promoting proline synthesis and accumulation, transcription of calmodulin protein, accelerated floral transition and seed maturation, and expression of ABA-drought responsive genes [199–201]. To achieve this, various drought-responsive genes and metabolic pathways are triggered leading to beneficial adjustments in the morphological structures and growth rate [86].

Various regulatory factors complement cell-to-cell or tissue-to-tissue communication and long-distance translocation from tissues for synthesis to target tissues. Recently, advances have been made in understanding the ABA regulatory mechanism at the molecular and cellular levels providing novel insights for biotechnology and agriculture merits for further research for crop breeding programs and crop innovation. Moreover, integrated phenomics will be key in addition to other high-throughput approaches to assess plant phenotype against drought stress. Similarly, artificial intelligence and machine learning are powerful techniques to understand these complex integrated systems in plant responses to drought.

Author Contributions: W.Y. and Q.Z. contributed in funding acquisition; M.M.A., M.W. and Q.Z. designed the manuscript; M.W., M.M.A., B.H.J. and E.J.O. wrote the first draft; M.M.A. and M.W. designed the diagrams; Q.Z., W.X., H.S.A.S., Z.L. and W.Y. reviewed and approved the final manuscript. All authors have read and agreed to the published version of the manuscript.

Funding: We are grateful for grants support from the Science and Technology Projects of China National Tobacco Corporation Fujian Company (2021350000240014), Fujian Province Natural Science Foundation (2020J01553) and the Education Department of Fujian Province (JAT190134).

Institutional Review Board Statement: Not applicable.

Informed Consent Statement: Not applicable.

Data Availability Statement: All data were available within the manuscript.

Conflicts of Interest: The authors declare no conflict of interest.

References

1. Rubin, R.L.; van Groenigen, K.J.; Hungate, B.A. Plant growth promoting rhizobacteria are more effective under drought: A meta-analysis. *Plant Soil* **2017**, *416*, 309–323. [CrossRef]
2. Ali, S.; Hayat, K.; Iqbal, A.; Xie, L. Implications of Abscisic Acid in the Drought Stress Tolerance of Plants. *Agronomy* **2020**, *10*, 1323. [CrossRef]
3. Müller, M. Foes or friends: ABA and ethylene interaction under abiotic stress. *Plants* **2021**, *10*, 448. [CrossRef] [PubMed]
4. Wani, S.H.; Kumar, V. Plant stress tolerance: Engineering ABA: A potent phytohormone. *Transcriptomics* **2015**, *3*, 1000113. [CrossRef]
5. Chaves, M.M.; Maroco, J.P.; Pereira, J.S. Understanding plant responses to drought-from genes to the whole plant. *Funct. Plant Biol. FPB* **2003**, *30*, 239–264. [CrossRef]
6. Gonzalez-Villagra, J.; Figueroa, C.; Luengo-Escobar, A.; Morales, M.; Inostroza-Blancheteau, C.; Reyes-Díaz, M. Abscisic Acid and Plant Response under Adverse Environmental Conditions. In *Plant Performance under Environmental Stress*; Springer: Berlin/Heidelberg, Germany, 2021; pp. 17–47.
7. Trivedi, D.K.; Gill, S.S.; Tuteja, N. Abscisic acid (ABA): Biosynthesis, regulation, and role in abiotic stress tolerance. *Abiotic Stress Response Plants* **2016**, *8*, 315–326.
8. Dejonghe, W.; Okamoto, M.; Cutler, S.R. Small molecule probes of ABA biosynthesis and signaling. *Plant Cell Physiol.* **2018**, *59*, 1490–1499. [CrossRef] [PubMed]
9. Vishwakarma, K.; Upadhyay, N.; Kumar, N.; Yadav, G.; Singh, J.; Mishra, R.K.; Kumar, V.; Verma, R.; Upadhyay, R.G.; Pandey, M.; et al. Abscisic Acid Signaling and Abiotic Stress Tolerance in Plants: A Review on Current Knowledge and Future Prospects. *Front. Plant Sci.* **2017**, *8*, 161. [CrossRef]
10. Kim, T.-H.; Böhmer, M.; Hu, H.; Nishimura, N.; Schroeder, J.I. Guard cell signal transduction network: Advances in understanding abscisic acid, CO2, and Ca2+ signaling. *Annu. Rev. Plant Biol.* **2010**, *61*, 561–591. [CrossRef]
11. Dar, N.A.; Amin, I.; Wani, W.; Wani, S.A.; Shikari, A.B.; Wani, S.H.; Masoodi, K.Z. Abscisic acid: A key regulator of abiotic stress tolerance in plants. *Plant Gene* **2017**, *11*, 106–111. [CrossRef]
12. Verma, V.; Ravindran, P.; Kumar, P.P. Plant hormone-mediated regulation of stress responses. *BMC Plant Biol.* **2016**, *16*, 1–10. [CrossRef] [PubMed]
13. Boominathan, P.; Shukla, R.; Kumar, A.; Manna, D.; Negi, D.; Verma, P.K.; Chattopadhyay, D. Long term transcript accumulation during the development of dehydration adaptation in Cicer arietinum. *Plant Physiol.* **2004**, *135*, 1608–1620. [CrossRef] [PubMed]
14. Zhu, J.-K. Salt and drought stress signal transduction in plants. *Annu. Rev. Plant Biol.* **2002**, *53*, 247–273. [CrossRef] [PubMed]
15. Zhang, H.; Zhu, H.; Pan, Y.; Yu, Y.; Luan, S.; Li, L. A DTX/MATE-type transporter facilitates abscisic acid efflux and modulates ABA sensitivity and drought tolerance in Arabidopsis. *Mol. Plant* **2014**, *7*, 1522–1532. [CrossRef] [PubMed]
16. Saavedra, X.; Modrego, A.; Rodri̇guez, D.; Gonzȧlez-Garċa, M.P.; Sanz, L.; Nicolȧs, G.; Lorenzo, O. The nuclear interactor PYL8/RCAR3 of Fagus sylvatica FsPP2C1 is a positive regulator of abscisic acid signaling in seeds and stress. *Plant Physiol.* **2010**, *152*, 133–150. [CrossRef] [PubMed]
17. Zhang, X.; Wollenweber, B.; Jiang, D.; Liu, F.; Zhao, J. Water deficits and heat shock effects on photosynthesis of a transgenic Arabidopsis thaliana constitutively expressing ABP9, a bZIP transcription factor. *J. Exp. Bot.* **2008**, *59*, 839–848. [CrossRef]
18. Sridha, S.; Wu, K. Identification of AtHD2C as a novel regulator of abscisic acid responses in Arabidopsis. *Plant J.* **2006**, *46*, 124–133. [CrossRef]

19. Hamel, L.-P.; Nicole, M.-C.; Duplessis, S.; Ellis, B.E. Mitogen-activated protein kinase signaling in plant-interacting fungi: Distinct messages from conserved messengers. *Plant Cell* **2012**, *24*, 1327–1351. [CrossRef]

20. Muchhal, U.S.; Raghothama, K.G. Transcriptional regulation of plant phosphate transporters. *Proc. Natl. Acad. Sci. USA* **1999**, *96*, 5868–5872. [CrossRef]

21. Daszkowska-Golec, A. The role of abscisic acid in drought stress: How aba helps plants to cope with drought stress. In *Drought Stress Tolerance in Plants*; Springer: Berlin/Heidelberg, Germany, 2016; Volume 2, pp. 123–151.

22. Fujii, H.; Zhu, J.-K. Arabidopsis mutant deficient in 3 abscisic acid-activated protein kinases reveals critical roles in growth, reproduction, and stress. *Proc. Natl. Acad. Sci. USA* **2009**, *106*, 8380–8385. [CrossRef]

23. Umezawa, T.; Sugiyama, N.; Mizoguchi, M.; Hayashi, S.; Myouga, F.; Yamaguchi-Shinozaki, K.; Ishihama, Y.; Hirayama, T.; Shinozaki, K. Type 2C protein phosphatases directly regulate abscisic acid-activated protein kinases in Arabidopsis. *Proc. Natl. Acad. Sci. USA* **2009**, *106*, 17588–17593. [CrossRef] [PubMed]

24. Vlad, F.; Rubio, S.; Rodrigues, A.; Sirichandra, C.; Belin, C.; Robert, N.; Leung, J.; Rodriguez, P.L.; Laurière, C.; Merlot, S. Protein phosphatases 2C regulate the activation of the Snf1-related kinase OST1 by abscisic acid in Arabidopsis. *Plant Cell* **2009**, *21*, 3170–3184. [CrossRef] [PubMed]

25. González-Guzmán, M.; Rodríguez, L.; Lorenzo-Orts, L.; Pons, C.; Sarrión-Perdigones, A.; Fernández, M.A.; Peirats-Llobet, M.; Forment, J.; Moreno-Alvero, M.; Cutler, S.R. Tomato PYR/PYL/RCAR abscisic acid receptors show high expression in root, differential sensitivity to the abscisic acid agonist quinabactin, and the capability to enhance plant drought resistance. *J. Exp. Bot.* **2014**, *65*, 4451–4464. [CrossRef] [PubMed]

26. Kim, H.; Lee, K.; Hwang, H.; Bhatnagar, N.; Kim, D.-Y.; Yoon, I.S.; Byun, M.-O.; Kim, S.T.; Jung, K.-H.; Kim, B.-G. Overexpression of PYL5 in rice enhances drought tolerance, inhibits growth, and modulates gene expression. *J. Exp. Bot.* **2014**, *65*, 453–464. [CrossRef] [PubMed]

27. Okamoto, M.; Peterson, F.C.; Defries, A.; Park, S.-Y.; Endo, A.; Nambara, E.; Volkman, B.F.; Cutler, S.R. Activation of dimeric ABA receptors elicits guard cell closure, ABA-regulated gene expression, and drought tolerance. *Proc. Natl. Acad. Sci. USA* **2013**, *110*, 12132–12137. [CrossRef]

28. Pizzio, G.A.; Rodriguez, L.; Antoni, R.; Gonzalez-Guzman, M.; Yunta, C.; Merilo, E.; Kollist, H.; Albert, A.; Rodriguez, P.L. The PYL4 A194T mutant uncovers a key role of PYR1-LIKE4/PROTEIN PHOSPHATASE 2CA interaction for abscisic acid signaling and plant drought resistance. *Plant Physiol.* **2013**, *163*, 441–455. [CrossRef]

29. Zhang, H.; Liu, D.; Yang, B.; Liu, W.Z.; Mu, B.; Song, H.; Chen, B.; Li, Y.; Ren, D.; Deng, H.; et al. Arabidopsis CPK6 positively regulates ABA signaling and drought tolerance through phosphorylating ABA-responsive element-binding factors. *J. Exp. Bot.* **2020**, *71*, 188–203. [CrossRef]

30. Park, H.-Y.; Seok, H.-Y.; Park, B.-K.; Kim, S.-H.; Goh, C.-H.; Lee, B.-h.; Lee, C.-H.; Moon, Y.-H. Overexpression of Arabidopsis ZEP enhances tolerance to osmotic stress. *Biochem. Biophys. Res. Commun.* **2008**, *375*, 80–85. [CrossRef]

31. Lu, G.; Gao, C.; Zheng, X.; Han, B. Identification of OsbZIP72 as a positive regulator of ABA response and drought tolerance in rice. *Planta* **2009**, *229*, 605–615. [CrossRef]

32. Kang, J.; Hwang, J.-U.; Lee, M.; Kim, Y.-Y.; Assmann, S.M.; Martinoia, E.; Lee, Y. PDR-type ABC transporter mediates cellular uptake of the phytohormone abscisic acid. *Proc. Natl. Acad. Sci. USA* **2010**, *107*, 2355–2360. [CrossRef]

33. Chen, K.; Li, G.J.; Bressan, R.A.; Song, C.P.; Zhu, J.K.; Zhao, Y. Abscisic acid dynamics, signaling, and functions in plants. *J. Integr. Plant Biol.* **2020**, *62*, 25–54. [CrossRef] [PubMed]

34. Yu, Y.; Wang, P.; Bai, Y.; Wang, Y.; Wan, H.; Liu, C.; Ni, Z. The soybean F-box protein GmFBX176 regulates ABA-mediated responses to drought and salt stress. *Environ. Exp. Bot.* **2020**, *176*, 104056. [CrossRef]

35. Zhang, J.; Jia, W.; Yang, J.; Ismail, A.M. Role of ABA in integrating plant responses to drought and salt stresses. *Field Crops Res.* **2006**, *97*, 111–119. [CrossRef]

36. Dossa, K.; Mmadi, M.A.; Zhou, R.; Liu, A.; Yang, Y.; Diouf, D.; You, J.; Zhang, X. Ectopic expression of the sesame MYB transcription factor SiMYB305 promotes root growth and modulates ABA-mediated tolerance to drought and salt stresses in Arabidopsis. *AoB Plants* **2019**, *12*, plz081. [CrossRef] [PubMed]

37. Wang, L.; Zhu, J.; Li, X.; Wang, S.; Wu, J. Salt and drought stress and ABA responses related to bZIP genes from V. radiata and V. angularis. *Gene* **2018**, *651*, 152–160. [CrossRef] [PubMed]

38. Baek, D.; Chun, H.J.; Kang, S.; Shin, G.; Park, S.J.; Hong, H.; Kim, C.; Kim, D.H.; Lee, S.Y.; Kim, M.C. A role for Arabidopsis miR399f in salt, drought, and ABA signaling. *Mol. Cells* **2016**, *39*, 111.

39. Seleiman, M.F.; Al-Suhaibani, N.; Ali, N.; Akmal, M.; Alotaibi, M.; Refay, Y.; Dindaroglu, T.; Abdul-Wajid, H.H.; Battaglia, M.L. Drought stress impacts on plants and different approaches to alleviate its adverse effects. *Plants* **2021**, *10*, 259. [CrossRef]

40. Sreenivasulu, N.; Harshavardhan, V.T.; Govind, G.; Seiler, C.; Kohli, A. Contrapuntal role of ABA: Does it mediate stress tolerance or plant growth retardation under long-term drought stress? *Gene* **2012**, *506*, 265–273. [CrossRef]

41. Cruz de Carvalho, M.H. Drought stress and reactive oxygen species: Production, scavenging and signaling. *Plant Signal. Behav.* **2008**, *3*, 156–165. [CrossRef]

42. Foyer, C.H.; Noctor, G. Redox homeostasis and antioxidant signaling: A metabolic interface between stress perception and physiological responses. *Plant Cell* **2005**, *17*, 1866–1875. [CrossRef]

43. Verslues, P.E.; Zhu, J.K. Before and beyond ABA: Upstream sensing and internal signals that determine ABA accumulation and response under abiotic stress. *Biochem. Soc. Trans.* **2005**, *33*, 375–379. [CrossRef] [PubMed]

44. Li, X.; Chen, L.; Forde, B.G.; Davies, W.J. The Biphasic Root Growth Response to Abscisic Acid in Arabidopsis Involves Interaction with Ethylene and Auxin Signalling Pathways. *Front. Plant Sci.* **2017**, *8*, 1493. [CrossRef] [PubMed]

45. Zhou, Y.; He, R.; Guo, Y.; Liu, K.; Huang, G.; Peng, C.; Liu, Y.; Zhang, M.; Li, Z.; Duan, L. A novel ABA functional analogue B2 enhances drought tolerance in wheat. *Sci. Rep.* **2019**, *9*, 2887. [CrossRef] [PubMed]

46. Sharp, R.E.; Poroyko, V.; Hejlek, L.G.; Spollen, W.G.; Springer, G.K.; Bohnert, H.J.; Nguyen, H.T. Root growth maintenance during water deficits: Physiology to functional genomics. *J Exp. Bot.* **2004**, *55*, 2343–2351. [CrossRef] [PubMed]

47. Wu, Y.; Thorne, E.T.; Sharp, R.E.; Cosgrove, D.J. Modification of expansin transcript levels in the maize primary root at low water potentials. *Plant Physiol.* **2001**, *126*, 1471–1479. [CrossRef]

48. Kwak, J.M.; Mori, I.C.; Pei, Z.M.; Leonhardt, N.; Torres, M.A.; Dangl, J.L.; Bloom, R.E.; Bodde, S.; Jones, J.D.; Schroeder, J.I. NADPH oxidase AtrbohD and AtrbohF genes function in ROS-dependent ABA signaling in Arabidopsis. *EMBO J.* **2003**, *22*, 2623–2633. [CrossRef]

49. Gilroy, S.; Read, N.; Trewavas, A.J. Elevation of cytoplasmic calcium by caged calcium or caged inositol trisphosphate initiates stomatal closure. *Nature* **1990**, *346*, 769–771. [CrossRef]

50. Aharoni, A.; Galili, G. Metabolic engineering of the plant primary–secondary metabolism interface. *Curr. Opin. Biotechnol.* **2011**, *22*, 239–244. [CrossRef]

51. Rausch, T.; Wachter, A. Sulfur metabolism: A versatile platform for launching defence operations. *Trends Plant Sci.* **2005**, *10*, 503–509. [CrossRef]

52. Kumar, M.; Patel, M.K.; Kumar, N.; Bajpai, A.B.; Siddique, K.H.M. Metabolomics and molecular approaches reveal drought stress tolerance in plants. *Int. J. Mol. Sci.* **2021**, *22*, 9108. [CrossRef]

53. Goodacre, R.; Vaidyanathan, S.; Dunn, W.B.; Harrigan, G.G.; Kell, D.B. Metabolomics by numbers: Acquiring and understanding global metabolite data. *Trends Biotechnol.* **2004**, *22*, 245–252. [CrossRef] [PubMed]

54. Zhu, M.; Assmann, S.M. Metabolic Signatures in Response to Abscisic Acid (ABA) Treatment in Brassica napus Guard Cells Revealed by Metabolomics. *Sci. Rep.* **2017**, *7*, 12875. [CrossRef] [PubMed]

55. Yoshida, T.; Obata, T.; Feil, R.; Lunn, J.E.; Fujita, Y.; Yamaguchi-Shinozaki, K.; Fernie, A.R. The role of abscisic acid signaling in maintaining the metabolic balance required for Arabidopsis growth under nonstress conditions. *Plant Cell* **2019**, *31*, 84–105. [CrossRef] [PubMed]

56. Dekkers, B.J.W.; Costa, M.C.D.; Maia, J.; Bentsink, L.; Ligterink, W.; Hilhorst, H.W.M. Acquisition and loss of desiccation tolerance in seeds: From experimental model to biological relevance. *Planta* **2015**, *241*, 563–577. [CrossRef]

57. Munemasa, S.; Hauser, F.; Park, J.; Waadt, R.; Brandt, B.; Schroeder, J.I. Mechanisms of abscisic acid-mediated control of stomatal aperture. *Curr. Opin. Plant Biol.* **2015**, *28*, 154–162. [CrossRef] [PubMed]

58. Nakashima, K.; Yamaguchi-Shinozaki, K. ABA signaling in stress-response and seed development. *Plant Cell Rep.* **2013**, *32*, 959–970. [CrossRef]

59. LeNoble, M.E.; Spollen, W.G.; Sharp, R.E. Maintenance of shoot growth by endogenous ABA: Genetic assessment of the involvement of ethylene suppression. *J. Exp. Bot.* **2004**, *55*, 237–245. [CrossRef]

60. Sharp, R.E.; LeNoble, M.E.; Else, M.A.; Thorne, E.T.; Gherardi, F. Endogenous ABA maintains shoot growth in tomato independently of effects on plant water balance: Evidence for an interaction with ethylene. *J. Exp. Bot.* **2000**, *51*, 1575–1584. [CrossRef]

61. Abdullah, H.M.; Rodriguez, J.; Salacup, J.M.; Castañeda, I.S.; Schnell, D.J.; Pareek, A.; Dhankher, O.P. Increased Cuticle Waxes by Overexpression of WSD1 Improves Osmotic Stress Tolerance in Arabidopsis thaliana and Camelina sativa. *Int. J. Mol. Sci.* **2021**, *22*, 5173. [CrossRef]

62. Zhu, M.; Simons, B.; Zhu, N.; Oppenheimer, D.G.; Chen, S. Analysis of abscisic acid responsive proteins in Brassica napus guard cells by multiplexed isobaric tagging. *J. Proteom.* **2010**, *73*, 790–805. [CrossRef]

63. Cramer, G.R.; Urano, K.; Delrot, S.; Pezzotti, M.; Shinozaki, K. Effects of abiotic stress on plants: A systems biology perspective. *BMC Plant Biol.* **2011**, *11*, 1–14. [CrossRef] [PubMed]

64. Sweetlove, L.J.; Fell, D.; Fernie, A.R. Getting to grips with the plant metabolic network. *Biochem. J.* **2008**, *409*, 27–41. [CrossRef] [PubMed]

65. Fiers, M.; Golemiec, E.; Xu, J.; van der Geest, L.; Heidstra, R.; Stiekema, W.; Liu, C.-M. The 14–amino acid CLV3, CLE19, and CLE40 peptides trigger consumption of the root meristem in Arabidopsis through a CLAVATA2-dependent pathway. *Plant Cell* **2005**, *17*, 2542–2553. [CrossRef] [PubMed]

66. Krasensky, J.; Jonak, C. Drought, salt, and temperature stress-induced metabolic rearrangements and regulatory networks. *J. Exp. Bot.* **2012**, *63*, 1593–1608. [CrossRef]

67. Dixon, R.A.; Paiva, N.L. Stress-induced phenylpropanoid metabolism. *Plant Cell* **1995**, *7*, 1085. [CrossRef]

68. Takahashi, F.; Kuromori, T.; Urano, K.; Yamaguchi-Shinozaki, K.; Shinozaki, K. Drought Stress Responses and Resistance in Plants: From Cellular Responses to Long-Distance Intercellular Communication. *Front. Plant Sci.* **2020**, *11*, 556972. [CrossRef]

69. Urano, K.; Maruyama, K.; Ogata, Y.; Morishita, Y.; Takeda, M.; Sakurai, N.; Suzuki, H.; Saito, K.; Shibata, D.; Kobayashi, M. Characterization of the ABA-regulated global responses to dehydration in Arabidopsis by metabolomics. *Plant J.* **2009**, *57*, 1065–1078. [CrossRef]

70. Pires, M.V.; Pereira Júnior, A.A.; Medeiros, D.B.; Daloso, D.M.; Pham, P.A.; Barros, K.A.; Engqvist, M.K.M.; Florian, A.; Krahnert, I.; Maurino, V.G. The influence of alternative pathways of respiration that utilize branched-chain amino acids following water shortage in Arabidopsis. *Plant Cell Environ.* **2016**, *39*, 1304–1319. [CrossRef]

71. Tarazona, P.; Feussner, K.; Feussner, I. An enhanced plant lipidomics method based on multiplexed liquid chromatography–mass spectrometry reveals additional insights into cold-and drought-induced membrane remodeling. *Plant J.* **2015**, *84*, 621–633. [CrossRef]

72. Nakabayashi, R.; Mori, T.; Saito, K. Alternation of flavonoid accumulation under drought stress in Arabidopsis thaliana. *Plant Signal. Behav.* **2014**, *9*, e29518. [CrossRef]

73. Ma, X.; Xia, H.; Liu, Y.; Wei, H.; Zheng, X.; Song, C.; Chen, L.; Liu, H.; Luo, L. Transcriptomic and metabolomic studies disclose key metabolism pathways contributing to well-maintained photosynthesis under the drought and the consequent drought-tolerance in rice. *Front. Plant Sci.* **2016**, *7*, 1886. [CrossRef] [PubMed]

74. Obata, T.; Witt, S.; Lisec, J.; Palacios-Rojas, N.; Florez-Sarasa, I.; Yousfi, S.; Araus, J.L.; Cairns, J.E.; Fernie, A.R. Metabolite profiles of maize leaves in drought, heat, and combined stress field trials reveal the relationship between metabolism and grain yield. *Plant Physiol.* **2015**, *169*, 2665–2683. [CrossRef] [PubMed]

75. Chmielewska, K.; Rodziewicz, P.; Swarcewicz, B.; Sawikowska, A.; Krajewski, P.; Marczak, Ł.; Ciesiołka, D.; Kuczyńska, A.; Mikołajczak, K.; Ogrodowicz, P. Analysis of drought-induced proteomic and metabolomic changes in barley (Hordeum vulgare L.) leaves and roots unravels some aspects of biochemical mechanisms involved in drought tolerance. *Front. Plant Sci.* **2016**, *7*, 1108. [CrossRef] [PubMed]

76. Liu, X.; Zhai, S.; Zhao, Y.; Sun, B.; Liu, C.; Yang, A.; Zhang, J. Overexpression of the phosphatidylinositol synthase gene (ZmPIS) conferring drought stress tolerance by altering membrane lipid composition and increasing ABA synthesis in maize. *Plant Cell Environ.* **2013**, *36*, 1037–1055. [CrossRef]

77. Nam, K.-H.; Kim, H.J.; Pack, I.-S.; Kim, H.J.; Chung, Y.S.; Kim, S.Y.; Kim, C.-G. Global metabolite profiling based on GC–MS and LC–MS/MS analyses in ABF3-overexpressing soybean with enhanced drought tolerance. *Appl. Biol. Chem.* **2019**, *62*, 1–9. [CrossRef]

78. Hong, J.H.; Seah, S.W.; Xu, J. The root of ABA action in environmental stress response. *Plant Cell Rep.* **2013**, *32*, 971–983. [CrossRef] [PubMed]

79. Benderradji, L.; Saibi, W.; Brini, F. Role of ABA in Overcoming Environmental Stress: Sensing, Signaling and Crosstalk. *Annu. Agric. Crop Sci.* **2021**, *6*, 1070.

80. Liang, Y.; Mitchell, D.M.; Harris, J.M. Abscisic acid rescues the root meristem defects of the Medicago truncatula latd mutant. *Dev. Biol.* **2007**, *304*, 297–307. [CrossRef]

81. De Smet, I.; Signora, L.; Beeckman, T.; Inzé, D.; Foyer, C.H.; Zhang, H. An abscisic acid-sensitive checkpoint in lateral root development of Arabidopsis. *Plant J.* **2003**, *33*, 543–555. [CrossRef]

82. Deak, K.I.; Malamy, J. Osmotic regulation of root system architecture. *Plant J.* **2005**, *43*, 17–28. [CrossRef]

83. Guo, D.; Liang, J.; Qiao, Y.; Yan, Y.; Li, L.; Dai, Y. Involvement of G1-to-S transition and AhAUX-dependent auxin transport in abscisic acid-induced inhibition of lateral root primodia initiation in *Arachis hypogaea* L. *J. Plant Physiol.* **2012**, *169*, 1102–1111. [CrossRef] [PubMed]

84. Quiroz-Figueroa, F.; Rodríguez-Acosta, A.; Salazar-Blas, A.; Hernández-Domínguez, E.; Campos, M.E.; Kitahata, N.; Asami, T.; Galaz-Avalos, R.M.; Cassab, G.I. Accumulation of high levels of ABA regulates the pleiotropic response of the nhr1 Arabidopsis mutant. *J. Plant Biol.* **2010**, *53*, 32–44. [CrossRef]

85. Zhang, C.; Bousquet, A.; Harris, J.M. Abscisic acid and lateral root organ defective/modulate root elongation via reactive oxygen species in Medicago truncatula. *Plant Physiol.* **2014**, *166*, 644–658. [CrossRef] [PubMed]

86. Fang, Y.; Xiong, L. General mechanisms of drought response and their application in drought resistance improvement in plants. *Cell. Mol. Life Sci.* **2015**, *72*, 673–689. [CrossRef]

87. Zhang, Q.; Yuan, W.; Wang, Q.; Cao, Y.; Xu, F.; Dodd, I.C.; Xu, W. ABA regulation of root growth during soil drying and recovery can involve auxin response. *Plant Cell Environ.* **2021**. [CrossRef]

88. Sharp, R.E.; LeNoble, M.E. ABA, ethylene and the control of shoot and root growth under water stress. *J. Exp. Bot.* **2002**, *53*, 33–37. [CrossRef]

89. Williams, A.; de Vries, F.T. Plant root exudation under drought: Implications for ecosystem functioning. *New Phytol.* **2020**, *225*, 1899–1905. [CrossRef]

90. Lynch, J.M.; Whipps, J.M. Substrate flow in the rhizosphere. *Plant Soil* **1990**, *129*, 1–10. [CrossRef]

91. Badri, D.V.; Vivanco, J.M. Regulation and function of root exudates. *Plant Cell Environ.* **2009**, *32*, 666–681. [CrossRef]

92. Baetz, U.; Martinoia, E. Root exudates: The hidden part of plant defense. *Trends Plant Sci.* **2014**, *19*, 90–98. [CrossRef]

93. Rudrappa, T.; Czymmek, K.J.; Paré, P.W.; Bais, H.P. Root-Secreted Malic Acid Recruits Beneficial Soil Bacteria. *Plant Physiol.* **2008**, *148*, 1547–1556. [CrossRef]

94. Iannucci, A.; Fragasso, M.; Platani, C.; Papa, R. Plant growth and phenolic compounds in the rhizosphere soil of wild oat (Avena fatua L.). *Front. Plant Sci.* **2013**, *4*, 509. [CrossRef]

95. Zhalnina, K.; Louie, K.B.; Hao, Z.; Mansoori, N.; da Rocha, U.N.; Shi, S.; Cho, H.; Karaoz, U.; Loqué, D.; Bowen, B.P. Dynamic root exudate chemistry and microbial substrate preferences drive patterns in rhizosphere microbial community assembly. *Nat. Microbiol.* **2018**, *3*, 470–480. [CrossRef]

96. Williams, A.; Langridge, H.; Straathof, A.L.; Muhamadali, H.; Hollywood, K.A.; Goodacre, R.; de Vries, F.T. Root functional traits explain root exudation rate and composition across a range of grassland species. *J. Ecol.* **2021**, *110*, 21–33. [CrossRef]

97. Dennis, P.G.; Miller, A.J.; Hirsch, P.R. Are root exudates more important than other sources of rhizodeposits in structuring rhizosphere bacterial communities? *FEMS Microbiol. Ecol.* **2010**, *72*, 313–327. [CrossRef]

98. Naylor, D.; Coleman-Derr, D. Drought stress and root-associated bacterial communities. *Front. Plant Sci.* **2018**, *8*, 2223. [CrossRef]

99. Bulgarelli, D.; Schlaeppi, K.; Spaepen, S.; Van Themaat, E.V.L.; Schulze-Lefert, P. Structure and functions of the bacterial microbiota of plants. *Annu. Rev. Plant Biol.* **2013**, *64*, 807–838. [CrossRef] [PubMed]

100. De Ollas, C.; Arbona, V.; GóMez-Cadenas, A. Jasmonoyl isoleucine accumulation is needed for abscisic acid build-up in roots of A rabidopsis under water stress conditions. *Plant Cell Environ.* **2015**, *38*, 2157–2170. [CrossRef]

101. Yang, J.; Zhang, J.; Wang, Z.; Zhu, Q.; Liu, L. Abscisic acid and cytokinins in the root exudates and leaves and their relationship to senescence and remobilization of carbon reserves in rice subjected to water stress during grain filling. *Planta* **2002**, *215*, 645–652. [CrossRef]

102. Song, F.; Han, X.; Zhu, X.; Herbert, S.J. Response to water stress of soil enzymes and root exudates from drought and non-drought tolerant corn hybrids at different growth stages. *Can. J. Soil Sci.* **2012**, *92*, 501–507. [CrossRef]

103. Gargallo-Garriga, A.; Preece, C.; Sardans, J.; Oravec, M.; Urban, O.; Peñuelas, J. Root exudate metabolomes change under drought and show limited capacity for recovery. *Sci. Rep.* **2018**, *8*, 12696. [CrossRef] [PubMed]

104. Porcel, R.; Zamarreño, Á.M.; García-Mina, J.M.; Aroca, R. Involvement of plant endogenous ABA in Bacillus megaterium PGPR activity in tomato plants. *BMC Plant Biol.* **2014**, *14*, 36. [CrossRef] [PubMed]

105. Karadeniz, A.; Topcuoğlu, Ş.; Inan, S. Auxin, gibberellin, cytokinin and abscisic acid production in some bacteria. *World J. Microbiol. Biotechnol.* **2006**, *22*, 1061–1064. [CrossRef]

106. Kai, M.; Vespermann, A.; Piechulla, B. The growth of fungi and Arabidopsis thaliana is influenced by bacterial volatiles. *Plant Signal. Behav.* **2008**, *3*, 482–484. [CrossRef] [PubMed]

107. Nakayama, T.; Homma, Y.; Hashidoko, Y.; Mizutani, J.; Tahara, S. Possible role of xanthobaccins produced by Stenotrophomonas sp. strain SB-K88 in suppression of sugar beet damping-off disease. *Appl. Environ. Microbiol.* **1999**, *65*, 4334–4339. [CrossRef] [PubMed]

108. Henry, A.; Doucette, W.; Norton, J.; Bugbee, B. Changes in crested wheatgrass root exudation caused by flood, drought, and nutrient stress. *J. Environ. Qual.* **2007**, *36*, 904–912. [CrossRef] [PubMed]

109. Gagné-Bourque, F.; Bertrand, A.; Claessens, A.; Aliferis, K.A.; Jabaji, S. Alleviation of drought stress and metabolic changes in timothy (Phleum pratense L.) colonized with Bacillus subtilis B26. *Front. Plant Sci.* **2016**, *7*, 584. [CrossRef]

110. Sánchez-Fernández, D.; Rizzo, V.; Bourdeau, C.; Cieslak, A.; Comas, J.; Faille, A.; Fresneda, J.; Lleopart, E.; Millán, A.; Montes, A. The deep subterranean environment as a potential model system in ecological, biogeographical and evolutionary research. *Subterr. Biol.* **2018**, *25*, 1–7. [CrossRef]

111. Dodd, I.C.; Zinovkina, N.Y.; Safronova, V.I.; Belimov, A.A. Rhizobacterial mediation of plant hormone status. *Ann. Appl. Biol.* **2010**, *157*, 361–379. [CrossRef]

112. Belimov, A.A.; Dodd, I.C.; Safronova, V.I.; Dumova, V.A.; Shaposhnikov, A.I.; Ladatko, A.G.; Davies, W.J. Abscisic acid metabolizing rhizobacteria decrease ABA concentrations in planta and alter plant growth. *Plant Physiol. Biochem.* **2014**, *74*, 84–91. [CrossRef]

113. Gowtham, H.G.; Duraivadivel, P.; Ayusman, S.; Sayani, D.; Gholap, S.L.; Niranjana, S.R.; Hariprasad, P. ABA analogue produced by Bacillus marisflavi modulates the physiological response of Brassica juncea L. under drought stress. *Appl. Soil Ecol.* **2021**, *159*, 103845. [CrossRef]

114. Luo, X.; Chen, Z.; Gao, J.; Gong, Z. Abscisic acid inhibits root growth in Arabidopsis through ethylene biosynthesis. *Plant J. Cell Mol. Biol.* **2014**, *79*, 44–55. [CrossRef]

115. Fujita, Y.; Nakashima, K.; Yoshida, T.; Katagiri, T.; Kidokoro, S.; Kanamori, N.; Umezawa, T.; Fujita, M.; Maruyama, K.; Ishiyama, K.; et al. Three SnRK2 protein kinases are the main positive regulators of abscisic acid signaling in response to water stress in Arabidopsis. *Plant Cell Physiol.* **2009**, *50*, 2123–2132. [CrossRef]

116. Liu, X.; Yue, Y.; Li, B.; Nie, Y.; Li, W.; Wu, W.H.; Ma, L. A G protein-coupled receptor is a plasma membrane receptor for the plant hormone abscisic acid. *Science* **2007**, *315*, 1712–1716. [CrossRef]

117. Hirayama, T.; Shinozaki, K. Perception and transduction of abscisic acid signals: Keys to the function of the versatile plant hormone ABA. *Trends Plant Sci.* **2007**, *12*, 343–351. [CrossRef]

118. Kuromori, T.; Seo, M.; Shinozaki, K. ABA Transport and Plant Water Stress Responses. *Trends Plant Sci.* **2018**, *23*, 513–522. [CrossRef]

119. Merilo, E.; Jalakas, P.; Laanemets, K.; Mohammadi, O.; Hõrak, H.; Kollist, H.; Brosché, M. Abscisic Acid Transport and Homeostasis in the Context of Stomatal Regulation. *Mol. Plant* **2015**, *8*, 1321–1333. [CrossRef]

120. Zhang, A.; Ji, Y.; Sun, M.; Lin, C.; Zhou, P.; Ren, J.; Luo, D.; Wang, X.; Ma, C.; Zhang, X.; et al. Research on the drought tolerance mechanism of Pennisetum glaucum (L.) in the root during the seedling stage. *BMC Genom.* **2021**, *22*, 568. [CrossRef]

121. Li, X.; Yu, B.; Wu, Q.; Min, Q.; Zeng, R.; Xie, Z.; Huang, J. OsMADS23 phosphorylated by SAPK9 confers drought and salt tolerance by regulating ABA biosynthesis in rice. *PLoS Genet.* **2021**, *17*, e1009699. [CrossRef]

122. Mathan, J.; Singh, A.; Ranjan, A. Sucrose transport in response to drought and salt stress involves ABA-mediated induction of OsSWEET13 and OsSWEET15 in rice. *Physiol. Plant.* **2021**, *171*, 620–637. [CrossRef]

123. Cao, L.; Lu, X.; Wang, G.; Zhang, Q.; Zhang, X.; Fan, Z.; Cao, Y.; Wei, L.; Wang, T.; Wang, Z. Maize ZmbZIP33 is involved in drought resistance and recovery ability through an abscisic acid-dependent signaling pathway. *Front. Plant Sci.* **2021**, *12*, 442. [CrossRef] [PubMed]

124. Aswath, C.R.; Kim, S.H.; Mo, S.Y.; Kim, D.H. Transgenic plants of creeping bent grass harboring the stress inducible gene, 9-cis-epoxycarotenoid dioxygenase, are highly tolerant to drought and NaCl stress. *Plant Growth Regul.* **2005**, *47*, 129–139. [CrossRef]

125. Zhang, Y.; Vasuki, H.; Liu, J.; Bar, H.; Lazary, S.; Egbaria, A.; Ripper, D.; Charrier, L.; Mussa, Z.; Wulff, N. ABA homeostasis and long-distance translocation is redundantly regulated by ABCG ABA importers. *bioRxiv* **2021**, *7*, eabf6069. [CrossRef] [PubMed]

126. Kanno, Y.; Hanada, A.; Chiba, Y.; Ichikawa, T.; Nakazawa, M.; Matsui, M.; Koshiba, T.; Kamiya, Y.; Seo, M. Identification of an abscisic acid transporter by functional screening using the receptor complex as a sensor. *Proc. Natl. Acad. Sci. USA* **2012**, *109*, 9653–9658. [CrossRef]

127. Huque, A.K.M.; So, W.; Noh, M.; You, M.K.; Shin, J.S. Overexpression of AtBBD1, Arabidopsis Bifunctional Nuclease, Confers Drought Tolerance by Enhancing the Expression of Regulatory Genes in ABA-Mediated Drought Stress Signaling. *Int. J. Mol. Sci.* **2021**, *22*, 2936. [CrossRef]

128. He, Y.; Liu, Y.; Li, M.; Lamin-Samu, A.T.; Yang, D.; Yu, X.; Izhar, M.; Jan, I.; Ali, M.; Lu, G. The Arabidopsis SMALL AUXIN UP RNA32 Protein Regulates ABA-Mediated Responses to Drought Stress. *Front. Plant Sci.* **2021**, *12*, 259. [CrossRef]

129. Le Hir, R.; Castelain, M.; Chakraborti, D.; Moritz, T.; Dinant, S.; Bellini, C. At bHLH68 transcription factor contributes to the regulation of ABA homeostasis and drought stress tolerance in Arabidopsis thaliana. *Physiol. Plant.* **2017**, *160*, 312–327. [CrossRef]

130. Baek, D.; Shin, G.; Kim, M.C.; Shen, M.; Lee, S.Y.; Yun, D.-J. Histone deacetylase HDA9 with ABI4 contributes to abscisic acid homeostasis in drought stress response. *Front. Plant Sci.* **2020**, *11*, 143. [CrossRef] [PubMed]

131. Kuromori, T.; Sugimoto, E.; Shinozaki, K. Brachypodium BdABCG25 is a homolog of Arabidopsis AtABCG25 involved in the transport of abscisic acid. *FEBS Lett.* **2021**, *595*, 954–959. [CrossRef]

132. Wang, Y.; Beaith, M.; Chalifoux, M.; Ying, J.; Uchacz, T.; Sarvas, C.; Griffiths, R.; Kuzma, M.; Wan, J.; Huang, Y. Shoot-specific down-regulation of protein farnesyltransferase (α-subunit) for yield protection against drought in canola. *Mol. Plant* **2009**, *2*, 191–200. [CrossRef]

133. Xu, M.; Li, H.; Liu, Z.-N.; Wang, X.-H.; Xu, P.; Dai, S.-J.; Cao, X.; Cui, X.-Y. The soybean CBL-interacting protein kinase, GmCIPK2, positively regulates drought tolerance and ABA signaling. *Plant Physiol. Biochem.* **2021**, *167*, 980–989. [CrossRef] [PubMed]

134. Zhang, Y.; Yang, J.; Lu, S.; Cai, J.; Guo, Z. Overexpressing SgNCED1 in tobacco increases ABA level, antioxidant enzyme activities, and stress tolerance. *J. Plant Growth Regul.* **2008**, *27*, 151–158. [CrossRef]

135. Zhang, S.; Haider, I.; Kohlen, W.; Jiang, L.; Bouwmeester, H.; Meijer, A.H.; Schluepmann, H.; Liu, C.-M.; Ouwerkerk, P.B.F. Function of the HD-Zip I gene Oshox22 in ABA-mediated drought and salt tolerances in rice. *Plant Mol. Biol.* **2012**, *80*, 571–585. [CrossRef] [PubMed]

136. Tang, N.; Zhang, H.; Li, X.; Xiao, J.; Xiong, L. Constitutive activation of transcription factor OsbZIP46 improves drought tolerance in rice. *Plant Physiol.* **2012**, *158*, 1755–1768. [CrossRef] [PubMed]

137. Yao, L.; Cheng, X.; Gu, Z.; Huang, W.; Li, S.; Wang, L.; Wang, Y.-F.; Xu, P.; Ma, H.; Ge, X. The AWPM-19 family protein OsPM1 mediates abscisic acid influx and drought response in rice. *Plant Cell* **2018**, *30*, 1258–1276. [CrossRef] [PubMed]

138. Estrada-Melo, A.C.; Chao, C.; Reid, M.S.; Jiang, C.-Z. Overexpression of an ABA biosynthesis gene using a stress-inducible promoter enhances drought resistance in petunia. *Hortic. Res.* **2015**, *2*, 15013. [CrossRef] [PubMed]

139. Li, C.; Yue, J.; Wu, X.; Xu, C.; Yu, J. An ABA-responsive DRE-binding protein gene from Setaria italica, SiARDP, the target gene of SiAREB, plays a critical role under drought stress. *J. Exp. Bot.* **2014**, *65*, 5415–5427. [CrossRef] [PubMed]

140. Liu, Y.; Wen, L.; Shi, Y.; Su, D.; Lu, W.; Cheng, Y.; Li, Z. Stress-responsive tomato gene SlGRAS4 function in drought stress and abscisic acid signaling. *Plant Sci.* **2021**, *304*, 110804. [CrossRef]

141. Gonzalez-Guzman, M.; Pizzio, G.A.; Antoni, R.; Vera-Sirera, F.; Merilo, E.; Bassel, G.W.; Fernández, M.A.; Holdsworth, M.J.; Perez-Amador, M.A.; Kollist, H. Arabidopsis PYR/PYL/RCAR receptors play a major role in quantitative regulation of stomatal aperture and transcriptional response to abscisic acid. *Plant Cell* **2012**, *24*, 2483–2496. [CrossRef]

142. Sivamani, E.; Bahieldin, A.; Wraith, J.M.; Al-Niemi, T.; Dyer, W.E.; Ho, T.-H.D.; Qu, R. Improved biomass productivity and water use efficiency under water deficit conditions in transgenic wheat constitutively expressing the barley HVA1 gene. *Plant Sci.* **2000**, *155*, 1–9. [CrossRef]

143. Yuasa, T.; Kubo, Y.; Fujimaki, W.; Ishii, T. Water Stress Activates The Expression of Abscisic Acid Transporter, VuABCG25, in Cowpea [*Vigna unguiculata* (L.) Walp.]. *Cryobiol. Cryotechnology* **2021**, *67*, 65–70.

144. Speirs, J.; Binney, A.; Collins, M.; Edwards, E.; Loveys, B. Expression of ABA synthesis and metabolism genes under different irrigation strategies and atmospheric VPDs is associated with stomatal conductance in grapevine (Vitis vinifera L. cv Cabernet Sauvignon). *J. Exp. Bot.* **2013**, *64*, 1907–1916. [CrossRef] [PubMed]

145. Soar, C.J.; Speirs, J.; Maffei, S.M.; Penrose, A.B.; McCarthy, M.G.; Loveys, B.R. Grape vine varieties Shiraz and Grenache differ in their stomatal response to VPD: Apparent links with ABA physiology and gene expression in leaf tissue. *Aust. J. Grape Wine Res.* **2006**, *12*, 2–12. [CrossRef]

146. Rossdeutsch, L.; Edwards, E.; Cookson, S.J.; Barrieu, F.; Gambetta, G.A.; Delrot, S.; Ollat, N. ABA-mediated responses to water deficit separate grapevine genotypes by their genetic background. *BMC Plant Biol.* **2016**, *16*, 1–15. [CrossRef]

147. Xiong, C.; Zhao, S.; Yu, X.; Sun, Y.; Li, H.; Ruan, C.; Li, J. Yellowhorn drought-induced transcription factor XsWRKY20 acts as a positive regulator in drought stress through ROS homeostasis and ABA signaling pathway. *Plant Physiol. Biochem.* **2020**, *155*, 187–195. [CrossRef]

148. Brugière, N.; Zhang, W.; Xu, Q.; Scolaro, E.J.; Lu, C.; Kahsay, R.Y.; Kise, R.; Trecker, L.; Williams, R.W.; Hakimi, S. Overexpression of RING domain E3 ligase ZmXerico1 confers drought tolerance through regulation of ABA homeostasis. *Plant Physiol.* **2017**, *175*, 1350–1369. [CrossRef]

149. Qian, D.; Zhang, Z.; He, J.; Zhang, P.; Ou, X.; Li, T.; Niu, L.; Nan, Q.; Niu, Y.; He, W. Arabidopsis ADF5 promotes stomatal closure by regulating actin cytoskeleton remodeling in response to ABA and drought stress. *J. Exp. Bot.* **2019**, *70*, 435–446. [CrossRef]

150. Banerjee, A.; Roychoudhury, A. Abscisic-acid-dependent basic leucine zipper (bZIP) transcription factors in plant abiotic stress. *Protoplasma* **2017**, *254*, 3–16. [CrossRef] [PubMed]

151. Kerr, T.C.C.; Abdel-Mageed, H.; Aleman, L.; Lee, J.; Payton, P.; Cryer, D.; Allen, R.D. Ectopic expression of two AREB/ABF orthologs increases drought tolerance in cotton (Gossypium hirsutum). *Plant Cell Environ.* **2018**, *41*, 898–907. [CrossRef]

152. Wang, Y.-H.; Que, F.; Li, T.; Zhang, R.-R.; Khadr, A.; Xu, Z.-S.; Tian, Y.-S.; Xiong, A.-S. DcABF3, an ABF transcription factor from carrot, alters stomatal density and reduces ABA sensitivity in transgenic Arabidopsis. *Plant Sci.* **2021**, *302*, 110699. [CrossRef]

153. Zou, M.; Guan, Y.; Ren, H.; Zhang, F.; Chen, F. A bZIP transcription factor, OsABI5, is involved in rice fertility and stress tolerance. *Plant Mol. Biol.* **2008**, *66*, 675–683. [CrossRef] [PubMed]

154. Casaretto, J.; Ho, T.-h.D. The transcription factors HvABI5 and HvVP1 are required for the abscisic acid induction of gene expression in barley aleurone cells. *Plant Cell* **2003**, *15*, 271–284. [CrossRef] [PubMed]

155. Casaretto, J.A.; Ho, T.-h.D. Transcriptional regulation by abscisic acid in barley (Hordeum vulgare L.) seeds involves autoregulation of the transcription factor HvABI5. *Plant Mol. Biol.* **2005**, *57*, 21–34. [CrossRef] [PubMed]

156. Cho, S.K.; Ryu, M.Y.; Song, C.; Kwak, J.M.; Kim, W.T. Arabidopsis PUB22 and PUB23 are homologous U-Box E3 ubiquitin ligases that play combinatory roles in response to drought stress. *Plant Cell* **2008**, *20*, 1899–1914. [CrossRef]

157. Liu, Y.-C.; Wu, Y.-R.; Huang, X.-H.; Sun, J.; Xie, Q. AtPUB19, a U-box E3 ubiquitin ligase, negatively regulates abscisic acid and drought responses in Arabidopsis thaliana. *Mol. Plant* **2011**, *4*, 938–946. [CrossRef]

158. Ryu, M.Y.; Cho, S.K.; Kim, W.T. The Arabidopsis C3H2C3-type RING E3 ubiquitin ligase AtAIRP1 is a positive regulator of an abscisic acid-dependent response to drought stress. *Plant Physiol.* **2010**, *154*, 1983–1997. [CrossRef]

159. Lee, H.K.; Cho, S.K.; Son, O.; Xu, Z.; Hwang, I.; Kim, W.T. Drought stress-induced Rma1H1, a RING membrane-anchor E3 ubiquitin ligase homolog, regulates aquaporin levels via ubiquitination in transgenic Arabidopsis plants. *Plant Cell* **2009**, *21*, 622–641. [CrossRef]

160. Lee, S.C.; Lan, W.; Buchanan, B.B.; Luan, S. A protein kinase-phosphatase pair interacts with an ion channel to regulate ABA signaling in plant guard cells. *Proc. Natl. Acad. Sci. USA* **2009**, *106*, 21419–21424. [CrossRef]

161. Zhang, H.; Cui, F.; Wu, Y.; Lou, L.; Liu, L.; Tian, M.; Ning, Y.; Shu, K.; Tang, S.; Xie, Q. The RING finger ubiquitin E3 ligase SDIR1 targets SDIR1-INTERACTING PROTEIN1 for degradation to modulate the salt stress response and ABA signaling in Arabidopsis. *Plant Cell* **2015**, *27*, 214–227. [CrossRef] [PubMed]

162. Ko, J.H.; Yang, S.H.; Han, K.H. Upregulation of an Arabidopsis RING-H2 gene, XERICO, confers drought tolerance through increased abscisic acid biosynthesis. *Plant J.* **2006**, *47*, 343–355. [CrossRef]

163. Zhang, Y.e.; Xu, W.; Li, Z.; Deng, X.W.; Wu, W.; Xue, Y. F-box protein DOR functions as a novel inhibitory factor for abscisic acid-induced stomatal closure under drought stress in Arabidopsis. *Plant Physiol.* **2008**, *148*, 2121–2133. [CrossRef] [PubMed]

164. Meng, X.-B.; Zhao, W.-S.; Lin, R.-M.; Wang, M.; Peng, Y.-L. Molecular cloning and characterization of a rice blast-inducible RING-H2 type Zinc finger gene: Full Length Research Paper. *DNA Seq.* **2006**, *17*, 41–48. [CrossRef] [PubMed]

165. Huai, J.; Zheng, J.; Wang, G. Overexpression of a new Cys 2/His 2 zinc finger protein ZmZF1 from maize confers salt and drought tolerance in transgenic Arabidopsis. *Plant Cell Tissue Organ Cult.* **2009**, *99*, 117–124. [CrossRef]

166. Sirko, A.; Wawrzyńska, A.; Brzywczy, J.; Sieńko, M. Control of ABA Signaling and Crosstalk with Other Hormones by the Selective Degradation of Pathway Components. *Int. J. Mol. Sci.* **2021**, *22*, 4638. [CrossRef]

167. De Ollas, C.; Dodd, I.C. Physiological impacts of ABA–JA interactions under water-limitation. *Plant Mol. Biol.* **2016**, *91*, 641–650. [CrossRef]

168. Seo, P.J.; Xiang, F.; Qiao, M.; Park, J.Y.; Lee, Y.N.; Kim, S.G.; Lee, Y.H.; Park, W.J.; Park, C.M. The MYB96 transcription factor mediates abscisic acid signaling during drought stress response in Arabidopsis. *Plant Physiol.* **2009**, *151*, 275–289. [CrossRef]

169. Lu, C.; Chen, M.X.; Liu, R.; Zhang, L.; Hou, X.; Liu, S.; Ding, X.; Jiang, Y.; Xu, J.; Zhang, J.; et al. Abscisic Acid Regulates Auxin Distribution to Mediate Maize Lateral Root Development Under Salt Stress. *Front. Plant Sci.* **2019**, *10*, 716. [CrossRef] [PubMed]

170. Xu, W.; Jia, L.; Shi, W.; Liang, J.; Zhou, F.; Li, Q.; Zhang, J. Abscisic acid accumulation modulates auxin transport in the root tip to enhance proton secretion for maintaining root growth under moderate water stress. *New Phytol.* **2013**, *197*, 139–150. [CrossRef]

171. Thole, J.M.; Beisner, E.R.; Liu, J.; Venkova, S.V.; Strader, L.C. Abscisic acid regulates root elongation through the activities of auxin and ethylene in Arabidopsis thaliana. *G3* **2014**, *4*, 1259–1274. [CrossRef]

172. Bandurska, H.; Stroiński, A.; Kubiś, J. The effect of jasmonic acid on the accumulation of ABA, proline and spermidine and its influence on membrane injury under water deficit in two barley genotypes. *Acta Physiol. Plant.* **2003**, *25*, 279–285. [CrossRef]

173. De Ollas, C.; Hernando, B.; Arbona, V.; Gómez-Cadenas, A. Jasmonic acid transient accumulation is needed for abscisic acid increase in citrus roots under drought stress conditions. *Physiol. Plant.* **2013**, *147*, 296–306. [CrossRef] [PubMed]

174. Puertolas, J.; Alcobendas, R.; Alarcón, J.J.; Dodd, I.C. Long-distance abscisic acid signalling under different vertical soil moisture gradients depends on bulk root water potential and average soil water content in the root zone. *Plant Cell Environ.* **2013**, *36*, 1465–1475. [CrossRef] [PubMed]

175. Liao, X.; Guo, X.; Wang, Q.; Wang, Y.; Zhao, D.; Yao, L.; Wang, S.; Liu, G.; Li, T. Overexpression of Ms DREB 6.2 results in cytokinin-deficient developmental phenotypes and enhances drought tolerance in transgenic apple plants. *Plant J.* **2017**, *89*, 510–526. [CrossRef]

176. Guo, Y.; Gan, S. AtMYB2 regulates whole plant senescence by inhibiting cytokinin-mediated branching at late stages of development in Arabidopsis. *Plant Physiol.* **2011**, *156*, 1612–1619. [CrossRef]

177. Huang, X.; Hou, L.; Meng, J.; You, H.; Li, Z.; Gong, Z.; Yang, S.; Shi, Y. The antagonistic action of abscisic acid and cytokinin signaling mediates drought stress response in Arabidopsis. *Mol. Plant* **2018**, *11*, 970–982. [CrossRef]

178. Gazzarrini, S.; McCourt, P. Cross-talk in plant hormone signalling: What Arabidopsis mutants are telling us. *Ann. Bot.* **2003**, *91*, 605–612. [CrossRef]

179. Leshem, Y.; Seri, L.; Levine, A. Induction of phosphatidylinositol 3-kinase-mediated endocytosis by salt stress leads to intracellular production of reactive oxygen species and salt tolerance. *Plant J.* **2007**, *51*, 185–197. [CrossRef]

180. Lorenzo, O.; Piqueras, R.; Sánchez-Serrano, J.J.; Solano, R. Ethylene response factor1 integrates signals from ethylene and jasmonate pathways in plant defense. *Plant Cell* **2003**, *15*, 165–178. [CrossRef]

181. Sharma, S.; Villamor, J.G.; Verslues, P.E. Essential role of tissue-specific proline synthesis and catabolism in growth and redox balance at low water potential. *Plant Physiol.* **2011**, *157*, 292–304. [CrossRef] [PubMed]

182. Müller, M.; Munné-Bosch, S. Ethylene response factors: A key regulatory hub in hormone and stress signaling. *Plant Physiol.* **2015**, *169*, 32–41. [CrossRef]

183. Hwang, K.; Susila, H.; Nasim, Z.; Jung, J.-Y.; Ahn, J.H. Arabidopsis ABF3 and ABF4 transcription factors act with the NF-YC complex to regulate SOC1 expression and mediate drought-accelerated flowering. *Mol. Plant* **2019**, *12*, 489–505. [CrossRef] [PubMed]

184. Yoshida, T.; Mogami, J.; Yamaguchi-Shinozaki, K. ABA-dependent and ABA-independent signaling in response to osmotic stress in plants. *Curr. Opin. Plant Biol.* **2014**, *21*, 133–139. [CrossRef] [PubMed]

185. Da Silva, E.C.; Nogueira, R.; da Silva, M.A.; de Albuquerque, M.B. Drought stress and plant nutrition. *Plant Stress* **2011**, *5*, 32–41.

186. Ahanger, M.A.; Morad-Talab, N.; Abd-Allah, E.F.; Ahmad, P.; Hajiboland, R. Plant growth under drought stress: Significance of mineral nutrients. *Water Stress Crop Plants A Sustain. Approach* **2016**, *2*, 649–668.

187. Wu, Z.Z.; Ying, Y.Q.; Zhang, Y.B.; Bi, Y.F.; Wang, A.K.; Du, X.H. Alleviation of drought stress in Phyllostachys edulis by N and P application. *Sci. Rep.* **2018**, *8*, 1–9. [CrossRef]

188. Hosseini, S.A.; Réthoré, E.; Pluchon, S.; Ali, N.; Billiot, B.; Yvin, J.-C. Calcium application enhances drought stress tolerance in sugar beet and promotes plant biomass and beetroot sucrose concentration. *Int. J. Mol. Sci.* **2019**, *20*, 3777. [CrossRef]

189. Zahoor, R.; Dong, H.; Abid, M.; Zhao, W.; Wang, Y.; Zhou, Z. Potassium fertilizer improves drought stress alleviation potential in cotton by enhancing photosynthesis and carbohydrate metabolism. *Environ. Exp. Bot.* **2017**, *137*, 73–83. [CrossRef]

190. Agami, R.A.; Alamri, S.A.; Abd El-Mageed, T.; Abousekken, M.; Hashem, M. Role of exogenous nitrogen supply in alleviating the deficit irrigation stress in wheat plants. *Agric. Water Manag.* **2018**, *210*, 261–270. [CrossRef]

191. El-Mageed, A.; Taia, A.; Shaaban, A.; El-Mageed, A.; Shimaa, A.; Semida, W.M.; Rady, M.O. Silicon defensive role in maize (Zea mays L.) against drought stress and metals-contaminated irrigation water. *Silicon* **2021**, *13*, 2165–2176. [CrossRef]

192. Nawaz, F.; Shehzad, M.A.; Majeed, S.; Ahmad, K.S.; Aqib, M.; Usmani, M.M.; Shabbir, R.N. *Role of Mineral Nutrition in Improving Drought and Salinity Tolerance in Field Crops*; Springer: Singapore, 2020; pp. 129–147.

193. Riboni, M.; Robustelli Test, A.; Galbiati, M.; Tonelli, C.; Conti, L. ABA-dependent control of GIGANTEA signalling enables drought escape via up-regulation of flowering locus T in Arabidopsis thaliana. *J. Exp. Bot.* **2016**, *67*, 6309–6322. [CrossRef] [PubMed]

194. Riboni, M.; Galbiati, M.; Tonelli, C.; Conti, L. Gigantea enables drought escape response via abscisic acid-dependent activation of the florigens and suppressor of overexpression of constans1. *Plant Physiol.* **2013**, *162*, 1706–1719. [CrossRef] [PubMed]

195. Du, H.; Huang, F.; Wu, N.; Li, X.; Hu, H.; Xiong, L. Integrative regulation of drought escape through ABA-dependent and -independent pathways in rice. *Mol. Plant* **2018**, *11*, 584–597. [CrossRef] [PubMed]

196. Yu, W.; Zhao, R.; Wang, L.; Zhang, S.; Li, R.; Sheng, J.; Shen, L. ABA signaling rather than ABA metabolism is involved in trehalose-induced drought tolerance in tomato plants. *Planta* **2019**, *250*, 643–655. [CrossRef]

197. Martignago, D.; Siemiatkowska, B.; Lombardi, A.; Conti, L. Abscisic Acid and Flowering Regulation: Many Targets, Different Places. *Int. J. Mol. Sci.* **2020**, *21*, 9700. [CrossRef] [PubMed]

198. Shavrukov, Y.; Kurishbayev, A.; Jatayev, S.; Shvidchenko, V.; Zotova, L.; Koekemoer, F.; de Groot, S.; Soole, K.; Langridge, P. Early flowering as a drought escape mechanism in plants: How can it aid wheat production? *Front. Plant Sci.* **2017**, *8*, 1950. [CrossRef]

199. Lim, C.W.; Baek, W.; Jung, J.; Kim, J.-H.; Lee, S.C. Function of ABA in stomatal defense against biotic and drought stresses. *Int. J. Mol. Sci.* **2015**, *16*, 15251–15270. [CrossRef] [PubMed]

200. Bhaskara, G.B.; Yang, T.-H.; Verslues, P.E. Dynamic proline metabolism: Importance and regulation in water limited environments. *Front. Plant Sci.* **2015**, *6*, 484. [CrossRef]

201. Ghate, T.; Barvkar, V.; Deshpande, S.; Bhargava, S. Role of ABA signaling in regulation of stem sugar metabolism and transport under post-flowering drought stress in sweet sorghum. *Plant Mol. Biol. Rep.* **2019**, *37*, 303–313. [CrossRef]

International Journal of
Molecular Sciences

MDPI

Review

Roles of Abscisic Acid and Gibberellins in Stem/Root Tuber Development

Peilei Chen [1], Ruixue Yang [1], Dorothea Bartels [2], Tianyu Dong [1] and Hongying Duan [1,*]

[1] College of Life Sciences, Henan Normal University, Xinxiang 453007, China; chenpl521x@gmail.com (P.C.); 18737590152@163.com (R.Y.); dongty536@163.com (T.D.)
[2] Institute of Molecular Physiology and Biotechnology of Plants (IMBIO), Faculty of Natural Sciences, University of Bonn, Kirschallee 1, D-53115 Bonn, Germany; dbartels@uni-bonn.de
* Correspondence: dxdhy@126.com

Abstract: Root and tuber crops are of great importance. They not only contribute to feeding the population but also provide raw material for medicine and small-scale industries. The yield of the root and tuber crops is subject to the development of stem/root tubers, which involves the initiation, expansion, and maturation of storage organs. The formation of the storage organ is a highly intricate process, regulated by multiple phytohormones. Gibberellins (GAs) and abscisic acid (ABA), as antagonists, are essential regulators during stem/root tuber development. This review summarizes the current knowledge of the roles of GA and ABA during stem/root tuber development in various tuber crops.

Keywords: root and tuber crops; stem/root tuber development; GA; ABA

Citation: Chen, P.; Yang, R.; Bartels, D.; Dong, T.; Duan, H. Roles of Abscisic Acid and Gibberellins in Stem/Root Tuber Development. *Int. J. Mol. Sci.* **2022**, *23*, 4955. https://doi.org/10.3390/ijms23094955

Academic Editor: Víctor Quesada

Received: 29 March 2022
Accepted: 27 April 2022
Published: 29 April 2022

Publisher's Note: MDPI stays neutral with regard to jurisdictional claims in published maps and institutional affiliations.

1. Introduction

Root and tuber crops are characterized by their underground storage organs. These subterranean structures are rich in water, carbohydrates, and a range of proteins and secondary metabolites [1,2]. Root and tuber crops are indispensable for human food due to the storage contents, of which many are healthy or useful for various industrial applications [1–4]. Potato (*Solanum tuberosum*), sweet potato (*Ipomoea batatas*), and cassava (*Manihot esculenta*) are among the most important root and tuber crops. They not only feed billions of people, especially in the humid regions, but also are used as fodder because of the high levels of carbohydrates and proteins, as well as minerals and vitamins [1,3–5]. Yams (*Dioscorea* spp.) are the fourth-most-important tuber crop, which are viewed as both famine food during the food scarcity periods and medicines with pharmacological effects [6]. Other root and tuber crops, such as Chinese arrowhead (*Sagittaria trifolia*) [7], Chinese foxglove (*Rehmannia glutinosa*) [2], yacon (*Smallanthus sonchifolius*) [8], and sugar beet (*Beta vulgaris* L.) [9], also provide vegetables, fruits, medicine, industry materials, and economical values.

Plant root system architecture shows high diversity. There are various kinds of modified underground stems/roots among the root and tuber crops. Potato and yam tubers are belowground modified stems originating from horizontally growing stolons and vertical hypocotyls, respectively [10–12]. The tuberous roots of sweet potato and cassava develop from adventitious roots [1]. The morphological diversity of some other root and tuber crops are summarized in Table 1. In general, the edible organs of the root and tuber crops arise from the underground swollen stems or roots. These stem/root tubers usually go through three development stages: initiation, expansion/enlargement, and maturation [2,12,13] (or partitioned into induction, initiation, and tuberization in some literatures [14,15]). Potato tuber initiation starts with the expansion of the central pith tissue in the hooked stolon [1,4]. Subsequently, the enlargement of the perimedullary region, comprised of xylem, phloem, parenchyma, and meristematic cells, accelerates the tuber expansion and finally leads to the maturation of stem tuber [1]. The growth of tuberous roots in cassava is a result of the enlargement and division of cells in the vascular cylinder, which eventually develops into a storage zone mainly containing the xylem vessels and starch-storing xylem parenchyma cells [1]. The mature tubers will reach the final size and accumulate maximum nutrients, along with the decay of the overground leaf canopy and the dormancy of the underground tuber [2,12]. All physiological and morphological developments of stem/root tuber crops are regulated by a series of environment factors [15,16]. Potato tuber organogenesis is initiated by a short photoperiod and negatively regulated by other environment perturbations, for example, high night temperatures or high nitrogen levels [16]. Chinese yam tuberization is also initiated by short-day photoperiods, but further enlargement is suppressed under short-day photoperiods after rapid growth [12]. The effects of the environmental cues on tuber development require the finely tuned modulation of endogenous phytohormones [15,16]. Gibberellins (GAs) and abscisic acid (ABA) are two important phytohormones that are involved in various developmental processes, including tuberization with antagonistic [17–19]. In this review, we summarize the recent advances in understanding the functions of GA and ABA in the stem/root tuber development of different tuber crops.

2. Gibberellins in the Development of Stem/Root Tubers

Gibberellin (GA) is an umbrella phytohormone term encompassing a large family of tetracyclic diterpenoid carboxylic acids with the *ent*-gibberellane or 20-nor-*ent*-gibberellane carbon skeletons [20]. To date, more than 130 GAs have been identified [21]. However, the major bioactive GAs are restricted to GA_1, GA_3, GA_4, and GA_7, while others are considered as the precursors or the deactivated forms of bioactive GAs [21]. GA has been implicated in multiple developmental processes via dominant reinforcement of cell elongation or sometimes cell division, as exemplified by tuber growth [1,17,18,20–23]. Potato tuber formation is retarded because of the stolon elongation promoted by the exogeneous bioactive $GA_{4/7}$ [18], while the yam tuber and bulbil yield are increased by the application of GA [23,24]. The opposite effects of GA treatments suggest the intricate roles of GA during tuber development in different stem/tuber root crops [23].

2.1. Gibberellin Biosynthesis and Deactivation

In plants, the common GA precursor is the biologically inactive GA_{12} [20,22,25]. It is transported through the plant vascular system as a mobile GA signal involved in plant developmental transitions [26]. The production of GA_{12} starts from *trans*-geranylgeranyl diphosphate (GGPP) in the proplastide [20,22,25]. GGPP is converted into the hydrocarbon intermediate *ent*-kaurene after catalyzation by *ent*-copalyl diphosphate synthase and *ent*-kaurene synthase (KS) [20,22,25] (Figure 1). In cassava, the transcript of the *KS* homolog was mainly detected in the cortex and parenchyma of fibrous roots and significantly downregulated as the storage root enlarged [27]. The conversion from *ent*-kaurene to GA_{12} occurs in the endoplasmic reticulum [20,22,25], which involves oxidation steps catalyzed by two membrane-associated cytochrome P450 mono-oxygenases: *ent*-kaurene oxidase (KO, localized in both plastid and endoplasmic reticulum), and *ent*-kaurenoic acid oxidase (KAO, localized in the endoplasmic reticulum) [20,22,25] (Figure 1). During corm formation in *S. trifolia*, the expression of *KS* and *KO* was enhanced in the process of stolon elongation and decreased when the corms swelled [28]. Both KS and KO are the key enzymes in the early steps of GA biosynthesis. Thus, the modification of *KS* and *KO* transcript expression indicated the reduced level of GA after tuber initiation. The observation is parallel to the report that the accumulated GA_1 during potato tuber initiation was decreased conspicuously in the enlarged potato tuber [18]. The biosynthesis of bioactive GAs follows the divergence of GA_{12} into two branches in the cytosol, one leading to the formation of GA_4 and the other to that of GA_1 and GA_3 [20,22,25]. GA_4 is produced through the successive oxidization of GA_{12} by two soluble oxoglutarate-dependent dioxygenases (ODDs), GA20ox and GA3ox [20,22,25]. The formation of GA_1 and GA_3 goes through similar pathways after the transformation of GA_{12} to GA_{53} catalyzed by 13-hydroxylases (GA13oxs) [20,22,25] (Figure 1). The potato GA 20-oxidase gene, *StGA20ox1*, is expressed mainly in leaves, but it still intervenes in potato tuberization via the biosynthesis of bioactive GAs [29,30]. Under short-day photoperiods, the potato plants over-expressing *StGA20ox1* turned out to be taller, with longer internodes and delayed time of tuberization [30]. The opposite phenotypes, shorter stems, shorter internodes, and earlier tuberization, were observed in plants expressing antisense copies [30]. Another potato ODD gene, *StGA3ox*, has partial analogy with *StGA20ox1* [31,32]. The silencing of *StGA3ox1* using RNAi also resulted in smaller plants with shorter internodes, but the time of tuber initiation and the tuber yield were not affected despite induction of more tubers [31]. The over-expression of *StGA3ox1* in tubers also gave rise to slightly delayed tuberization, whereas earlier tuberization and taller plants were observed as *StGA3ox1* was over-expressed in leaves [32]. In the tuberous root of *R. glutinosa*, the diminution of *GA20ox* and *GA3ox* expression was consistent with the decline in the GA_3 levels during root development [33]. These results concerning the two ODD genes demonstrate the negative effects of GA biosynthesis on tuber growth. GA may be involved in triggering similar downstream actions on sink tissue development regardless of tubers or tuberous roots originating from different organs [33].

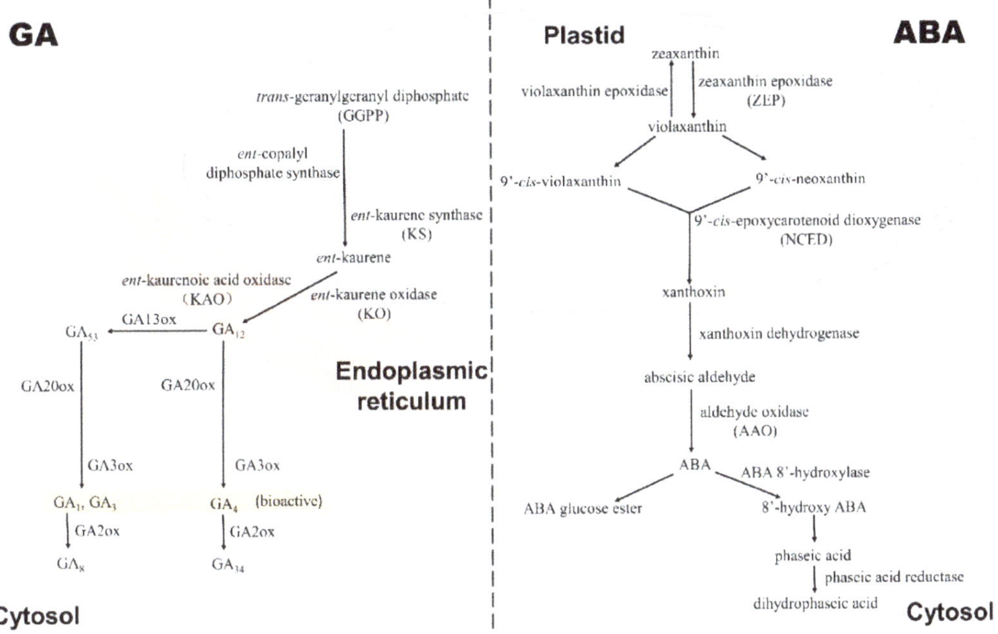

Figure 1. The major GA and ABA metabolism in tuber and root crops. The major GA and ABA metabolism (biosynthesis and catabolism) is predicated according to the expression patterns of genes for the key enzymes during stem/root tuber development and the detection of GA or ABA metabolites in stem/root tubers.

GA is an important phytohormone implicated in multiple biological processes so that the concentration of the bioactive GAs is finely tuned. The tight regulation depends on both the synthesis and deactivation of bioactive GAs [22,25]. The classical bioactive deactivation was catalyzed by two major groups of ODDs. GA 2-oxidases (GA2oxs) are capable of converting the bioactive GA_1 and GA_4 into inactive GA_8 and GA_{34}, respectively, and transform some other GA intermediates [22,25] (Figure 1). In potato, the upregulation of the *StGA2ox1* transcript level was detected in the subapical zone of stolon and growing tubers before tuber expansion [34]. The reduction and over-expression of *StGA2ox1* in transformants caused late and earlier tuberization, respectively, by modifying the GA levels [34]. The phenotypes of the transformants showed that the implication of GA2oxs in GA metabolism exerts influence on tuber development as well. The regulation of *ODD* expression involving GA biosynthesis and deactivation has also been identified in turnip, carrot, and yam [23,35,36] (Table 1). Similar to potato, GA plays a negative role in taproot development as well because the swelling of the taproot in turnip or carrot is inhibited by exogenous GA [35,37,38]. The early development of taproot is associated with a decline in the content of endogenous active GA (GA_1, GA_3, GA_4) and GA metabolites (GA_{53}, GA_{12}, and so on) [35]. Liu et al. ascribed the decrease in the GA content in turnip to GA biosynthesis but not GA bioactivation in the light of the downregulation of both biosynthesis ODDs (*BrrGA20ox*, *BrrGA3ox*, and *BrrGA13ox*) and the bioactivation ODD (most *BrrGA2ox* except for *BrrGA2ox8-4*) [35]. A similar expression trend of *ODDs* was detected in carrot [36]. The function of GA in the development of yam tubers is different to that in potato. The yam tuber yield was increased by treatment with GA, and the GA homeostasis in yam after treatment with GA_3 was maintained as a result of the downregulation of *DoGA20ox1* and *DoGA3ox1* and the upregulation of *DoGA2ox3* and *DoGA2ox4* [23]. Besides the catalyzation of ODDs, there are two other GA deactivation mechanisms reported [22,25]. One rests with a cytochrome P450 monooxygenase, named EUI in rice, epoxidizing the 16,17-double bond of non13-hydroxylated GAs, and another one depends on the methylation of the C-6 carboxyl group of GAs with gibberellin methyltransferases [22,25]. So far, these genes are still not identified in stem/root tuber crops.

2.2. Gibberellin Signaling Pathway

It is acknowledged that the GA signaling pathway (Figure 2) centers on DELLA proteins, which repress GA-dependent biological processes [39,40]. As the hub of GA signaling, DELLAs behave in a manner similar to the lock for the GA responsive genes, while GA acts as the pivotal part of the assembled key. The key can trigger the degradation of "the lock" DELLA and the transcriptional activation of GA responsive genes. Another key component of the unassembled key is the soluble nuclear GA receptor gibberellin-insensitive dwarf 1 (GID1) [39,40]. GID1 has the potential to bind GA molecules with a GA-binding pocket [39]. The binding between GID1 and bioactive GAs leads to a conformational change, and thus the assembled molecule becomes capable of turning on the expression of GA-responsive genes "locked" by DELLAs through forming the GA–GID1–DELLA complex [39,40]. Subsequently, the degradation of DELLAs is triggered by the SCF (SKP1, CULLIN, F-BOX) E3 ubiquitin–ligase complexes [39,40]. In rice and Arabidopsis, the F-box protein components of SCF were identified as GID2 and SLEEPY1 (SLY1), respectively, presiding over labeling DELLA using polyubiquitin chains for subsequent degradation by the 26S proteasome [39,40]. Ultimately, the GA downstream signaling cascade is stimulated along with the decrease in DELLAs [39,40]. This describes the upstream of the DELLA-dependent GA signaling pathway. However, DELLAs can exert negative impacts on GA signaling without degradation [39]. In the absence of $SCF^{SLY1/GID2}$ activity, DELLAs also serve as transactivation factors whose transcriptional activity is inhibited by the GA–GID1 complex [39]. DELLAs are highly conserved [39]. DELLAs also have a major role in GA-induced development in stem/root tuber crops. The GA–GID1–DELLA regulatory module was identified in yam [23]. The interaction between DoGID1s and DoDELLAs in yeast was in a GA-mediated manner [23]. Moreover, the expression patterns of *DoGID1s* and *DoDELLAs* during tuber development and their responses to exogenous GA and the GA biosynthesis inhibitor paclobutrazol further substantiated the involvement of DELLA-dependent GA signaling pathways in yam tuber growth [23]. Tomato/potato heterografting (potato as rootstocks) caused a decreased potato tuber number [41]. The up-regulated expression of GA networking regulatory genes, *StDELLA* and *StGID1*, in rootstock was a hint for the conclusion that changes in DELLA-dependent GA signaling contribute to potato phenotypic modification [41]. DELLA-family proteins are versatile [39]. Their various negative regulatory functions are consistent with the roles of their target genes [39,40]. DELLAs are involved in flowering time regulation because of their interaction with *FLOWERING LOCUS T* activators, such as CONSTANS and PHYTOCHROME INTERACTING FACTOR4, and *FLOWERING LOCUS T* suppressors [17,39]. The expression of these genes was significantly regulated during both rhizome development and early flowering in lotus [42]. Cao et al. transferred *CONSTANS-LIKE 5* of lotus (*Nelumbo nucifera*) into potato and therefore achieved increased weights of potato tubers and starch content [43]. These results suggest that DELLA-mediated flowering signaling might regulate tuber growth as well. Except for the flowering regulation factors, other transcriptional factors associated with DELLAs were detected in stem/root tuber crops. In turnip, the interactions between DELLAs and NAC transcription factors, related to lignification of the secondary cell wall, resulted in GA-induced xylem lignification and thus the inhibition of taproot formation [35]. In Arabidopsis, the GA-biosynthetic enzyme GA 20-oxidase is also one of DELLAs' target genes [44]. Therefore, DELLAs are major components of GA feedback regulation through interacting with another transcription factor GAF1 [44]. The GA feedback regulation of *GA20oxs* also exists in potato [45], but the involvement of DELLAs still needs to be verified.

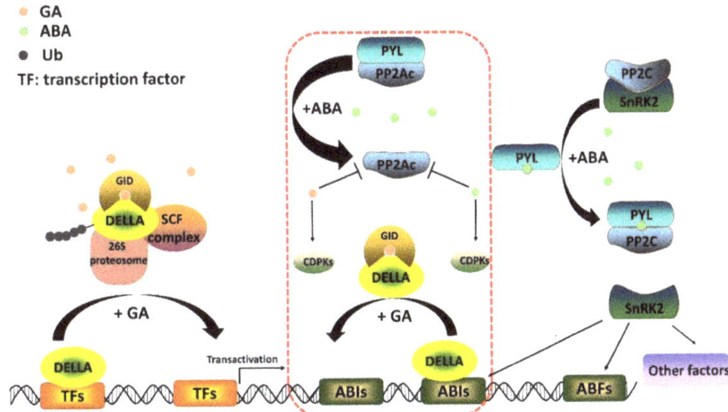

Figure 2. The potential GA and ABA signaling during stem/root tuber development. Arrows and linkers represent positive and negative effects, respectively. The components encircled by a red dotted rectangle line indicate the potential protein factors in both core GA and ABA signaling.

Except for the DELLA-dependent GA signaling pathway, some studies have indicated alternative GA signaling without hinging on DELLA [46,47]. One of the putative DELLA-independent GA signals is involved in the mediation of cytosolic Ca^{2+} ($[Ca^{2+}]_{cyt}$) [46,47]. GA application can trigger an increase in the second messenger Ca^{2+} in the cytoplasm, even for *della* pentuple mutant plants [46,48]. The DELLA-independent increase in $[Ca^{2+}]_{cyt}$ may be induced through GID1 or some other membrane-localized GA receptor [46]. Ca^{2+}-dependent protein kinases (CDPKs) are an important group of Ser/Thr protein kinases conveying signals to physiological responses. The expression of CDPKs was detected in tuberizing potato stolons and upregulated after the application of exogenous GA_3 [49,50]. In cassava, the KS was viewed as involving the induction of CDPK expression according to similar expression patterns in the storage root initiation and early developmental stages [27]. In addition to CDPKs, another calcium-dependent protein, annexin, belonging to calcium-dependent phospholipid-binding proteins, was upregulated as a consequence of treatment with exogenous GA_3 in potato [49]. The expression of the calcium-dependent proteins by the application of GA in potato suggests the GA-induced calcium-dependent signaling impacts on the inhibition of tuberization [49].

Both GA metabolism and signal transduction can affect plant development, spanning the entire plant life. The effects of GA metabolism depend on the regulation of the bioactive

GA content level, which then further regulates the expression of target genes through DELLA-dependent and DELLA-independent signaling pathways. Since the last century, ample evidence exists from stem/root tuber crop studies that gibberellins are a negative regulator of tuber formation (Table 1). The inhibitory role of GA in tuber formation correlates with facilitating cell elongation. Over-expression of GA biosynthetic or metabolic enzymes tends to result in taller mutants with delayed tuberization (*GA20ox* and *GA3ox*) or dwarf plants with earlier tuberization (*GA2ox*) [30–32,34,40]. Moreover, the stimulation of root xylem development and lignification by GA is another reason for the inhibition of tuber expansion [35,37,38,51], as normally the downregulation of lignin biosynthesis genes and continuous decreasing lignin content take place at an early stage of tuber formation [52,53]. Nevertheless, GA is essential at some stages of tuberization. The cumulative bioactive GA in yam may induce cell expansion during early tuberization [23,34]. Cell enlargement during the later developmental stages of sugar beet taproot correlates with increased gibberellin levels resulting from the preferential expression of *GA20ox* and *GA3ox* during late development stages [54–56].

3. Abscisic Acid in the Development of Stem/Root Tubers

Abscisic acid (ABA) is an important sesquiterpenoid phytohormone. It is commonly considered as a stress hormone because of its crucial role in regulating adaptive responses, such as suberin biosynthesis, stomatal movement, and osmotic modification, in response to various stresses. However, basal ABA plays a significant role during plant growth and development as well, such as in bud and tuber dormancy and seed germination [57–59]. About twofold downregulation of ABA signaling pathway genes under drought stress in potato tubers indicates the important role of ABA in normal tuber development [60]. ABA was first found as a growth-inhibiting substance from peels of resting potatoes [57]. Afterward, retarded tuberization was observed in ABA-deficient potatoes [61]. A decreased endogenous ABA level was identified during potato stolon and tuber development, which counteracted GA and assisted in tuber development [18]. Stringent regulation of endogenous ABA levels involving ABA metabolism and signaling pathways has been well characterized, but how it exerts actions on tuber development is still not clear. Hitherto, there has been a paucity of studies focusing on the function of ABA in tuberization besides some omics analysis concerning tuber development. In this part, ABA metabolism and signaling implicated in tuber development will be summarized and discussed.

3.1. ABA Biosynthesis, Catabolism, and Conjunction

ABA belongs to terpenoid metabolites and contains 15 carbon atoms [57,62], yet its biosynthesis requires the cleavage of C_{40} carotenoid (indirect pathway) but not the C_{15} sesquiterpene precursor (direct pathway) in higher plants [57,62]. The first C_{40} precursor is the oxygenated C_{40} carotenoids, called zeaxanthin [63]. The conversion from zeaxanthin to the first C_{15} precursor xanthoxin in the de novo biosynthesis pathway occurs in plastids, involving several enzymes [57,62–65] (Figure 1). Zeaxanthin epoxidase (ZEP) catalyzes the conversion from zeaxanthin to violaxanthin by introducing oxygen into zeaxanthin [63]. This reaction can be reversed by violaxanthin epoxidase [63]. Violaxanthin is then isomerized into 9′-*cis*-violaxanthin by an unknown isomerase or 9′-*cis*-neoxanthin with the catalyzation of neoxanthin synthase and some isomerase [57,62–65]. Subsequently, the key rate-limiting enzyme NCEDs (9′-*cis*-epoxycarotenoid dioxygenases) oxidatively cleaves the two *cis*-xanthophylls into the C_{15} precursor xanthoxin [57,62–65]. In sugar beet, *ZEP* was preferentially expressed during the late development of beet according to the transcript profiles [55]. Similarly, the expression of *ZEPs* and *NECDs* was elevated at the final stages of yam aerial tuber (bulbils) development as well. The expression pattern conformed to the endogenous ABA content change in bulbil tissue [24]. The genes encoding key ABA biosynthetic enzymes StZEP and StNCED have been identified in potato tubers, but only their roles in tuber dormancy have been meticulously analyzed [66,67]. Römer et al. succeeded in improving the carotenoid content level in potato tuber by inhibiting the synthesis of violaxanthin in transformants using the sense and antisense constructs encoding *ZEP* [68]. However, no growth or tuber yield difference was observed between wild-type and transgenic plants [68]. This result may be attributed to the unchanged ABA content in transformants, which suggests the possibility of ZEP isoforms and redundancy of ABA biosynthesis pathways. Except for potato, the *ZEP* gene was also cloned and characterized in *Pseudostellaria heterophylla* [69]. It was upregulated along with the development of tuberous roots after flowering and responded to fluridone (the inhibitor of ABA biosynthesis) [69]. ABA biosynthesis moves to the cytosol accompanied by the transportation of the C_{15} precursor xanthoxin. In cytosol, xanthoxin is converted into abscisic aldehyde by xanthoxin dehydrogenase and finally oxidized to ABA by aldehyde oxidase (AAO) (Figure 1). In *R. glutinosa*, the increased endogenous ABA level during tuberous root development was correlated with the expression of ABA biosynthesis genes encoding ZEP, NCED, and AAO [33]. The similar overall enhanced expression of ABA biosynthesis genes was also detected in the development of carrot roots [36]. However, in sweet potato the expression of the *AAO* gene was enhanced at the early stage of storage root development but then sharply decreased until the final stage of storage root development [70].

The modulation of active ABA homoeostasis is reminiscent of GA, requiring not only biosynthesis but catabolism and conjunction as well [57,62–65] (Figure 1). Irreversible catabolism is the hydroxylation of ABA at three different methyl groups: C-7′, C-8′ (primary catabolic route), and C-9′ [57,63]. Among the three hydroxylation products, 8′- and 9′-hydroxy ABA, generated by the catalyzation of the CYP707A subfamily of P450 monooxygenases, are then isomerized spontaneously into phaseic acid and neophaseic acid respectively [63]. Phaseic acid with partial biological activity is finally converted into the inactive dihydrophaseic acid by phaseic acid reductase [63]. In line with the expression of the ABA biosynthesis genes in *R. glutinosa* and carrot, the ABA catabolism genes encoding ABA 8′-hydroxylase were concomitantly downregulated [33,36]. The expression patterns carried a potential for triggering the elevation of endogenous ABA levels during storage organ development [33,36]. The expression of *ABA 8-hydroxylase* in *S. trifolia* was also decreased during corm formation, but the downregulation only took place at the initial swelling stage but not during later stages of stolon development [28]. The modulated expression of ABA biosynthesis and catabolism genes in various species indicates the potential function of ABA in stem/root tuber development. Another pathway involved in the dynamic equilibrium of ABA is the reversible conjugation of ABA with glucose. This pathway generates inactive ABA glucosyl esters by the catalyzing addition of glucose moieties via uridine diphosphate glucosyl transferases [62,63]. The product can sequester ABA and thus serves as a transport form or as a storage reservoir [62,63]. Although the intermediate products, dihydrophaseic acid, phaseic acid, and ABA glucose ester, have been detected in root/stem tubers [66,71], genes involving ABA catabolism and conjugation, except for *ABA 8-hydroxylase*, have not been identified in the development of stem/root tubers yet.

3.2. Core ABA Signaling Pathway

The core components of the ABA signaling pathway encompass the intracellular receptor, pyrabactin resistance-like proteins (PYLs), the co-receptor, clade A protein phosphatases of type 2Cs (PP2Cs), and sucrose non-fermenting-1 (SNF1)-related protein kinase 2s (SnRK2s) [57,58,63–65] (Figure 2). The soluble receptor PYLs are localized in the cytosol and nucleus [64]. The direct binding of PYLs to ABA elicits conformational changes in PYLs, which gives rise to the interface in PLYs for interacting with PP2Cs [64]. The expression of group A PP2Cs can be intensively induced by the application of exogenous ABA [72]. The increased endogenous ABA level and its concomitant upregulation of PP2Cs were observed in the early microtuber formation of *Dioscorea opposite* [73]. The accumulation of transcripts of PLYs and PP2Cs was also detected during radish taproot thickening and *R. glutinosa* tuberous root formation [74,75]. The interaction between PYLs and PP2Cs releases SnRK2s due to the occupation of the enzymatic active site of PP2Cs, and then SnRK2s are activated by autophosphorylation [57,58,64]. SnRK2s are the positive regulators in ABA signaling. Many downstream components, including membrane channels, transcription factors, and transporters, are controlled by SnRK2s [57]. When SnRK2s are released from PP2Cs, the activity of ABA-responsive element binding proteins/factors (ABEBs/ABFs) and ABA INSENSITIVE 5 (ABI5), belonging to the basic leucine zipper domain (bZIP) transcription factors, is stimulated and thereby the ABA-responsive gene transcription is actuated [76,77]. The expression of yam *ABF3* at the final stage of bulbil growth was three times higher than during early stages, suggesting enhanced ABA signaling at later stages of yam bulbil growth [24]. The constitutive expression of *AtABF4* in potato not only led to the increased number of tubers per plant and improved tuber yield under non-stress conditions but also induced enhanced salt and drought tolerance [78]. However, this result could not be repeated using another bZIP transcription factor of hot pepper (*Capsicum annuum*) [79]. The heterologous expression of *CaBZ1* in potato only improved the tuber drought tolerance, while no phenotypic change was observed under non-stress conditions [79]. The different results suggest that not all bZIP transcription factors involving ABA signaling are associated with tuber development. Transcriptomic analyses of the rhizome formation in lotus and taproot thickening in *Panax notoginseng* and carrot show that numerous ABA core signaling component genes are regulated differently during tuber development [36,42,80] (Table 1). Nonetheless, the genes related to ABA signaling in some species exhibited a consistent expression trend, i.e., significant upregulation in the early and late root development stages in *P. notoginseng* and downregulation in carrot roots after 40 days of growing [36,80]. The different expression patterns in different species present the different ABA signaling pathways in modulating tuber development in various species.

It is obvious from the forementioned studies that the ABA metabolism and signaling pathway participates in stem/root tuber formation. ABA plays a positive role in the process of tuber development. In sweet potato, much more endogenous ABA was determined in the storage roots compared to that in the non-storage roots and in the heavier tuberous roots of cultivars compared to that in the lighter tuberous roots of wild species [70,81]. Tuberous root development displayed a positive correlation with ABA contents. The facilitated yam microtuber formation by the application of ABA and the significantly arrested microtuber development by ABA inhibitors further confirm the positive function of ABA [73]. In addition, ABA content increases along with the development of yam bulbs [24]. These results further suggest that ABA plays a major role in the maturation and storability of bulbs or tubers. The positive influence of ABA on tuber development may be associated with its role in sugar metabolism [33], as ABA implies a potential for stimulating transport of sugars derived from source tissues to sink tissues, for instance, the storage stem/root tubers [82].

4. Conclusions: The Antagonistic Role of GA and ABA in the Development of Stem/Root Tubers

GA and ABA play antagonistic roles during tuber formation. GA represses tuber swelling, while ABA stimulates tuber formation (Figure 3). Furthermore, the death of tubers in an ABA-deficient droopy potato mutant was stopped by the application of the GA inhibitor tetcyclacis [18,61]. The observation demonstrates that ABA exerts positive effects on tuberization by counteracting the negative effect of GA [18,61]. The profiles of endogenous ABA and GA levels during the process of tuberization in various species are different (Table 1). In potato, the endogenous GA_1 reached a peak during stolon elongation when the ABA level decreased and then GA_1 decreased during tuber swelling when a little more ABA was accumulated [18]. The ABA level in *R. glutinosa* exhibited a continuous increase during tuberous root development, yet GA was reduced after 45 days of growing [33]. However, in sweet potato, continuous increase was detected for GA, and a strikingly increased ABA level was observed at the early stage, which then decreased along with storage root development, with a minor rise in the late development stage [70]. The diverse species-specific profiles of ABA and GA content during tuber formation imply the complexity and diversity of the mechanisms underlying stem/root tuber development.

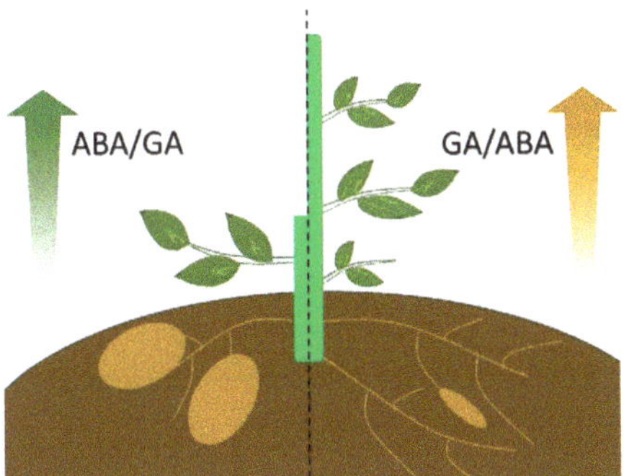

Figure 3. A model showing the antagonistic role of GA and ABA during tuber development. A higher ABA/GA ratio leads to the tuberization, while a higher GA/ABA ratio results in longer stems and delayed tuberization.

The antagonistic regulatory roles of GA and ABA are implicated in various stages of plant development, i.e., during tuber formation and also during seed maturation and dormancy, root growth, and flowering [19]. Numerous factors are relevant to the ABA and GA antagonism function in diverse biological processes. In the model plant *Arabidopsis thaliana*, during post-embryonic root development, a C2H2-type zinc finger, GA-AND ABA-RESPONSIVE ZINC FINGER, is regulated by both GA and ABA, which in turn intervenes in the transcriptional regulation of ABA and GA homeostasis [83]. A C3H-type zinc finger SOMNUS, associated with seed germination, is controlled by the two phytohormones as well, whose promoter is targeted by both GA and ABA signaling components, DELLAs and ABIs [84]. Liu et al. summarized the antagonistic interaction between core GA- and ABA-signaling components [85]. However, the specific factors involving the antagonism between ABA and GA are still not identified in the regulation of tuber formation, although many important common factors in the phytohormone signaling are associated with the process (Figure 2). In potato, *StMYBs* and *StCDPK1* are induced by both GA and ABA and their expression is regulated during potato tuber development [50,86]. The protein phosphatase type 2A (PP2A) catalytic subunits downregulated by GA can positively affect tuber formation in potato [87,88]. Meanwhile, the new ABA PYLs-PP2A signaling modulating root architecture was identified recently, in which ABA binds to PYLs and thus inhibits PP2A activity [89]. The signaling components collectively responsive to both GA and ABA offer a possible entry point into the antagonistic regulation of ABA and GA in tuber formation.

Despite the important antagonistic roles of ABA and GA in the development of stem/root tubers, tubers cannot be formed without the contributions of multiple phytohormones. Auxin is an essential hormone in all the developmental events involving the formation and maintenance of meristem [90]. Before tuber initiation, the auxin content remains relatively low [90]. However, along with stolon elongation, the auxin content increases, while GA goes down. When tubers start swelling, the content level of auxin keeps increasing and then slowly decreases after reaching a peak [90]. Cytokinin is another important phytohormone during plant development. Cytokinin and auxin play either a synergistical or an antagonistic role in meristem biology [91]. Although cytokinin is upregulated during tuberous root development, the auxin/cytokinin ratio may be the key factor for tuberization [92]. Recently, the positive role of another phytohormone, jasmonic acid, was further identified in potato with the investigation of a suppressor of JA signaling [93]. JA is increased in stolon in the initial stage of tuber formation, during which, it exerts a negative role as the inhibitor of GA synthesis [94]. In addition, there is a spectrum of other phytohormones involved in tuber formation. Hormone profiling during tuberization has been performed in various tuber crops, such as potato [94–96], cassava [92], yam [24,97], *R. glutinosa* [33], and sweet potato [70].

Table 1. The GA-and ABA-related gene expression patterns and GA and ABA profiles during stem/root tuber development in different tuber and root crops described in the article.

Species	Storage Organ	GA-and ABA-Related Gene Expression	GA and ABA Profile	Ref.
Potato (*Solanum tuberosum*)	Tuber (developing from the stolon)	*GA20ox1*: Expressed mainly in leaves. *GA2ox1*: Upregulated in the subapical zone of the stolon and growing tuber before tuber expansion. *CDPK*: Mainly located in the plasma membrane of swelling stolons and sprouting tubers. *GA3ox2*: Increased in the aerial parts and repressed in the stolons under the short-day condition.	GA_3: The content in the tuber dropped by around 30%.	[18,29–34,41–45,49,50,86,98]
			ABA: Around 30% increase detected in the tuber.	[18,41,61,78,86–88]
Chinese arrowhead *Sagittaria trifolia*	Corm (developing from the tips of the stolon)	*KS* and *KO*: Enhanced during stolon elongation and decreased when the corm swells. *CDPK*: Upregulated.		[28]
Lotus (*Nelumbo nucifera*)	Rhizome (developing from the stem)	*ABA 8-hydroxylase*: Decreased at the initial swelling stage. *GAI*: Downregulated during rhizome development.		[28]
		ABFs, *PP2C*, *PYL*, and *SnRK2*: Different expression patterns, with some upregulated, some downregulated, and two *PP2Cs* expressed high in the middle stage.		[42]
Yams (*Dioscorea opposita*)	Underground tuber (developing from the hypocotyl); and aerial tubers (bulbils)	*GA20ox1*, *GA3ox1*, and four *GA2oxs*: Significantly abundant in the early expansion stage and gradually declined along with tuber growth. Three *GIDs* and three *DELLAs*: Different expression patterns in the early expansion stage and gradually declined along with tuber growth.	GA_3 and GA_4: Reached a peak of around 150 ng/g at 90 days after field planting and then decreased.	[42]
				[23,97]
		NCED, *ZEP*, and *ABF*: Increased at the final stage. *PP2Cs*: Upregulated in the early microtuber formation stage.	ABA (during tuber development): Reached a maximum of over 600 ng/g at 90 days after field planting and then decreased. ABA (in bulbil tissues): Continuously increased by around 22-fold.	[24,73,97]
Turnip (*Brassica rapa* var. *rapa*)	Taproot (developing from the hypocotyl and a part of the root)	*GA20ox*, *GA3ox*, and *GA13ox*: Increased in the early growth stage but declined during the late developmental stage. *KS*: Decreased.	GA: Most abundant active GA is GA3, and all the active GAs decreased by more than 50%.	[35]
Sugar beet (*Beta vulgaris* L.)	Taproot	*GA2oxs* (except for *GA2ox8-4*): Downregulated. *GA20-ox* and *GA3-ox*: Preferentially expressed during late development. *GA2-ox*: Preferentially expressed during early development. *ZEP*: Preferentially expressed during late development.	GA/ABA ratio: Decreased from 1358.5 to 18.8.	[54–56]
		GA20oxs and *GA3oxs*: Different expression patterns, with some upregulated in the middle stage and then downregulated and some downregulated from the early development stage.		[55,56]
Carrot (*Daucus carota* L.)	Taproot	*GA2ox*, *KO*, and *KS*: Upregulated and then downregulated. *KAO*: Downregulated and then upregulated. *NCED* and *AAO*: Upregulated in the final growth stage. *ABAH*, *PYL* and *SnRK2*: Decreased.		[36–38]
		PP2c: Upregulated in the early stage and then downregulated.		[36]
Radish (*Raphanus sativus* L.)	Taproot	*PLYs* and *PP2Cs*: Upregulated.		[74]
San qi (*Panax notoginseng*)	Taproot	*GAI* and *GID*: Downregulated and then upregulated. *PYL*: Downregulated and then upregulated. *PP2Cs*: Different expression patterns, with some downregulated and then upregulated and some upregulated at the final stage. Most of *SnRK2*: Upregulated.		[80]
				[80]
Sweet potato (*Ipomoea batatas*)	Tuberous root (developing from the adventitious root)	*GA3ox*: Downregulated.	GAs: Around 2.5-fold decrease in the storage root.	[51,52,70]
Cassava (*Manihot esculenta*)	Tuberous root	*AAO*: Enhanced in the early stage and then sharply decreased until the final stage.	ABA: Increased to 6.5 nmol/g in the early stage and finally decreased to around 3 nmol/g.	[70,81]
		KS: Mainly detected in the cortex and parenchyma of fibrous root and significantly downregulated. *CDPK1*: Upregulated during the initial stage and gradually downregulated.	ABA: Decreased in fibrous roots compared with the pretuberous roots.	[27,92]
Chinese foxglove *Rehmannia glutinosa*	Tuberous root	*GA20ox* and *GA 3-beta-dioxygenase*: Downregulated. Most of GA 2-beta-dioxygenase: Upregulated and then downregulated. *ZEP*, *AAO*, *PYL*, and most of the *NCEDs*: Upregulated. Most of *ABA 8'-hydroxylase 3*: Downregulated.	GA: Decreased by approximately 30%.	[33,75]
Tai zi shen *Pseudostellaria heterophylla*	Tuberous root	*ZEP*: Upregulated.	ABA: Increased 2-fold.	[33,75]
				[69]

Root and tuber crops are a great source of nutrition for humans. Improvement in the nutritional or mechanical properties of tuber crops requires a clear understanding of tuber organogenesis. The regulation of tuber development is influenced by multiple environmental cues depending on the regulation of hormone signaling crosstalk. The antagonism between ABA and GA can be considered as a part of the hormone signaling crosstalk. Therefore, the antagonistic role of ABA and GA during tuber development needs to be further dissected. As the model plant *A. thaliana* does not have stem/root tubers, potato becomes an important material for investigating the mechanism of tuber organogenesis. However, tuber crops include a multitude of species that may achieve tuber organogenesis through different mechanisms. Thereby, studies on the tuber organogenesis of tuber crops will be a big challenge.

Author Contributions: P.C. and R.Y. wrote the manuscript together. P.C. and T.D. prepared the table and figures. D.B. and H.D. corrected the manuscript and supervised the work. All authors have read and agreed to the published version of the manuscript.

Funding: This work was supported by the National Science Foundation of China (No. 31870312) and Talent Program of Henan Normal University, China (No. qd20035).

Institutional Review Board Statement: Not applicable.

Informed Consent Statement: Not applicable.

Data Availability Statement: Not applicable.

Conflicts of Interest: The authors declare that there is no conflict of interest.

Abbreviations

AAO	Aldehyde oxidase
ABA	Abscisic acid
ABEBs/ABFs	ABA-responsive element binding proteins/factors
ABI5	ABA INSENSITIVE 5
bZIP	Basic leucine zipper domain
CDPKs	Ca^{2+}-dependent protein kinases
GA	Gibberellin
GGPP	*trans*-Geranylgeranyl diphosphate
GID1	Gibberellin-insensitive dwarf 1
KAO	*ent*-Kaurenoic acid oxidase
KO	*ent*-Kaurene oxidase
KS	*ent*-Kaurene synthase
NCEDs	$9'$-cis-Epoxycarotenoid dioxygenases
ODD	Oxoglutarate-dependent dioxygenase
PP2C	Clade A protein phosphatases of type 2C
PYL	Pyrabactin resistance-like protein
SLY1	F-box proteins SLEEPY1
SnRK2	Sucrose non-fermenting-1 (SNF1)-related protein kinase 2
ZEP	Zeaxanthin epoxidase

References

1. Zierer, W.; Ruscher, D.; Sonnewald, U.; Sonnewald, S. Tuber and Tuberous Root Development. *Annu. Rev. Plant Biol.* **2021**, *72*, 551–580. [CrossRef] [PubMed]
2. Chen, P.; Wei, X.; Qi, Q.; Jia, W.; Zhao, M.; Wang, H.; Zhou, Y.; Duan, H. Study of Terpenoid Synthesis and Prenyltransferase in Roots of Rehmannia glutinosa Based on iTRAQ Quantitative Proteomics. *Front. Plant Sci.* **2021**, *12*, 693758. [CrossRef] [PubMed]
3. Scott, G.J. A review of root, tuber and banana crops in developing countries: Past, present and future. *Int. J. Food Sci. Technol.* **2021**, *56*, 1093–1114. [CrossRef] [PubMed]
4. Liu, Q.; Liu, J.; Zhang, P.; He, S. Root and Tuber Crops. In *Encyclopedia of Agriculture and Food Systems*; Academic Press: Cambridge, MA, USA, 2014; pp. 46–61.
5. Villordon, A.Q.; Ginzberg, I.; Firon, N. Root architecture and root and tuber crop productivity. *Trends Plant Sci.* **2014**, *19*, 419–425. [CrossRef]

6. Padhan, B.; Panda, D. Potential of Neglected and Underutilized Yams (Dioscorea spp.) for Improving Nutritional Security and Health Benefits. *Front. Pharmacol.* **2020**, *11*, 496. [CrossRef] [PubMed]

7. Zhang, Y.; Yang, G.; Wang, X.; Ni, G.; Cui, Z.; Yan, Z. Sagittaria trifolia tuber: Bioconstituents, processing, products, and health benefits. *J. Sci. Food Agric.* **2021**, *101*, 3085–3098. [CrossRef]

8. Caetano, B.F.; de Moura, N.A.; Almeida, A.P.; Dias, M.C.; Sivieri, K.; Barbisan, L.F. Yacon (Smallanthus sonchifolius) as a Food Supplement: Health-Promoting Benefits of Fructooligosaccharides. *Nutrients* **2016**, *8*, 436. [CrossRef]

9. Yu, B.; Chen, M.; Grin, I.; Ma, C. Mechanisms of Sugar Beet Response to Biotic and Abiotic Stresses. *Adv. Exp. Med. Biol.* **2020**, *1241*, 167–194. [CrossRef]

10. Gregory, P.J.; Wojciechowski, T. Root systems of major tropical root and tuber crops: Root architecture, size, and growth and initiation of storage organs. *Adv. Agron.* **2020**, *161*, 1–25.

11. Kondhare, K.R.; Natarajan, B.; Banerjee, A.K. Molecular signals that govern tuber development in potato. *Int. J. Dev. Biol.* **2020**, *64*, 133–140. [CrossRef]

12. Epping, J.; Laibach, N. An underutilized orphan tuber crop-Chinese yam: A review. *Planta* **2020**, *252*, 58. [CrossRef] [PubMed]

13. Zhou, Y.; Luo, S.; Hameed, S.; Xiao, D.; Zhan, J.; Wang, A.; He, L. Integrated mRNA and miRNA transcriptome analysis reveals a regulatory network for tuber expansion in Chinese yam (Dioscorea opposita). *BMC Genom.* **2020**, *21*, 117. [CrossRef] [PubMed]

14. Ewing, E.E.; Struik, P.C. Tuber formation in the potato: Induction, initiation and growth. In *Horticultural Reviews*; Janick, J., Ed.; Wiley: New York, NY, USA, 1992; Volume 14, pp. 89–198.

15. Liu, L.; Huang, Y.; Huang, X.; Yang, J.; Wu, W.; Xu, Y.; Cong, Z.; Xie, J.; Xia, W.; Huang, D. Characterization of the Dioscorin Gene Family in Dioscorea alata Reveals a Role in Tuber Development and Environmental Response. *Int. J. Mol. Sci.* **2017**, *18*, 1579. [CrossRef] [PubMed]

16. Roumeliotis, E.; Kloosterman, B.; Oortwijn, M.; Kohlen, W.; Bouwmeester, H.J.; Visser, R.G.; Bachem, C.W. The effects of auxin and strigolactones on tuber initiation and stolon architecture in potato. *J. Exp. Bot.* **2012**, *63*, 4539–4547. [CrossRef] [PubMed]

17. Bao, S.; Hua, C.; Shen, L.; Yu, H. New insights into gibberellin signaling in regulating flowering in Arabidopsis. *J. Integr. Plant Biol.* **2020**, *62*, 118–131. [CrossRef] [PubMed]

18. Xu, X.; van Lammeren, A.A.; Vermeer, E.; Vreugdenhil, D. The role of gibberellin, abscisic acid, and sucrose in the regulation of potato tuber formation in vitro. *Plant Physiol.* **1998**, *117*, 575–584. [CrossRef] [PubMed]

19. Shu, K.; Zhou, W.; Chen, F.; Luo, X.; Yang, W. Abscisic Acid and Gibberellins Antagonistically Mediate Plant Development and Abiotic Stress Responses. *Front. Plant Sci.* **2018**, *9*, 416. [CrossRef]

20. Salazar-Cerezo, S.; Martinez-Montiel, N.; Garcia-Sanchez, J.; Perez, Y.T.R.; Martinez-Contreras, R.D. Gibberellin biosynthesis and metabolism: A convergent route for plants, fungi and bacteria. *Microbiol. Res.* **2018**, *208*, 85–98. [CrossRef]

21. Binenbaum, J.; Weinstain, R.; Shani, E. Gibberellin Localization and Transport in Plants. *Trends Plant Sci.* **2018**, *23*, 410–421. [CrossRef]

22. Hedden, P.; Thomas, S.G. Gibberellin biosynthesis and its regulation. *Biochem. J.* **2012**, *444*, 11–25. [CrossRef]

23. Zhou, Y.; Li, Y.; Gong, M.; Qin, F.; Xiao, D.; Zhan, J.; Wang, A.; He, L. Regulatory mechanism of GA3 on tuber growth by DELLA-dependent pathway in yam (Dioscorea opposita). *Plant Mol. Biol.* **2021**, *106*, 433–448. [CrossRef] [PubMed]

24. Wu, Z.G.; Jiang, W.; Tao, Z.M.; Pan, X.J.; Yu, W.H.; Huang, H.L. Morphological and stage-specific transcriptome analyses reveal distinct regulatory programs underlying yam (Dioscorea alata L.) bulbil growth. *J. Exp. Bot.* **2020**, *71*, 1899–1914. [CrossRef] [PubMed]

25. Yamaguchi, S. Gibberellin metabolism and its regulation. *Annu. Rev. Plant Biol.* **2008**, *59*, 225–251. [CrossRef] [PubMed]

26. Regnault, T.; Davière, J.M.; Wild, M.; Sakvarelidze-Achard, L.; Heintz, D.; Carrera Bergua, E.; Lopez Diaz, I.; Gong, F.; Hedden, P.; Achard, P. The gibberellin precursor GA12 acts as a long-distance growth signal in Arabidopsis. *Nat. Plants* **2015**, *1*, 15073. [CrossRef] [PubMed]

27. Sojikul, P.; Kongsawadworakul, P.; Viboonjun, U.; Thaiprasit, J.; Intawong, B.; Narangajavana, J.; Svasti, M.R. AFLP-based transcript profiling for cassava genome-wide expression analysis in the onset of storage root formation. *Physiol. Plant* **2010**, *140*, 189–198. [CrossRef]

28. Cheng, L.; Li, S.; Xu, X.; Hussain, J.; Yin, J.; Zhang, Y.; Li, L.; Chen, X. Identification of differentially expressed genes relevant to corm formation in Sagittaria trifolia. *PLoS ONE* **2013**, *8*, e54573. [CrossRef]

29. Jackson, S.D.; James, P.E.; Carrera, E.; Prat, S.; Thomas, B. Regulation of transcript levels of a potato gibberellin 20-oxidase gene by light and phytochrome B. *Plant Physiol.* **2000**, *124*, 423–430. [CrossRef]

30. Carrera, E.; Bou, J.; García-Martínez, J.L.; Prat, S. Changes in GA 20-oxidase gene expression strongly affect stem length, tuber induction and tuber yield of potato plants. *Plant J.* **2000**, *22*, 247–256. [CrossRef]

31. Roumeliotis, E.; Kloosterman, B.; Oortwijn, M.; Lange, T.; Visser, R.G.; Bachem, C.W. Down regulation of StGA3ox genes in potato results in altered GA content and affect plant and tuber growth characteristics. *J. Plant Physiol.* **2013**, *170*, 1228–1234. [CrossRef]

32. Bou-Torrent, J.; Martinez-Garcia, J.F.; Garcia-Martinez, J.L.; Prat, S. Gibberellin A1 metabolism contributes to the control of photoperiod-mediated tuberization in potato. *PLoS ONE* **2011**, *6*, e24458. [CrossRef]

33. Sun, P.; Xiao, X.; Duan, L.; Guo, Y.; Qi, J.; Liao, D.; Zhao, C.; Liu, Y.; Zhou, L.; Li, X. Dynamic transcriptional profiling provides insights into tuberous root development in Rehmannia glutinosa. *Front. Plant Sci.* **2015**, *6*, 396. [CrossRef] [PubMed]

34. Kloosterman, B.; Navarro, C.; Bijsterbosch, G.; Lange, T.; Prat, S.; Visser, R.G.; Bachem, C.W. StGA2ox1 is induced prior to stolon swelling and controls GA levels during potato tuber development. *Plant J.* **2007**, *52*, 362–373. [CrossRef] [PubMed]

35. Liu, Y.; Wen, J.; Ke, X.; Zhang, J.; Sun, X.; Wang, C.; Yang, Y. Gibberellin inhibition of taproot formation by modulation of DELLA-NAC complex activity in turnip (Brassica rapa var. rapa). *Protoplasma* **2021**, *258*, 925–934. [CrossRef] [PubMed]

36. Wang, G.L.; Jia, X.L.; Xu, Z.S.; Wang, F.; Xiong, A.S. Sequencing, assembly, annotation, and gene expression: Novel insights into the hormonal control of carrot root development revealed by a high-throughput transcriptome. *Mol. Genet. Genom.* **2015**, *290*, 1379–1391. [CrossRef]

37. Wang, G.L.; Que, F.; Xu, Z.S.; Wang, F.; Xiong, A.S. Exogenous gibberellin altered morphology, anatomic and transcriptional regulatory networks of hormones in carrot root and shoot. *BMC Plant Biol.* **2015**, *15*, 290. [CrossRef]

38. Wang, G.L.; Que, F.; Xu, Z.S.; Wang, F.; Xiong, A.S. Exogenous gibberellin enhances secondary xylem development and lignification in carrot taproot. *Protoplasma* **2017**, *254*, 839–848. [CrossRef]

39. Daviere, J.M.; Achard, P. Gibberellin signaling in plants. *Development* **2013**, *140*, 1147–1151. [CrossRef]

40. Daviere, J.M.; Achard, P. A Pivotal Role of DELLAs in Regulating Multiple Hormone Signals. *Mol. Plant* **2016**, *9*, 10–20. [CrossRef]

41. Zhang, G.; Mao, Z.; Wang, Q.; Song, J.; Nie, X.; Wang, T.; Zhang, H.; Guo, H. Comprehensive transcriptome profiling and phenotyping of rootstock and scion in a tomato/potato heterografting system. *Physiol. Plant* **2019**, *166*, 833–847. [CrossRef]

42. Yang, M.; Zhu, L.; Pan, C.; Xu, L.; Liu, Y.; Ke, W.; Yang, P. Transcriptomic Analysis of the Regulation of Rhizome Formation in Temperate and Tropical Lotus (Nelumbo nucifera). *Sci. Rep.* **2015**, *5*, 13059. [CrossRef]

43. Cao, D.; Lin, Z.; Huang, L.; Damaris, R.N.; Li, M.; Yang, P. A CONSTANS-LIKE gene of Nelumbo nucifera could promote potato tuberization. *Planta* **2021**, *253*, 65. [CrossRef] [PubMed]

44. Fukazawa, J.; Mori, M.; Watanabe, S.; Miyamoto, C.; Ito, T.; Takahashi, Y. DELLA-GAF1 Complex Is a Main Component in Gibberellin Feedback Regulation of GA20 Oxidase 2. *Plant Physiol.* **2017**, *175*, 1395–1406. [CrossRef]

45. Carrera, E.; Jackson, S.D.; Prat, S. Feedback control and diurnal regulation of gibberellin 20-oxidase transcript levels in potato. *Plant Physiol.* **1999**, *119*, 765–774. [CrossRef] [PubMed]

46. Ito, T.; Okada, K.; Fukazawa, J.; Takahashi, Y. DELLA-dependent and -independent gibberellin signaling. *Plant Signal. Behav.* **2018**, *13*, e1445933. [CrossRef] [PubMed]

47. Wu, K.; Xu, H.; Gao, X.; Fu, X. New insights into gibberellin signaling in regulating plant growth-metabolic coordination. *Curr. Opin. Plant Biol.* **2021**, *63*, 102074. [CrossRef]

48. Gilroy, S.; Jones, R.L. Gibberellic acid and abscisic acid coordinately regulate cytoplasmic calcium and secretory activity in barley aleurone protoplasts. *Proc. Natl. Acad. Sci. USA* **1992**, *89*, 3591–3595. [CrossRef]

49. Cheng, L.; Wang, Y.; Liu, Y.; Zhang, Q.; Gao, H.; Zhang, F. Comparative proteomics illustrates the molecular mechanism of potato (Solanum tuberosum L.) tuberization inhibited by exogenous gibberellins in vitro. *Physiol. Plant* **2018**, *163*, 103–123. [CrossRef]

50. Gargantini, P.R.; Giammaria, V.; Grandellis, C.; Feingold, S.E.; Maldonado, S.; Ulloa, R.M. Genomic and functional characterization of StCDPK1. *Plant Mol. Biol.* **2009**, *70*, 153–172. [CrossRef]

51. Singh, V.; Sergeeva, L.; Ligterink, W.; Aloni, R.; Zemach, H.; Doron-Faigenboim, A.; Yang, J.; Zhang, P.; Shabtai, S.; Firon, N. Gibberellin Promotes Sweetpotato Root Vascular Lignification and Reduces Storage-Root Formation. *Front. Plant Sci.* **2019**, *10*, 1320. [CrossRef]

52. Firon, N.; LaBonte, D.; Villordon, A.; Kfir, Y.; Solis, J.; Lapis, E.; Perlman, T.S.; Doron-Faigenboim, A.; Hetzroni, A.; Althan, L.; et al. Transcriptional profiling of sweetpotato (Ipomoea batatas) roots indicates down-regulation of lignin biosynthesis and up-regulation of starch biosynthesis at an early stage of storage root formation. *BMC Genom.* **2013**, *14*, 460. [CrossRef]

53. Wang, G.L.; Huang, Y.; Zhang, X.Y.; Xu, Z.S.; Wang, F.; Xiong, A.S. Transcriptome-based identification of genes revealed differential expression profiles and lignin accumulation during root development in cultivated and wild carrots. *Plant Cell Rep.* **2016**, *35*, 1743–1755. [CrossRef] [PubMed]

54. Zhang, Y.F.; Li, G.L.; Wang, X.F.; Sun, Y.Q.; Zhang, S.Y. Transcriptomic profiling of taproot growth and sucrose accumulation in sugar beet (Beta vulgaris L.) at different developmental stages. *PLoS ONE* **2017**, *12*, e0175454. [CrossRef] [PubMed]

55. Bellin, D.; Schulz, B.; Soerensen, T.R.; Salamini, F.; Schneider, K. Transcript profiles at different growth stages and tap-root zones identify correlated developmental and metabolic pathways of sugar beet. *J. Exp. Bot.* **2007**, *58*, 699–715. [CrossRef] [PubMed]

56. Ozolina, N.V.; Pradedova, E.V.; Saliaev, R.K. The dynamics of hormonal status of developing red beet root (*Beta vulgaris* L.) in correlation with the dynamics of sugar accumulation. *Biol. Bull.* **2005**, *32*, 22–26. [CrossRef]

57. Chen, K.; Li, G.J.; Bressan, R.A.; Song, C.P.; Zhu, J.K.; Zhao, Y. Abscisic acid dynamics, signaling, and functions in plants. *J. Integr. Plant Biol.* **2020**, *62*, 25–54. [CrossRef]

58. Yoshida, T.; Christmann, A.; Yamaguchi-Shinozaki, K.; Grill, E.; Fernie, A.R. Revisiting the Basal Role of ABA-Roles Outside of Stress. *Trends Plant Sci.* **2019**, *24*, 625–635. [CrossRef]

59. Simko, I.; McMurry, S.; Yang, H.M.; Manschot, A.; Davies, P.J.; Ewing, E.E. Evidence from Polygene Mapping for a Causal Relationship between Potato Tuber Dormancy and Abscisic Acid Content. *Plant Physiol.* **1997**, *115*, 1453–1459. [CrossRef]

60. Da Ros, L.; Elferjani, R.; Soolanayakanahally, R.; Kagale, S.; Pahari, S.; Kulkarni, M.; Wahab, J.; Bizimungu, B. Drought-Induced Regulatory Cascades and Their Effects on the Nutritional Quality of Developing Potato Tubers. *Genes* **2020**, *11*, 864. [CrossRef]

61. Vreugdenhil, D.; Bindels, P.; Reinhoud, P.; Hendriks, K.T. Use of the growth retardant tetcyclacis for potato tuber formation in vitro. *Plant Growth Regul.* **1994**, *14*, 257–265. [CrossRef]

62. Nambara, E.; Marion-Poll, A. Abscisic acid biosynthesis and catabolism. *Annu. Rev. Plant Biol.* **2005**, *56*, 165–185. [CrossRef]

63. Sano, N.; Marion-Poll, A. ABA Metabolism and Homeostasis in Seed Dormancy and Germination. *Int. J. Mol. Sci.* **2021**, *22*, 5069. [CrossRef] [PubMed]

64. Dong, T.; Park, Y.; Hwang, I. Abscisic acid: Biosynthesis, inactivation, homoeostasis and signalling. *Essays Biochem.* **2015**, *58*, 29–48. [CrossRef] [PubMed]

65. Ma, Y.; Cao, J.; He, J.; Chen, Q.; Li, X.; Yang, Y. Molecular Mechanism for the Regulation of ABA Homeostasis During Plant Development and Stress Responses. *Int. J. Mol. Sci.* **2018**, *19*, 3643. [CrossRef] [PubMed]

66. Destefano-Beltran, L.; Knauber, D.; Huckle, L.; Suttle, J.C. Effects of postharvest storage and dormancy status on ABA content, metabolism, and expression of genes involved in ABA biosynthesis and metabolism in potato tuber tissues. *Plant Mol. Biol.* **2006**, *61*, 687–697. [CrossRef] [PubMed]

67. Destefano-Beltran, L.; Knauber, D.; Huckle, L.; Suttle, J. Chemically forced dormancy termination mimics natural dormancy progression in potato tuber meristems by reducing ABA content and modifying expression of genes involved in regulating ABA synthesis and metabolism. *J. Exp. Bot.* **2006**, *57*, 2879–2886. [CrossRef]

68. Römer, S.; Lübeck, J.; Kauder, F.; Steiger, S.; Adomat, C.; Sandmann, G. Genetic engineering of a zeaxanthin-rich potato by antisense inactivation and co-suppression of carotenoid epoxidation. *Metab. Eng.* **2002**, *4*, 263–272. [CrossRef]

69. Zheng, W.; Zhou, T.; Li, J.; Long, D.K.; Jiang, W.K.; Ding, L. Cloning and expression analysis on full length CDS of zeaxanthin epoxidase gene in Pseudostellaria heterophylla. *Zhongguo Zhong Yao Za Zhi* **2017**, *42*, 669–674. [CrossRef]

70. Dong, T.; Zhu, M.; Yu, J.; Han, R.; Tang, C.; Xu, T.; Liu, J.; Li, Z. RNA-Seq and iTRAQ reveal multiple pathways involved in storage root formation and development in sweet potato (Ipomoea batatas L.). *BMC Plant Biol.* **2019**, *19*, 136. [CrossRef]

71. Suttle, J.C.; Abrams, S.R.; De Stefano-Beltran, L.; Huckle, L.L. Chemical inhibition of potato ABA-8′-hydroxylase activity alters in vitro and in vivo ABA metabolism and endogenous ABA levels but does not affect potato microtuber dormancy duration. *J. Exp. Bot.* **2012**, *63*, 5717–5725. [CrossRef]

72. Wang, X.; Guo, C.; Peng, J.; Li, C.; Wan, F.; Zhang, S.; Zhou, Y.; Yan, Y.; Qi, L.; Sun, K.; et al. ABRE-BINDING FACTORS play a role in the feedback regulation of ABA signaling by mediating rapid ABA induction of ABA co-receptor genes. *New Phytol.* **2019**, *221*, 341–355. [CrossRef]

73. Li, J.; Zhao, X.; Dong, Y.; Li, S.; Yuan, J.; Li, C.; Zhang, X.; Li, M. Transcriptome Analysis Reveals Key Pathways and Hormone Activities Involved in Early Microtuber Formation of Dioscorea opposita. *Biomed. Res. Int.* **2020**, *2020*, 8057929. [CrossRef] [PubMed]

74. Xie, Y.; Xu, L.; Wang, Y.; Fan, L.; Chen, Y.; Tang, M.; Luo, X.; Liu, L. Comparative proteomic analysis provides insight into a complex regulatory network of taproot formation in radish (Raphanus sativus L.). *Hortic. Res.* **2018**, *5*, 51. [CrossRef] [PubMed]

75. Li, M.; Yang, Y.; Li, X.; Gu, L.; Wang, F.; Feng, F.; Tian, Y.; Wang, F.; Wang, X.; Lin, W.; et al. Analysis of integrated multiple 'omics' datasets reveals the mechanisms of initiation and determination in the formation of tuberous roots in Rehmannia glutinosa. *J. Exp. Bot.* **2015**, *66*, 5837–5851. [CrossRef] [PubMed]

76. Jung, C.; Nguyen, N.H.; Cheong, J.J. Transcriptional Regulation of Protein Phosphatase 2C Genes to Modulate Abscisic Acid Signaling. *Int. J. Mol. Sci.* **2020**, *21*, 9517. [CrossRef]

77. Collin, A.; Daszkowska-Golec, A.; Szarejko, I. Updates on the Role of ABSCISIC ACID INSENSITIVE 5 (ABI5) and ABSCISIC ACID-RESPONSIVE ELEMENT BINDING FACTORs (ABFs) in ABA Signaling in Different Developmental Stages in Plants. *Cells* **2021**, *10*, 1996. [CrossRef]

78. Muñiz García, M.N.; Cortelezzi, J.I.; Fumagalli, M.; Capiati, D.A. Expression of the Arabidopsis ABF4 gene in potato increases tuber yield, improves tuber quality and enhances salt and drought tolerance. *Plant Mol. Biol.* **2018**, *98*, 137–152. [CrossRef]

79. Moon, S.J.; Han, S.Y.; Kim, D.Y.; Yoon, I.S.; Shin, D.; Byun, M.O.; Kwon, H.B.; Kim, B.G. Ectopic expression of a hot pepper bZIP-like transcription factor in potato enhances drought tolerance without decreasing tuber yield. *Plant Mol. Biol.* **2015**, *89*, 421–431. [CrossRef]

80. Li, X.J.; Yang, J.L.; Hao, B.; Lu, Y.C.; Qian, Z.L.; Li, Y.; Ye, S.; Tang, J.R.; Chen, M.; Long, G.Q.; et al. Comparative transcriptome and metabolome analyses provide new insights into the molecular mechanisms underlying taproot thickening in Panax notoginseng. *BMC Plant Biol.* **2019**, *19*, 451. [CrossRef]

81. Nakatani, M.; Komeichi, M. Changes in the Endogenous Level of Zeatin Riboside, Abscisic Acid and Indole Acetic Acid during Formation and Thickening of Tuberous Roots in Sweet Potato. *Jpn. J. Crop Sci.* **2008**, *60*, 91–100. [CrossRef]

82. Thalmann, M.; Pazmino, D.; Seung, D.; Horrer, D.; Nigro, A.; Meier, T.; Kölling, K.; Pfeifhofer, H.W.; Zeeman, S.C.; Santelia, D. Regulation of Leaf Starch Degradation by Abscisic Acid Is Important for Osmotic Stress Tolerance in Plants. *Plant Cell* **2016**, *28*, 1860–1878. [CrossRef]

83. Lee, S.A.; Jang, S.; Yoon, E.K.; Heo, J.O.; Chang, K.S.; Choi, J.W.; Dhar, S.; Kim, G.; Choe, J.E.; Heo, J.B.; et al. Interplay between ABA and GA Modulates the Timing of Asymmetric Cell Divisions in the Arabidopsis Root Ground Tissue. *Mol. Plant* **2016**, *9*, 870–884. [CrossRef] [PubMed]

84. Lim, S.; Park, J.; Lee, N.; Jeong, J.; Toh, S.; Watanabe, A.; Kim, J.; Kang, H.; Kim, D.H.; Kawakami, N.; et al. ABA-insensitive3, ABA-insensitive5, and DELLAs Interact to activate the expression of SOMNUS and other high-temperature-inducible genes in imbibed seeds in Arabidopsis. *Plant Cell* **2013**, *25*, 4863–4878. [CrossRef] [PubMed]

85. Liu, X.; Hou, X. Antagonistic Regulation of ABA and GA in Metabolism and Signaling Pathways. *Front. Plant Sci.* **2018**, *9*, 251. [CrossRef] [PubMed]

86. Sun, W.; Ma, Z.; Chen, H.; Liu, M. MYB Gene Family in Potato (Solanum tuberosum L.): Genome-Wide Identification of Hormone-Responsive Reveals Their Potential Functions in Growth and Development. *Int. J. Mol. Sci.* **2019**, *20*, 4847. [CrossRef] [PubMed]

87. País, S.M.; García, M.N.; Téllez-Iñón, M.T.; Capiati, D.A. Protein phosphatases type 2A mediate tuberization signaling in Solanum tuberosum L. leaves. *Planta* **2010**, *232*, 37–49. [CrossRef]
88. Muñiz García, M.N.; Muro, M.C.; Mazzocchi, L.C.; País, S.M.; Stritzler, M.; Schlesinger, M.; Capiati, D.A. The protein phosphatase 2A catalytic subunit StPP2Ac2b acts as a positive regulator of tuberization induction in *Solanum tuberosum* L. *Plant Mol. Biol.* **2017**, *93*, 227–245. [CrossRef]
89. Li, Y.; Wang, Y.; Tan, S.; Li, Z.; Yuan, Z.; Glanc, M.; Domjan, D.; Wang, K.; Xuan, W.; Guo, Y.; et al. Root Growth Adaptation is Mediated by PYLs ABA Receptor-PP2A Protein Phosphatase Complex. *Adv. Sci.* **2020**, *7*, 1901455. [CrossRef]
90. Roumeliotis, E.; Visser, R.G.; Bachem, C.W. A crosstalk of auxin and GA during tuber development. *Plant Signal. Behav.* **2012**, *7*, 1360–1363. [CrossRef]
91. Su, Y.H.; Liu, Y.B.; Zhang, X.S. Auxin-cytokinin interaction regulates meristem development. *Mol. Plant* **2011**, *4*, 616–625. [CrossRef]
92. Utsumi, Y.; Tanaka, M.; Utsumi, C.; Takahashi, S.; Matsui, A.; Fukushima, A.; Kobayashi, M.; Sasaki, R.; Oikawa, A.; Kusano, M.; et al. Integrative omics approaches revealed a crosstalk among phytohormones during tuberous root development in cassava. *Plant Mol. Biol.* **2020**, *104*, 1–21. [CrossRef]
93. Begum, S.; Jing, S.; Yu, L.; Sun, X.; Wang, E.; Abu Kawochar, M.; Qin, J.; Liu, J.; Song, B. Modulation of JA signalling reveals the influence of StJAZ1-like on tuber initiation and tuber bulking in potato. *Plant J.* **2022**, *109*, 952–964. [CrossRef] [PubMed]
94. Aksenova, N.; Konstantinova, T.; Golyanovskaya, S.; Sergeeva, L.; Romanov, G. Hormonal regulation of tuber formation in potato plants. *Russ. J. Plant Physiol.* **2012**, *59*, 451–466. [CrossRef]
95. Kolachevskaya, O.O.; Sergeeva, L.I.; Getman, I.A.; Lomin, S.N.; Savelieva, E.M.; Romanov, G.A. Core features of the hormonal status in in vitro grown potato plants. *Plant Signal. Behav.* **2018**, *13*, e1467697. [CrossRef] [PubMed]
96. Saidi, A.; Hajibarat, Z. Phytohormones: Plant switchers in developmental and growth stages in potato. *J. Genet. Eng. Biotechnol.* **2021**, *19*, 89. [CrossRef]
97. Gong, M.; Luo, H.; Wang, A.; Zhou, Y.; Huang, W.; Zhu, P.; He, L. Phytohormone Profiling During Tuber Development of Chinese Yam by Ultra-high performance Liquid Chromatography–Triple Quadrupole Tandem Mass Spectrometry. *J. Plant Growth Regul.* **2016**, *36*, 362–373. [CrossRef]
98. Chen, H.; Banerjee, A.K.; Hannapel, D.J. The tandem complex of BEL and KNOX partners is required for transcriptional repression of ga20ox1. *Plant J.* **2004**, *38*, 276–284. [CrossRef]

International Journal of
Molecular Sciences

Review

Molecular Aspects of Seed Development Controlled by Gibberellins and Abscisic Acids

Akiko Kozaki [1,2,3,*] and Takuya Aoyanagi [1]

1   Graduate School of Science and Technology, Shizuoka University, Ohya 836, Suruga-ku,
Shizuoka 422-8021, Japan; aoyanagi.takuya.17@shizuoka.ac.jp
2   Course of Bioscience, Department of Science, Graduate School of Integrated Science and Technology,
Shizuoka University, Ohya 836, Suruga-ku, Shizuoka 422-8021, Japan
3   Department of Biological Science, Faculty of Science, Shizuoka University, Ohya 836, Suruga-ku,
Shizuoka 422-8021, Japan
*   Correspondence: kozaki.akiko@shizuoka.ac.jp

Abstract: Plants have evolved seeds to permit the survival and dispersion of their lineages by providing nutrition for embryo growth and resistance to unfavorable environmental conditions. Seed formation is a complicated process that can be roughly divided into embryogenesis and the maturation phase, characterized by accumulation of storage compound, acquisition of desiccation tolerance, arrest of growth, and acquisition of dormancy. Concerted regulation of several signaling pathways, including hormonal and metabolic signals and gene networks, is required to accomplish seed formation. Recent studies have identified the major network of genes and hormonal signals in seed development, mainly in maturation. Gibberellin (GA) and abscisic acids (ABA) are recognized as the main hormones that antagonistically regulate seed development and germination. Especially, knowledge of the molecular mechanism of ABA regulation of seed maturation, including regulation of dormancy, accumulation of storage compounds, and desiccation tolerance, has been accumulated. However, the function of ABA and GA during embryogenesis still remains elusive. In this review, we summarize the current understanding of the sophisticated molecular networks of genes and signaling of GA and ABA in the regulation of seed development from embryogenesis to maturation.

Keywords: gibberellin (GA); abscisic acid (ABA); seed development; seed maturation

Citation: Kozaki, A.; Aoyanagi, T. Molecular Aspects of Seed Development Controlled by Gibberellins and Abscisic Acids. *Int. J. Mol. Sci.* **2022**, *23*, 1876. https://doi.org/10.3390/ijms23031876

Academic Editor: Víctor Quesada

Received: 27 December 2021
Accepted: 2 February 2022
Published: 7 February 2022

Publisher's Note: MDPI stays neutral with regard to jurisdictional claims in published maps and institutional affiliations.

1. Introduction

Seeds are the products of the evolution of spermatophytes that enable the maintenance and spread of their lineages by providing nutrition for embryo growth and resistance to unfavorable environmental conditions through the state of dormancy. Until the environment is suitable for germination, seeds spend variable lengths of time in the dormancy stage.

The seed development process can be divided into two main phases, embryogenesis (cell division and morphogenesis) and maturation. Embryogenesis includes the formation and structural development of the mature seeds consisting of an embryo, endosperm, and maternal seed coat.

As a consequence of complex developmental processes that start from the end of embryogenesis and terminate with the state of dormancy, seed maturation occurs. The maturation stage is characterized by accumulation of storage compound, acquisition of desiccation tolerance, arrest of growth, and entry into dormancy. Seeds can germinate under favorable environmental conditions only after dormancy is broken.

Complex gene networks regulate seed development and germination, and diverse phytohormones are involved in these processes [1,2]. Gibberellin (GA) and abscisic acids (ABA) are recognized as primary hormones that antagonistically regulate seed development (including dormancy) and germination [2,3]. In early embryogenesis, auxin plays a major role in establishing the embryonic body plan via the effects of apical-basal polarity/pattern formation and vascular development. Together with auxin, cytokinins are linked to growth promotion by cell division, development, and differentiation. Brassinosteroids regulate the ovule number and size and shape of seeds, and also participate in seed germination by antagonizing the inhibitory effect of ABA [1,4].

Until now, impressive progress has been achieved in the understanding of the molecular network regulating the seed development, metabolism, and signaling pathways of ABA and GA in seed maturation and germination. However, the function of ABA and GA during embryogenesis still remains elusive.

In this review, we summarize the mechanism underlying the regulation of seed development (from embryogenesis to maturation) and the function of the phytohormones GA and ABA in seed development. Since there are several reviews on the function of other phytohormones in seed development in the literature, we focus on these two hormones [1,4–6].

2. The Level of ABA and GA during Seed Development

2.1. ABA Level during Seed Development

In seed development in *Arabidopsis*, a peak of ABA level in the whole silique is observed in the middle of development (around nine days after flowering (DAF)), and after 12 DAF, ABA increases until late stage of development (21 DAF) [7–9]. ABA was detected mostly in the seeds during the middle stage and in the envelopes during the late stage of maturation [7] (Figure 1).

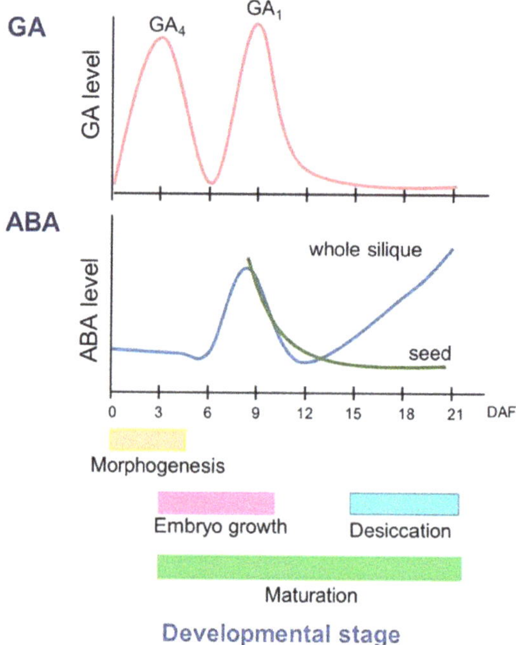

Figure 1. The level of GA and ABA during seed development of *Arabidopsis*. Schematic trend of hormone accumulation during seed development (Based on [7–10]). DAF: day after flowering.

It has been demonstrated that nine-cis epoxycarotenoid dioxygenase (NCED) is the key regulatory enzyme in the ABA biosynthetic pathway [11]. Among five *NCED* genes in *Arabidopsis*, *AtNCED6* and *AtNCED9* contribute to a high level of ABA at mid-seed development, while *AtNCED2* and *AtNCED3* contribute to the accumulation of ABA in the later stages of whole silique [9,12].

ABA accumulated in both phases is synthesized mainly in zygotic tissues. However, when zygotic tissues, but not maternal tissues, are deficient in ABA, ABA synthesized in maternal tissue is translocated into the embryos of zygotic tissues [7]. The main role of ABA synthesized in zygotic tissues is the induction and/or maintenance of seed dormancy [7,13,14]. On the other hand, maternal ABA affects the thickness of the mucilage layer released from mature seeds on imbibition in *Arabidopsis* [13].

In the seed development of wheat, there are two peaks of ABA level [15,16]. The ABA synthesized during the late phase of seed development (about 35–40 days after pollination (DAP)) is associated with the level of dormancy [15]. On the other hand, rice and triticale have one peak of ABA level in their seed development. In rice seeds, the accumulation of ABA involved in the induction of dormancy occurs during the early and middle stages of seed development (10–20 DAP), earlier than in wheat [17,18]. In triticale grains, peak ABA accumulation was around 35 DAP, before a significant loss of water [19].

Catabolism of ABA occurs by conversion from ABA to phaseic acid (PA), which is catalyzed by a cytochrome P450 monooxygenase (P450) encoded by *CYP707As* [20].

2.2. ABA Signaling

Three major components are involved in ABA signaling: pyrabactin resistance 1/pyrabactin-like/regulatory components of ABA receptors (PYR/PYL/RCAR), protein phosphatase 2Cs (PP2Cs), and SNF1-related protein kinase 2s (SnRK2s). In the absence of ABA, the activities of SnRK2s is inhibited by PP2Cs through dephosphorylation of their kinase activation loops, while in the presence of ABA, the ABA receptors PYR/PYL/RCAR form a complex with PP2C, which inhibits the phosphatase activity of PP2C and, as a result, SnRK2 is activated [21,22]. The activated form of SnRK2 subsequently activates ABRE-binding protein/ABRE-binding factor (AREB/ABF) transcription factors, which subsequently activate the transcription of ABA-responsive genes [22]. The ubiquitin-proteosome system (UPS) is also involved in ABA signaling. In the absence of ABA, ABA receptors PYR/PYL/RCAR, SnRK2s, and ABREB/ABF transcription factors are degraded via the UPS, which secures the inhibition of the ABA response. On the other hand, PP2C is degraded via the UPS in the presence of ABA leading to the enhancement of the ABA response [23,24].

2.3. GA Level during Seed Development

Among more than 130 GAs identified in plants, fungi, and bacteria, only four of them, GA1, GA3, GA4, and GA7, are thought to function as bioactive hormones. And among them, GA1 and GA4 are the major bioactive GAs in many plants including *Arabidopsis*. GA1and GA4 are synthesized via the 13-hydroxy pathway and the non-13-hydroxy pathway, respectively. The latter is the predominant pathway in *Arabidopsis* [25,26].

In *Arabidopsis*, GA4 and GA1 was accumulated in flower buds, flowers, and early developing silique (3 DAF), and in the mid-seed development (around 9 DAF), respectively [7,10,27] (Figure 1).

The conversion of intermediates to bioactive GAs is catalyzed by two enzymes, GA20-oxidase (GA20ox) and GA3ox in the last steps of the GA biosynthesis. Another enzyme, GA 2-oxidase (GA2ox), catalyzes the conversion of bioactive GAs to inactive catabolites [28]. The level of bioactive GA is controlled primarily by these three enzymes. Bioactive GAs are synthesized in developing seeds by all four *AtGA3ox* and by *AtGA3ox1* in replums and funiculi in developing *Arabidopsis* siliques [27]. In developing pea seeds, *PsGA20oxs* and *PsGA3oxs* were involved in the synthesis of bioactive Gas [29].

2.4. GA Signaling

GA signaling in plants is induced when bioactive GA is perceived by its receptor GIBBERELLIN INSENSITIVE DWARF1 (GID1) [30,31]. DELLA proteins are negative regulators of GA signaling [32]. When GA binds to GID1, the formation of the GA-GID1-DELLA complex is promoted, and the complex is associated with F-box protein, the central component of SCF$^{SLY1/GID2}$ E3 ubiquitin ligase, which leads to DELLA degradation via the ubiquitin 26S proteasome pathway [33–35]. As a result, GA response genes are activated.

3. Function of ABA and GA in Seed Development

3.1. Function of GA and ABA in Embryogenesis

Embryogenesis starts from a single cell zygote and ends when all embryo structures have been formed. In *Arabidopsis*, embryo development is divided into three phases: the earliest proembryo stage, is characterized by embryo polarity establishment; early embryogenesis is characterized by the embryo morphology shifting from the early globular stage to the heart stage (most of the structures have formed at this stage); and late embryogenesis is characterized by embryo expansion (elongation of cotyledon and axis) and maturation (storage compound accumulation, desiccation, and dormancy) [36]. The last phase, late embryogenesis, corresponds to the early stage of the maturation phase of seed development.

Several essential genes for embryogenesis, including *YUCCA* (*YUC*) family members, which are auxin biosynthesis genes, and *LEAFY COTYLEDON* genes (*LEC1, LEC2,* and *FUSCA3*), have been identified [37–41]. These *LEAFY COTYLEDON* genes also function in the seed maturation stage (described later).

For normal seed development, GAs are required. The evidence that GAs are necessary for seed development has been provided by the analysis of GA-deficient mutant in pea [42,43]. Overexpression of the gene for GA 2-oxidase (GA2ox) from pea in *Arabidopsis* seeds caused seed abortion and inhibition of pollen tube growth, demonstrating that active GAs in the endosperm are essential for normal seed development [44,45]. Similarly, overexpression of GA2ox from tomato in tomato fruit led to the reduction of fruit weight, seed number, and germination rate [46].

The maternal tissues, especially the seed coat, play an important role in embryonic development [47–49]. The plant proembryo is composed of an embryo-proper domain and a suspensor domain, and the suspensor is the major channel for maternal-to-proembryo communication. The transport of nutrients and signals from the mother to embryo is essential for embryonic development and plant fertility [50,51]. However, the degeneration of the suspensor through programmed cell death (PCD) occurs at a very early stage of embryonic development in plants [52]. In tobacco plant (*Nicotiana tabacum*), the suspensor PCD is established by the antagonistic action of two proteins; a protease inhibitor, cystatin NtCYS, and its target, cathepsin H-like protease NtCP14 [52]. Recently, it has been reported that a DELLA protein, NtCRF (NtCYS regulative factor 1) regulates suspensor PCD in tobacco by promoting the expression of *NtCYS*. GA generated in the micropylar endothelium trigger the suspensor PCD by suppression of *NtCYS* expression via degradation of NtCRF [53].

On the other hand, maternal ABA plays a significant role in embryo development and seed maturation in tobacco (*Nicotiana plumbaginifolia*), although it does not affect dormancy induction [54].

3.2. Gene Networks in the Maturation Phase

Following the embryogenesis phase, the maturation phase begins. In *Arabidopsis*, the embryo growth phase starts at the torpedo stage (around 7 DAF) and ends when the seed sac is filled with a mature embryo. During the growth phase, the volume ratio of embryo and endosperm is reversed. At the end of the growth phase, the endosperm is reduced to one cell layer while the embryo volume is increased. The cell division of the embryo increases at the beginning of the growth phase and then is arrested by the end of the phase [55]. During the maturation phase, accumulation of seed reserves, maturation and degradation of chloroplasts, acquisition of desiccation tolerance, and dormancy occur before water content decreases and the embryo enters a quiescent state.

In the case of cereals, the endosperm continues to increase the volume to accumulate storage materials and cover other important roles in embryo development and seed organization [1].

A complex network of transcription factors regulates seed maturation. Among these, the LAFL regulatory network is the central network. The LAFL genes include the AFL clade of B3 domain plant-specific transcription factors (ALF-B3), FUSCA3 (FUS3), ABA INSENSITIVE 3 (ABI3), LEAFY COTYLEDON 2 (LEC2) [38,39,56], and the HAP3 subunit of the CCAAT-binding transcription factors (CBF or NF-Y), LEC1, and LEC1-LIKE (L1L) [57,58]. Mutation of the LAFL genes affects many aspects of seed maturation: decreased dormancy at maturation [55], reduced expression of seed storage materials [59], reduced desiccation tolerance, and a low level of ABA content [60,61]. The LAFL network regulates several genes involved in modulation of various aspects of plant development besides seed development: genes for zinc finger factor PEI1, APETALA2 (AP2) family factor BABY BOOM (BBM), NAC factor CUP-SHAPED COTYLEDON1 (CUC1), and MADS box factor FLOWERING LOCUS C (FLC) [61].

AFL factors activate the target genes through the RY *cis*-element that is recognized by the B3-DNA binding domain [62–66]. LEC1 and L1L bind to the CCAAT DNA motif as a subunit of the NF-Y complex [57,67]. Genome-wide analysis of LEC1 binding sites in the upstream region of target genes in *Arabidopsis* and soybean revealed that, besides the CCAAT motif, G-Box, ABA-responsive promoter element (ABRE)-like, RY, and BPC1 *cis*-elements were enriched in the promoters of genes regulated during seed maturation, indicating that LEC1 regulates the target genes by interacting with several other kinds of transcription factors [68–70].

Genetic analysis shows that the LAFL genes organize a network with complex mutual interactions among LAFL genes (Figure 2). LEC1 can activate *ABI3*, *FUS3*, and *LEC2* expression, while ectopic expression of *LEC2* can up-regulate *LEC1*, *ABI3*, and *FUS3* [69,71,72]. *ABI3* and *FUS3* positively regulate each other and their own expression [69,73]. Moreover, *L1L* is regulated by FUS3 [74]. A recent ChIP analysis indicated that LEC1 regulates *L1L* [75], whereas FUS3 regulates *LEC1*, *FUS3*, and *ABI3* [76].

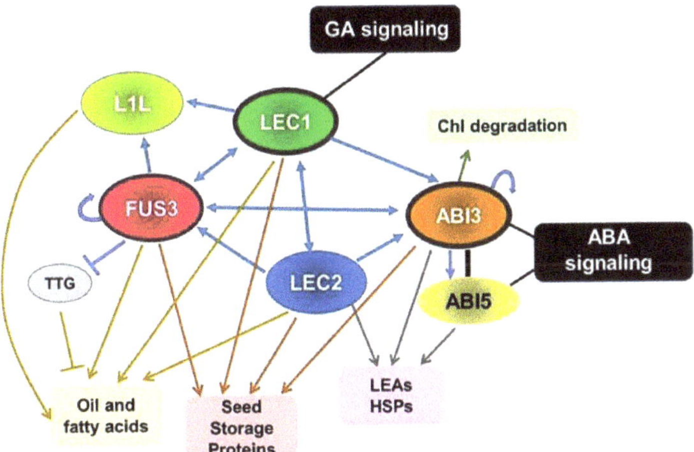

Figure 2. LAFL network regulates seed development. Arrows and blunted lines indicate activation and repression, respectively. Black line between ABI3 and ABI5 indicate the interaction of these proteins. LEC1, LEC2, and FUS3 (surrounded by the thick black line) are involved in acquisition of DT and all LAFL proteins are involved in the regulation of dormancy. LEC1 is related to GA signaling and ABI3 and ABI5 are related to ABA signaling.

In addition to the LAFL genes, ABI5 and ABI5-related bZIP transcription factors (bZIP), which bind to ABRE, are involved in the regulation of seed maturation. ABI5 is a key player in ABA signaling [77]. An important subset of LAFL-regulated genes during seed maturation includes *LATE EMBRYOGENESIS ABUNDANT* (*LEA*) genes, which have both RY and ABRE motifs in their promoters and are regulated by a combination of ABI3 and ABI5-related bZIP transcription factors [78,79]. Therefore, ABA signaling is integrated into the LAFL network by ABI5 and its related bZIP factors via physical interaction with the N-terminal COAR (co-activator/co-repressor) domain of ABI3 [78,79]. ABREs are also found in the promoters of target genes of other LAFL factors, suggesting that other components of the LAFL are potentially co-regulated by ABA [73,75,76].

In *Arabidopsis*, *FUS3* expression is increased by exogenously-introduced ABA [72], and FUS3 induces the increase of ABA [8]. Thus, FUS3 and ABA are positive regulators of each other [41]. Furthermore, the expression of FUS3 was found to be able to be positively regulated by auxin [8].

During seed maturation, GA's level should be down-regulated. GA's level is regulated by FUS3 and LEC2, which repress the enzymes involved in bioactive GA synthesis [8,80].

As mentioned above, LAFL genes play important roles in embryogenesis [70,81]. Recent research showed that GA signaling facilitates embryo development by promoting auxin accumulation in late embryogenesis via *LEC1* in *Arabidopsis*. The GA signaling repressors, DELLAs, interact with LEC1, which promotes the expression of the *YUC* gene that facilitate embryogenesis by promoting the accumulation of auxin. GA triggers the degradation of DELLAs to relieve their repression of LEC1, leading to the activation of genes essential genes for embryogenesis [10].

3.3. Accumulation of Seed Storage Products

During seed maturation, seed storage compounds needed for germination and initial seedling growth and development, such as seed storage proteins (SSP), lipids, and carbohydrates, are accumulated, and ABA is involved in this process [21,82]. Mutations in ABA signaling, such as *PYL* and *SnRK2*, often exhibit reduced seed storage products [83–86]. Inactivation of *SnRK2.6* results in reduction of seed oil content, while overexpression of *SnRK2.6* increases overall seed products [84]. *SnRK2s* triple mutant (*snrk2,2/3/6*) and *pyl* duodecuple mutant exhibited lower levels of seed storage products such as 12S globulin [83,85]. The starch biosynthesis in maize and rice is regulated synergistically by sucrose and ABA [87–89].

The *LAFL* genes are involved in the regulation of storage material accumulation [90]. LEC1 and FUS3 control the accumulation of ABI3 and function with each other to regulate the accumulation of storage proteins (including *Arabidopsis* 2S albumin storage protein 3 (At2S3) and Cruciferin C (CRC)), anthocyanins synthesis, and accumulation of chlorophyll and lipid during maturation in an ABA-dependent manner [68,72,91–93]. LEC1 activates *CRC* as well, via a direct interaction with bZIP67 [74].

FUS3 negatively regulates the expression of *TRANSPARENT TESTA GLABRA1* (*TTG1*), which encodes a transcription factor that suppresses the accumulation of seed storage proteins and oils in *Arabidopsis* [94]. A mutant of *ttg1* is characterized by a dramatic increase in storage reserves, such as oil and SSP [95]. FUS3 may lead to the accumulation of storage reserves by suppressing *TTG1* [94]. FUS3, in combination with LEC2, also induces the expression of *WRINKLED 1* (*WRI1*), which encodes AP2 transcription factor and plays roles in the regulation of sugars and oil content in seeds by increasing the gene expression for fatty acid synthesis and sugar degradation [74,96]. Together with repressing *TTG1* expression and enhancing *WRI1* expression, FUS3 promotes the accumulation of storage oils. This storage oil accumulation is also regulated by *LEC1* and *AFL* genes through activation of *WRI1* [93]. LEC2 regulates oil and protein accumulation by activating the expression of *OLE1*, encoding oleosin and genes encoding 2S and 12S storage proteins [62,63,97,98].

Other factors than *LAFL* genes are also involved in the accumulation of storage materials. bZIP67, together with L1L and NUCLEAR FACTOR-YC2 (NF-YC2), regulate *FATTY ACID DESATURASE 3* (*FAD3*), which functions in the storage of omega-3 fatty acid during maturation [99]. The DOG1-LIKE4 (DOGL4) gene, whose expression is induced by ABA, regulates the expression of some seed storage proteins including CRC, albumins, and oleosins during seed maturation [100].

3.4. Desiccation Tolerance and De-Greening

Desiccation tolerance (DT) is an important trait that seeds have to survive prolonged periods until favorable conditions for germination are present for. In many plants, the DT process during seed maturation is intricately linked to loss of chlorophyll (chl), namely de-greening. In terms of commercial products, the presence of chlorophyll in mature seeds can be an undesirable characteristic that can affect seed maturation and quality [60,82].

In *Arabidopsis*, the *abi3–6* mutant shows a lack of de-greening, and ABI3 was found to control embryo de-greening through regulating the expression of *STAY GREEN (SGR)* genes (*AtSGR1* and *AtSGR2*), which are orthologs of the *SGR* gene encoded by Mendel's *I* locus [101–103]. The seeds of the triple mutant *snrk2.2/3/6* also have greenish-brown seed coats, which indicate that ABA signaling is involved in the de-greening process [83].

LAFL genes play important roles in the DT acquisition process. A mutation in *LEC1*, *ABI3*, or *FUS3* drastically affects DT, indicating that all three of these regulators are required to activate DT [104], while a mutation in *LEC2* does not show this effect [69,97].

To acquire DT, a set of genes, including genes encoding protective proteins such as LEA [105,106] and HEAT SHOCK PROTEINs (HSPs) [107], and other protective enzymes, compounds, and antioxidants are required [3,108–111]. LEA proteins are highly hydrophilic glycine-rich proteins that display antioxidant, metal ion binding, membrane and protein stabilization, hydration buffering, and DNA and RNA interaction properties [112–115].

The expression of the LEA gene is regulated by ABI3 and ABI5 [116–119]. ABI3 also regulates the expression of seed-specific heat shock factor HSFA9 [120]. LEA and HSP gene expression is increased by DELAY OF GERMINATION (DOG1) through ABI5/ABI3, and enhances the storage of N-rich compounds in the seed, which promotes the seed's dormancy and viability [117–119,121].

Although a *lec2* mutant did not show DT reduction [97,122], *LEC2* is involved in DT establishment. LEC2 affects the expression of *LEA*, *EM1*, and *EM6* genes by induction of the expression of the gene for ENHANCED EM LEVEL (EEL) bZIP transcription factor [62], which is a negative regulator of those EM proteins in *Arabidopsis*. EEL competes with a positive regulator of EMS, ABI5, by competing for their promoter sites [118].

In *Medicago truncatula* and pea (*Pisum sativum*), ABI3, ABI4, and ABI5 were identified as major hubs to regulate DT acquisition to control genes involved in raffinose family oligosaccharide (RFO) metabolism, LEA proteins synthesis, and photosynthesis associated nuclear genes [106,109,123]. *ABI5* also regulates de-greening and seed longevity in legumes [123].

3.5. Induction and Maintenance of Primary Seed Dormancy

Dormancy, a temporary quiescent state, is the important characteristic of seeds of wild plant species to avoid germination under unfavorable environmental conditions and ensure the initiation of a next generation. Whereas in the case of domesticated species, seeds with fast and uniform germination have been selected for rapid growth to achieve good crop yield. On the other hand, lack of seed dormancy is undesirable because it may cause preharvest sprouting (PHS), a serious problem in cereal crops, and non-dormant mutants can have reduced seed longevity [16,124].

At the end of seed maturation after storage products are synthesized, dehydration starts, and de novo ABA is stored, seed dormancy is achieved [55]. Several pieces of evidence have established that ABA is a key regulator in this process [3,14,124]. Mutation in ABA biosynthesis, sensing, and signaling affect seed dormancy [12,83,85,125,126].

In *Arabidopsis*, mutants of *AtNCED6* and *AtNCED9* show decreased ABA levels and dormancy in mature, dry seeds [12]. Other ABA-deficient mutants, such as *aba1* and *aba2/3*, also show reduced dormancy levels [82,125,126]. In wheat, mutations in the two homologs of *TaABA8′ OH1* (*TaABA8′OH1A* and *TaABA8′OH1D*; *AtCYP707* homolog) resulted in an increase of ABA and an enhanced degree of dormancy [127]. *TsNCED1* is also related to a higher ABA content and higher resistance to PHS [19].

In *Arabidopsis*, AtMYB96 directly activates ABA synthesis genes (*AtNCED2,5,6*, and *9*) and inactivates GA biosynthesis genes (*AtGA3ox1* and *AtGA20ox1*) to induce primary seed dormancy [128]. AtABI4 deepens seed dormancy through direct interaction with promoter regions of *AtNECD6* to increase ABA biosynthesis and, with promoter regions of *AtGA2ox7*, a GA inactivation gene, to inhibit GA accumulation [129,130].

A mutation in ABA signaling, such as in the rice *ospyl* septuple and *snrk2.2/3/6* triple mutant, also leads to premature germination in rice and *Arabidopsis* [83,131].

Members of LAFL genes are involved in the achievement of dormancy. Growth arrest of embryo in mature seeds is controlled by *FUS3*, *LEC1*, and *LEC2*, whose mutants all fail in complete cessation of embryo growth and exhibit premature germination [39,98,132].

The maize *Viviparous 1* (*Vp1*) gene, an ortholog of the *ABI3* of *Arabidopsis*, was one of the key ABA signaling components first identified and characterized. A mutation in *Vp1* leads to PHS and disruption of embryo maturation in maize [133,134]. *Vp1* genes of wheat, rice, and sorghum are also associated with the level of dormancy and sensitivity to ABA and PHS [135–137]. Members of *LAFL* genes are regulated by the *VP8* encoding of a putative peptidase in maize [138]. The mutations in the *VP8* homolog gene *PLAS-TOCHRON3/COLIATH* (*PLA3/GO*) in rice and *ALTERDMERISTEM PROGRAM 1* (*AMP1*) in *Arabidopsis* show a reduced dormancy phenotype [139].

ABI5 is also important for the induction of dormancy during wheat and pea seed maturation [123,140,141]. In sorghum bicolor, SbABI4 and SbABI5 enhance the transcription of *SbGA2ox3* through directly binding to its promoter, and accordingly prolong seed dormancy [142].

Two major dormancy genes, *DOG1* and *REDUCED DORMANCY 5* (*RDO5*), have been identified that seem to function independently of the plant hormones, including ABA [122,143]. RDO5 is a member of the PP2C protein phosphatase family, but does not show phosphatase activity [143], while DOG1 is a protein of unknown function [122]. Mutations in *DOG1* and *RDO5* completely abolish or reduce seed dormancy, respectively [122,143]. Genetic analysis revealed *DOG1* and ABA are both required for normal seed dormancy [82,122,144].

DOG1 interacts with four phosphatases and two of them are belong to clade A of type 2C protein phosphatase, ABA-HYPERSENSITIVE GERMINATION 1 (AHG1), and AHG3. The ABA and DOG1 pathways converge at the level of PP2C phosphatases: DOG1 inhibits AHG1 and AHG3, while ABA inhibits other PP2CAs and AHG3. By inhibiting PP2C phosphatases, ABA and DOG1 promote and maintain dormancy [145,146]. DOG1 is also required for multiple aspects of seed maturation, partially by interfering with ABA signaling components [121].

SEED DORMANCY 4 (OsSDR4) is considered as a regulator involved in seed dormancy with an unknown function in rice [147]. In *Arabidopsis*, SDR4-LIKE (AtSDR4L) regulates dormancy release and germination through regulation of *DOG1* and *RGA-LIKE2* (*RGL2* encoding DELLA protein) in the GA pathway [148]. A recent study speculated that AtODR1 (for reversal of rdo5), an ortholog of OsSDR4, acts together with bHLH57 and functions upstream of *AtNCED6* and *AtNCED9* to control ABA synthesis and seed dormancy in *Arabidopsis* [149].

In addition to gene regulation networks, other regulation, such as protein phosphorylation and chromatin remodeling, is involved in the regulation of dormancy. RAF-like MAPKKKs, RAF10/11 can phosphorylate SnRK2 and ABFs (ABRE binding factors) to influence seed dormancy [150,151]. A member of the histone deacetylation complex in *Arabidopsis*, SIN3-like 1 (SNL1), interacts with HISTONE DEACETYLASE 19 (HDA19) to modulate the ABA signaling pathway to promote seed dormancy [152]. Several regulators, including HISTONE MONOUBIQUITINATION (HUB1: C3HC4-RING finger protein) and REDUCED DORMANCY 2 (RDO2: transcription elongation factor TFIIS), are involved in the regulation of seed dormancy [153,154].

4. Conclusions and Future Perspectives

ABA and GA play important roles in seed development and germination. Most attention has been paid to the functions of these hormones in the induction, maintenance, and breaking of dormancy and germination [2,3,16,155]. In *Arabidopsis* seed, one of the GA level peaks is at the late stage of embryogenesis as well that of ABA level, during the growth phase of maturation, indicating GA plays an important role at this stage [7,10,27]. However, the detailed function of GA in embryogenesis has remained elusive. A recent finding shed light on the mechanism of GA signaling in the regulation of embryogenesis: GA signaling regulates late embryogenesis via *LEC1* activation [10]. Moreover, GA has been revealed to be a maternal-to-proembryo communication signal to control the embryonic suspensor PCD [53].

These recent findings show that DELLA proteins play important roles in the integration of the GA signal with other signals, such as PCD or auxin synthesis [10]. Because DELLA proteins interact with many kinds of proteins and are involved in the various aspects of signal transduction [156,157], novel functions of GA signaling through DELLA proteins in seed development might be found in the future.

Although the importance of maternal ABA in embryogenesis in tobacco was reported [54], the detailed function of ABA in embryogenesis has not been clarified. It has been reported that ABA is required for formation of the somatic embryo, induced by auxin [158]. Auxin promotes the expression of *ABI3*, which induces embryo identity genes through *AUXIN RESPONSE FACTOR* (*ARF*) genes activation [158]. Similarly, auxin controls seed dormancy through stimulation of ABA signaling by inducing *ABI3* expression [159]. Interaction between ABA and auxin signaling has functions in many aspects of plant development [160]. Further research will reveal the detailed functions of ABA in embryogenesis.

In this review, we focus on the function of ABA and GA in seed development. However, besides GA and ABA, there is elaborate crosstalk among phytohormone signaling during seed development. Further research on the crosstalk among signaling of ABA, GA, and other hormones will provide a more complete mechanism of the regulation of seed development.

Funding: This research received no external funding.

Institutional Review Board Statement: Not applicable.

Informed Consent Statement: Not applicable.

Data Availability Statement: Not applicable.

Conflicts of Interest: The authors declare that there is no conflict of interest.

References

1. Locascio, A.; Roig-Villanova, I.; Bernardi, J.; Varotto, S. Current perspectives on the hormonal control of seed development in Arabidopsis and maize: A focus on auxin. *Front. Plant Sci.* **2014**, *5*. [CrossRef] [PubMed]
2. Shu, K.; Liu, X.-D.; Xie, Q.; He, Z.-H. Two Faces of One Seed: Hormonal Regulation of Dormancy and Germination. *Mol. Plant* **2016**, *9*, 34–45. [CrossRef] [PubMed]
3. Finch-Savage, W.E.; Leubner Metzger, G. Seed dormancy and the control of germination. *New Phytol.* **2006**, *171*, 501–523. [CrossRef] [PubMed]
4. Matilla, A.J. Auxin: Hormonal Signal Required for Seed Development and Dormancy. *Plants* **2020**, *9*, 705. [CrossRef]
5. Figueiredo, D.D.; Köhler, C. Auxin: A molecular trigger of seed development. *Genes Dev.* **2018**, *32*, 479–490. [CrossRef]
6. Cao, J.; Li, G.; Qu, D.; Li, X.; Wang, Y. Into the Seed: Auxin Controls Seed Development and Grain Yield. *Int. J. Mol. Sci.* **2020**, *21*, 1662. [CrossRef]
7. Kanno, Y.; Jikumaru, Y.; Hanada, A.; Nambara, E.; Abrams, S.R.; Kamiya, Y.; Seo, M. Comprehensive hormone profiling in developing Arabidopsis seeds: Examination of the site of ABA biosynthesis, ABA transport and hormone interactions. *Plant Cell Physiol.* **2010**, *51*, 1988–2001. [CrossRef]
8. Gazzarrini, S.; Tsuchiya, Y.; Lumba, S.; Okamoto, M.; McCourt, P. The transcription factor FUSCA3 controls developmental timing in Arabidopsis through the hormones gibberellin and abscisic acid. *Dev. Cell* **2004**, *7*, 373–385. [CrossRef]

9. Okamoto, M.; Kuwahara, A.; Seo, M.; Kushiro, T.; Asami, T.; Hirai, N.; Kamiya, Y.; Koshiba, T.; Nambara, E. CYP707A1 and CYP707A2, which encode abscisic acid 8′-hydroxylases, are indispensable for proper control of seed dormancy and germination in Arabidopsis. *Plant Physiol.* **2006**, *141*, 97–107. [CrossRef]

10. Hu, Y.; Zhou, L.; Huang, M.; He, X.; Yang, Y.; Liu, X.; Li, Y.; Hou, X. Gibberellins play an essential role in late embryogenesis of Arabidopsis. *Nat. Plants* **2018**, *4*, 289–298. [CrossRef]

11. Nambara, E.; Marion-Poll, A. Abscisic acid biosynthesis and catabolism. *Annu. Rev. Plant Biol.* **2005**, *56*, 165–185. [CrossRef] [PubMed]

12. Lefebvre, V.; North, H.; Frey, A.; Sotta, B.; Seo, M.; Okamoto, M.; Nambara, E.; Marion-Poll, A. Functional analysis of Arabidopsis NCED6 and NCED9 genes indicates that ABA synthesized in the endosperm is involved in the induction of seed dormancy. *Plant J.* **2006**, *45*, 309–319. [CrossRef] [PubMed]

13. Karssen, C.; Brinkhorst-Van der Swan, D.; Breekland, A.; Koornneef, M. Induction of dormancy during seed development by endogenous abscisic acid: Studies on abscisic acid deficient genotypes of *Arabidopsis thaliana* (L.) Heynh. *Planta* **1983**, *157*, 158–165. [CrossRef] [PubMed]

14. Nambara, E.; Okamoto, M.; Tatematsu, K.; Yano, R.; Seo, M.; Kamiya, Y. Abscisic acid and the control of seed dormancy and germination. *Seed Sci. Res.* **2010**, *20*, 55–67. [CrossRef]

15. Suzuki, T.; Matsuura, T.; Kawakami, N.; Noda, K. Accumulation and leakage of abscisic acid during embryo development and seed dormancy in wheat. *Plant Growth Regul.* **2000**, *30*, 253–260. [CrossRef]

16. Tuan, P.A.; Kumar, R.; Rehal, P.K.; Toora, P.K.; Ayele, B.T. Molecular Mechanisms Underlying Abscisic Acid/Gibberellin Balance in the Control of Seed Dormancy and Germination in Cereals. *Front. Plant Sci.* **2018**, *9*. [CrossRef]

17. Liu, Y.; Fang, J.; Xu, F.; Chu, J.; Yan, C.; Schläppi, M.R.; Wang, Y.; Chu, C. Expression patterns of ABA and GA metabolism genes and hormone levels during rice seed development and imbibition: A comparison of dormant and non-dormant rice cultivars. *J. Genet. Genom.* **2014**, *41*, 327–338. [CrossRef]

18. Gu, X.Y.; Foley, M.E.; Horvath, D.P.; Anderson, J.V.; Feng, J.; Zhang, L.; Mowry, C.R.; Ye, H.; Suttle, J.C.; Kadowaki, K.; et al. Association between seed dormancy and pericarp color is controlled by a pleiotropic gene that regulates abscisic acid and flavonoid synthesis in weedy red rice. *Genetics* **2011**, *189*, 1515–1524. [CrossRef] [PubMed]

19. Fidler, J.; Zdunek-Zastocka, E.; Prabucka, B.; Bielawski, W. Abscisic acid content and the expression of genes related to its metabolism during maturation of triticale grains of cultivars differing in pre-harvest sprouting susceptibility. *J. Plant Physiol.* **2016**, *207*, 1–9. [CrossRef]

20. Kushiro, T.; Okamoto, M.; Nakabayashi, K.; Yamagishi, K.; Kitamura, S.; Asami, T.; Hirai, N.; Koshiba, T.; Kamiya, Y.; Nambara, E. The Arabidopsis cytochrome P450 CYP707A encodes ABA 8′-hydroxylases: Key enzymes in ABA catabolism. *Embo J.* **2004**, *23*, 1647–1656. [CrossRef]

21. Finkelstein, R. Abscisic Acid synthesis and response. *Arab. Book* **2013**, *11*, e0166. [CrossRef] [PubMed]

22. Ng, L.M.; Melcher, K.; Teh, B.T.; Xu, H.E. Abscisic acid perception and signaling: Structural mechanisms and applications. *Acta Pharmacol. Sin.* **2014**, *35*, 567–584. [CrossRef] [PubMed]

23. Su, T.; Yang, M.; Wang, P.; Zhao, Y.; Ma, C. Interplay between the Ubiquitin Proteasome System and Ubiquitin-Mediated Autophagy in Plants. *Cells* **2020**, *9*, 2219. [CrossRef] [PubMed]

24. Kumar, M.; Kesawat, M.S.; Ali, A.; Lee, S.C.; Gill, S.S.; Kim, A.H.U. Integration of Abscisic Acid Signaling with Other Signaling Pathways in Plant Stress Responses and Development. *Plants* **2019**, *8*, 592. [CrossRef] [PubMed]

25. Yamaguchi, S. Gibberellin metabolism and its regulation. *Annu. Rev. Plant Biol.* **2008**, *59*, 225–251. [CrossRef]

26. Binenbaum, J.; Weinstain, R.; Shani, E. Gibberellin Localization and Transport in Plants. *Trends Plant Sci.* **2018**, *23*, 410–421. [CrossRef]

27. Hu, J.; Mitchum, M.G.; Barnaby, N.; Ayele, B.T.; Ogawa, M.; Nam, E.; Lai, W.-C.; Hanada, A.; Alonso, J.M.; Ecker, J.R.; et al. Potential Sites of Bioactive Gibberellin Production during Reproductive Growth in Arabidopsis. *Plant Cell* **2008**, *20*, 320–336. [CrossRef]

28. Sun, T.-P. Gibberellin metabolism, perception and signaling pathways in Arabidopsis. *Arab. Book* **2008**, *6*, e0103. [CrossRef]

29. Nadeau, C.D.; Ozga, J.A.; Kurepin, L.V.; Jin, A.; Pharis, R.P.; Reinecke, D.M. Tissue-specific regulation of gibberellin biosynthesis in developing pea seeds. *Plant Physiol.* **2011**, *156*, 897–912. [CrossRef]

30. Ueguchi-Tanaka, M.; Ashikari, M.; Nakajima, M.; Itoh, H.; Katoh, E.; Kobayashi, M.; Chow, T.Y.; Hsing, Y.I.; Kitano, H.; Yamaguchi, I.; et al. Gibberellin Insensitive DWARF1 encodes a soluble receptor for gibberellin. *Nature* **2005**, *437*, 693–698. [CrossRef]

31. Griffiths, J.; Murase, K.; Rieu, I.; Zentella, R.; Zhang, Z.-L.; Powers, S.J.; Gong, F.; Phillips, A.L.; Hedden, P.; Sun, T.-p.; et al. Genetic Characterization and Functional Analysis of the GID1 Gibberellin Receptors in Arabidopsis. *Plant Cell* **2006**, *18*, 3399–3414. [CrossRef] [PubMed]

32. Sun, T.P. The molecular mechanism and evolution of the GA-GID1-DELLA signaling module in plants. *Curr. Biol.* **2011**, *21*, R338–R345. [CrossRef] [PubMed]

33. Murase, K.; Hirano, Y.; Sun, T.P.; Hakoshima, T. Gibberellin-induced DELLA recognition by the gibberellin receptor GID1. *Nature* **2008**, *456*, 459–463. [CrossRef] [PubMed]

34. McGinnis, K.M.; Thomas, S.G.; Soule, J.D.; Strader, L.C.; Zale, J.M.; Sun, T.P.; Steber, C.M. The Arabidopsis SLEEPY1 gene encodes a putative F-box subunit of an SCF E3 ubiquitin ligase. *Plant Cell* **2003**, *15*, 1120–1130. [CrossRef] [PubMed]

35. Sasaki, A.; Itoh, H.; Gomi, K.; Ueguchi-Tanaka, M.; Ishiyama, K.; Kobayashi, M.; Jeong, D.H.; An, G.; Kitano, H.; Ashikari, M.; et al. Accumulation of phosphorylated repressor for gibberellin signaling in an F-box mutant. *Science* **2003**, *299*, 1896–1898. [CrossRef]

36. Mayer, U.; Ruiz, R.A.T.; Berleth, T.; Miséra, S.; Jürgens, G. Mutations affecting body organization in the Arabidopsis embryo. *Nature* **1991**, *353*, 402–407. [CrossRef]

37. Cheng, Y.; Dai, X.; Zhao, Y. Auxin synthesized by the YUCCA flavin monooxygenases is essential for embryogenesis and leaf formation in Arabidopsis. *Plant Cell* **2007**, *19*, 2430–2439. [CrossRef]

38. Luerssen, H.; Kirik, V.; Herrmann, P.; Miséra, S. FUSCA3 encodes a protein with a conserved VP1/AB13-like B3 domain which is of functional importance for the regulation of seed maturation in Arabidopsis thaliana. *Plant J.* **1998**, *15*, 755–764. [CrossRef]

39. Stone, S.L.; Kwong, L.W.; Yee, K.M.; Pelletier, J.; Lepiniec, L.; Fischer, R.L.; Goldberg, R.B.; Harada, J.J. Leafy Cotyledon2 encodes a B3 domain transcription factor that induces embryo development. *Proc. Natl. Acad. Sci. USA* **2001**, *98*, 11806–11811. [CrossRef]

40. Lee, H.; Fischer, R.L.; Goldberg, R.B.; Harada, J.J. Arabidopsis LEAFY COTYLEDON1 represents a functionally specialized subunit of the CCAAT binding transcription factor. *Proc. Natl. Acad. Sci. USA* **2003**, *100*, 2152–2156. [CrossRef]

41. Braybrook, S.A.; Harada, J.J. LECs go crazy in embryo development. *Trends Plant Sci.* **2008**, *13*, 624–630. [CrossRef] [PubMed]

42. Swain, S.M.; Reid, J.B.; Kamiya, Y. Gibberellins are required for embryo growth and seed development in pea. *Plant J.* **1997**, *12*, 1329–1338. [CrossRef]

43. Groot, S.P.C.; Bruinsma, J.; Karssen, C.M. The role of endogenous gibberellin in seed and fruit development of tomato: Studies with a gibberellin-deficient mutant. *Physiol. Plant.* **1987**, *71*, 184–190. [CrossRef]

44. Singh, D.P.; Filardo, F.F.; Storey, R.; Jermakow, A.M.; Yamaguchi, S.; Swain, S.M. Overexpression of a gibberellin inactivation gene alters seed development, KNOX gene expression, and plant development in Arabidopsis. *Physiol. Plant* **2010**, *138*, 74–90. [CrossRef]

45. Singh, D.P.; Jermakow, A.M.; Swain, S.M. Gibberellins are required for seed development and pollen tube growth in Arabidopsis. *Plant Cell* **2002**, *14*, 3133–3147. [CrossRef]

46. Chen, S.; Wang, X.; Zhang, L.; Lin, S.; Liu, D.; Wang, Q.; Cai, S.; El-Tanbouly, R.; Gan, L.; Wu, H.; et al. Identification and characterization of tomato gibberellin 2-oxidases (GA2oxs) and effects of fruit-specific SlGA2ox1 overexpression on fruit and seed growth and development. *Hortic. Res.* **2016**, *3*, 16059. [CrossRef]

47. Friml, J.; Vieten, A.; Sauer, M.; Weijers, D.; Schwarz, H.; Hamann, T.; Offringa, R.; Jürgens, G. Efflux-dependent auxin gradients establish the apical–basal axis of Arabidopsis. *Nature* **2003**, *426*, 147–153. [CrossRef]

48. Liu, Y.; Li, X.; Zhao, J.; Tang, X.; Tian, S.; Chen, J.; Shi, C.; Wang, W.; Zhang, L.; Feng, X. Direct evidence that suspensor cells have embryogenic potential that is suppressed by the embryo proper during normal embryogenesis. *Proc. Natl. Acad. Sci. USA* **2015**, *112*, 12432–12437. [CrossRef]

49. Robert, H.S.; Park, C.; Gutièrrez, C.L.; Wójcikowska, B.; Pěnčík, A.; Novák, O.; Chen, J.; Grunewald, W.; Dresselhaus, T.; Friml, J. Maternal auxin supply contributes to early embryo patterning in Arabidopsis. *Nat. Plants* **2018**, *4*, 548–553. [CrossRef]

50. Kawashima, T.; Goldberg, R.B. The suspensor: Not just suspending the embryo. *Trends Plant Sci.* **2010**, *15*, 23–30. [CrossRef]

51. Yeung, E.C.; Meinke, D.W. Embryogenesis in angiosperms: Development of the suspensor. *Plant Cell* **1993**, *5*, 1371. [CrossRef] [PubMed]

52. Zhao, P.; Zhou, X.-M.; Zhang, L.-Y.; Wang, W.; Ma, L.-G.; Yang, L.-B.; Peng, X.-B.; Bozhkov, P.V.; Sun, M.-X. A bipartite molecular module controls cell death activation in the basal cell lineage of plant embryos. *PLoS Biol.* **2013**, *11*, e1001655. [CrossRef] [PubMed]

53. Shi, C.; Luo, P.; Du, Y.-T.; Chen, H.; Huang, X.; Cheng, T.-H.; Luo, A.; Li, H.-J.; Yang, W.-C.; Zhao, P.; et al. Maternal control of suspensor programmed cell death via gibberellin signaling. *Nat. Commun.* **2019**, *10*, 3484. [CrossRef] [PubMed]

54. Frey, A.; Godin, B.; Bonnet, M.; Sotta, B.; Marion-Poll, A. Maternal synthesis of abscisic acid controls seed development and yield in Nicotiana plumbaginifolia. *Planta* **2004**, *218*, 958–964. [CrossRef] [PubMed]

55. Raz, V.; Bergervoet, J.H.; Koornneef, M. Sequential steps for developmental arrest in Arabidopsis seeds. *Development* **2001**, *128*, 243–252. [CrossRef] [PubMed]

56. Giraudat, J.; Hauge, B.M.; Valon, C.; Smalle, J.; Parcy, F.; Goodman, H.M. Isolation of the Arabidopsis ABI3 gene by positional cloning. *Plant Cell* **1992**, *4*, 1251–1261. [CrossRef] [PubMed]

57. Lotan, T.; Ohto, M.; Yee, K.M.; West, M.A.; Lo, R.; Kwong, R.W.; Yamagishi, K.; Fischer, R.L.; Goldberg, R.B.; Harada, J.J. Arabidopsis Leafy Cotyledon1 is sufficient to induce embryo development in vegetative cells. *Cell* **1998**, *93*, 1195–1205. [CrossRef]

58. Kwong, R.W.; Bui, A.Q.; Lee, H.; Kwong, L.W.; Fischer, R.L.; Goldberg, R.B.; Harada, J.J. Leafy Cotyledon1-Like defines a class of regulators essential for embryo development. *Plant Cell* **2003**, *15*, 5–18. [CrossRef]

59. Gutierrez, L.; Van Wuytswinkel, O.; Castelain, M.; Bellini, C. Combined networks regulating seed maturation. *Trends Plant Sci.* **2007**, *12*, 294–300. [CrossRef]

60. Holdsworth, M.J.; Bentsink, L.; Soppe, W.J.J. Molecular networks regulating Arabidopsis seed maturation, after-ripening, dormancy and germination. *New Phytol.* **2008**, *179*, 33–54. [CrossRef]

61. Jia, H.; Suzuki, M.; McCarty, D.R. Regulation of the seed to seedling developmental phase transition by the LAFL and VAL transcription factor networks. *Wiley Interdiscip. Rev. Dev. Biol.* **2014**, *3*, 135–145. [CrossRef] [PubMed]

62. Braybrook, S.A.; Stone, S.L.; Park, S.; Bui, A.Q.; Le, B.H.; Fischer, R.L.; Goldberg, R.B.; Harada, J.J. Genes directly regulated by Leafy Cotyledon2 provide insight into the control of embryo maturation and somatic embryogenesis. *Proc. Natl. Acad. Sci. USA* **2006**, *103*, 3468–3473. [CrossRef] [PubMed]

63. Kroj, T.; Savino, G.; Valon, C.; Giraudat, J.; Parcy, F. Regulation of storage protein gene expression in Arabidopsis. *Development* **2003**, *130*, 6065–6073. [CrossRef] [PubMed]
64. Mönke, G.; Altschmied, L.; Tewes, A.; Reidt, W.; Mock, H.P.; Bäumlein, H.; Conrad, U. Seed-specific transcription factors ABI3 and FUS3: Molecular interaction with DNA. *Planta* **2004**, *219*, 158–166. [CrossRef]
65. Reidt, W.; Wohlfarth, T.; Ellerström, M.; Czihal, A.; Tewes, A.; Ezcurra, I.; Rask, L.; Bäumlein, H. Gene regulation during late embryogenesis: The RY motif of maturation-specific gene promoters is a direct target of the FUS3 gene product. *Plant J.* **2000**, *21*, 401–408. [CrossRef]
66. Suzuki, M.; Kao, C.Y.; McCarty, D.R. The conserved B3 domain of VIVIPAROUS1 has a cooperative DNA binding activity. *Plant Cell* **1997**, *9*, 799–807. [CrossRef]
67. Miller, M. Interactions of CCAAT/enhancer-binding protein β with transcriptional coregulators. *Postepy. Biochem.* **2016**, *62*, 343–348. [CrossRef]
68. Parcy, F.; Valon, C.; Kohara, A.; Miséra, S.; Giraudat, J. The ABSCISIC ACID-INSENSITIVE3, FUSCA3, and LEAFY COTYLEDON1 loci act in concert to control multiple aspects of Arabidopsis seed development. *Plant Cell* **1997**, *9*, 1265–1277.
69. To, A.; Valon, C.; Savino, G.; Guilleminot, J.; Devic, M.; Giraudat, J.; Parcy, F. A network of local and redundant gene regulation governs Arabidopsis seed maturation. *Plant Cell* **2006**, *18*, 1642–1651. [CrossRef]
70. Jo, L.; Pelletier, J.M.; Harada, J.J. Central role of the LEAFY COTYLEDON1 transcription factor in seed development. *J. Integr. Plant Biol.* **2019**, *61*, 564–580. [CrossRef]
71. Stone, S.L.; Braybrook, S.A.; Paula, S.L.; Kwong, L.W.; Meuser, J.; Pelletier, J.; Hsieh, T.-F.; Fischer, R.L.; Goldberg, R.B.; Harada, J.J. *Arabidopsis* LEAFY COTYLEDON2 induces maturation traits and auxin activity: Implications for somatic embryogenesis. *Proc. Natl. Acad. Sci. USA* **2008**, *105*, 3151–3156. [CrossRef] [PubMed]
72. Kagaya, Y.; Okuda, R.; Ban, A.; Toyoshima, R.; Tsutsumida, K.; Usui, H.; Yamamoto, A.; Hattori, T. Indirect ABA-dependent regulation of seed storage protein genes by FUSCA3 transcription factor in Arabidopsis. *Plant Cell Physiol.* **2005**, *46*, 300–311. [CrossRef] [PubMed]
73. Mönke, G.; Seifert, M.; Keilwagen, J.; Mohr, M.; Grosse, I.; Hähnel, U.; Junker, A.; Weisshaar, B.; Conrad, U.; Bäumlein, H.; et al. Toward the identification and regulation of the Arabidopsis thaliana ABI3 regulon. *Nucleic Acids Res.* **2012**, *40*, 8240–8254. [CrossRef] [PubMed]
74. Yamamoto, A.; Kagaya, Y.; Usui, H.; Hobo, T.; Takeda, S.; Hattori, T. Diverse roles and mechanisms of gene regulation by the Arabidopsis seed maturation master regulator FUS3 revealed by microarray analysis. *Plant Cell Physiol.* **2010**, *51*, 2031–2046. [CrossRef] [PubMed]
75. Junker, A.; Mönke, G.; Rutten, T.; Keilwagen, J.; Seifert, M.; Thi, T.M.; Renou, J.P.; Balzergue, S.; Viehöver, P.; Hähnel, U.; et al. Elongation-related functions of LEAFY COTYLEDON1 during the development of Arabidopsis thaliana. *Plant J.* **2012**, *71*, 427–442. [CrossRef] [PubMed]
76. Wang, F.; Perry, S.E. Identification of direct targets of FUSCA3, a key regulator of Arabidopsis seed development. *Plant Physiol.* **2013**, *161*, 1251–1264. [CrossRef] [PubMed]
77. Collin, A.; Daszkowska-Golec, A.; Szarejko, I. Updates on the Role of ABSCISIC ACID INSENSITIVE 5 (ABI5) and ABSCISIC ACID-RESPONSIVE ELEMENT BINDING FACTORs (ABFs) in ABA Signaling in Different Developmental Stages in Plants. *Cells* **2021**, *10*, 1996. [CrossRef]
78. Nakamura, S.; Lynch, T.J.; Finkelstein, R.R. Physical interactions between ABA response loci of Arabidopsis. *Plant J.* **2001**, *26*, 627–635. [CrossRef]
79. Alonso, R.; Oñate-Sánchez, L.; Weltmeier, F.; Ehlert, A.; Diaz, I.; Dietrich, K.; Vicente-Carbajosa, J.; Dröge-Laser, W. A pivotal role of the basic leucine zipper transcription factor bZIP53 in the regulation of Arabidopsis seed maturation gene expression based on heterodimerization and protein complex formation. *Plant Cell* **2009**, *21*, 1747–1761. [CrossRef]
80. Curaba, J.; Moritz, T.; Blervaque, R.; Parcy, F.O.; Raz, V.; Herzog, M.; Vachon, G. AtGA3ox2, a Key Gene Responsible for Bioactive Gibberellin Biosynthesis, Is Regulated during Embryogenesis by LEAFY COTYLEDON2 and FUSCA3 in Arabidopsis. *Plant Physiol.* **2004**, *136*, 3660–3669. [CrossRef]
81. Jia, H.; McCarty, D.R.; Suzuki, M. Distinct roles of LAFL network genes in promoting the embryonic seedling fate in the absence of VAL repression. *Plant Physiol.* **2013**, *163*, 1293–1305. [CrossRef] [PubMed]
82. Alonso-Blanco, C.; Bentsink, L.; Hanhart, C.J.; Vries, H.B.-d.; Koornneef, M. Analysis of Natural Allelic Variation at Seed Dormancy Loci of Arabidopsis thaliana. *Genetics* **2003**, *164*, 711–729. [CrossRef] [PubMed]
83. Nakashima, K.; Fujita, Y.; Kanamori, N.; Katagiri, T.; Umezawa, T.; Kidokoro, S.; Maruyama, K.; Yoshida, T.; Ishiyama, K.; Kobayashi, M.; et al. Three Arabidopsis SnRK2 protein kinases, SRK2D/SnRK2.2, SRK2E/SnRK2.6/OST1 and SRK2I/SnRK2.3, involved in ABA signaling are essential for the control of seed development and dormancy. *Plant Cell Physiol.* **2009**, *50*, 1345–1363. [CrossRef] [PubMed]
84. Zheng, Z.; Xu, X.; Crosley, R.A.; Greenwalt, S.A.; Sun, Y.; Blakeslee, B.; Wang, L.; Ni, W.; Sopko, M.S.; Yao, C. The protein kinase SnRK2. 6 mediates the regulation of sucrose metabolism and plant growth in Arabidopsis. *Plant Physiol.* **2010**, *153*, 99–113. [CrossRef] [PubMed]
85. Zhao, Y.; Zhang, Z.; Gao, J.; Wang, P.; Hu, T.; Wang, Z.; Hou, Y.-J.; Wan, Y.; Liu, W.; Xie, S. Arabidopsis duodecuple mutant of PYL ABA receptors reveals PYL repression of ABA-independent SnRK2 activity. *Cell Rep.* **2018**, *23*, 3340–3351.e3345. [CrossRef] [PubMed]

86. Gonzalez-Guzman, M.; Pizzio, G.A.; Antoni, R.; Vera-Sirera, F.; Merilo, E.; Bassel, G.W.; Fernández, M.A.; Holdsworth, M.J.; Perez-Amador, M.A.; Kollist, H.; et al. Arabidopsis PYR/PYL/RCAR receptors play a major role in quantitative regulation of stomatal aperture and transcriptional response to abscisic acid. *Plant Cell* **2012**, *24*, 2483–2496. [CrossRef] [PubMed]

87. Huang, H.; Xie, S.; Xiao, Q.; Wei, B.; Zheng, L.; Wang, Y.; Cao, Y.; Zhang, X.; Long, T.; Li, Y.; et al. Sucrose and ABA regulate starch biosynthesis in maize through a novel transcription factor, ZmEREB156. *Sci. Rep.* **2016**, *6*, 27590. [CrossRef] [PubMed]

88. Hu, Y.-F.; Li, Y.-P.; Zhang, J.; Liu, H.; Tian, M.; Huang, Y. Binding of ABI4 to a CACCG motif mediates the ABA-induced expression of the ZmSSI gene in maize (*Zea mays* L.) endosperm. *J. Exp. Bot.* **2012**, *63*, 5979–5989. [CrossRef]

89. Chen, T.; Li, G.; Islam, M.R.; Fu, W.; Feng, B.; Tao, L.; Fu, G. Abscisic acid synergizes with sucrose to enhance grain yield and quality of rice by improving the source-sink relationship. *BMC Plant Biol.* **2019**, *19*, 525. [CrossRef]

90. Finkelstein, R.R.; Gampala, S.S.; Rock, C.D. Abscisic acid signaling in seeds and seedlings. *Plant Cell* **2002**, *14*, S15–S45. [CrossRef]

91. Zhang, Y.-Q.; Lu, X.; Zhao, F.-Y.; Li, Q.-T.; Niu, S.-L.; Wei, W.; Zhang, W.-K.; Ma, B.; Chen, S.-Y.; Zhang, J.-S. Soybean GmDREBL increases lipid content in seeds of transgenic Arabidopsis. *Sci. Rep.* **2016**, *6*, 1–13. [CrossRef] [PubMed]

92. Kagaya, Y.; Toyoshima, R.; Okuda, R.; Usui, H.; Yamamoto, A.; Hattori, T. LEAFY COTYLEDON1 controls seed storage protein genes through its regulation of FUSCA3 and ABSCISIC ACID INSENSITIVE3. *Plant Cell Physiol.* **2005**, *46*, 399–406. [CrossRef] [PubMed]

93. Mu, J.; Tan, H.; Zheng, Q.; Fu, F.; Liang, Y.; Zhang, J.; Yang, X.; Wang, T.; Chong, K.; Wang, X.-J. LEAFY COTYLEDON1 is a key regulator of fatty acid biosynthesis in Arabidopsis. *Plant Physiol.* **2008**, *148*, 1042–1054. [CrossRef] [PubMed]

94. Chen, M.; Zhang, B.; Li, C.; Kulaveerasingam, H.; Chew, F.T.; Yu, H. TRANSPARENT TESTA GLABRA1 regulates the accumulation of seed storage reserves in Arabidopsis. *Plant Physiol.* **2015**, *169*, 391–402. [CrossRef] [PubMed]

95. Baud, S.; Dubreucq, B.; Miquel, M.; Rochat, C.; Lepiniec, L. Storage reserve accumulation in Arabidopsis: Metabolic and developmental control of seed filling. *Arab. Book Am. Soc. Plant Biol.* **2008**, *6*, e0113. [CrossRef]

96. To, A.; Joubès, J.; Barthole, G.; Lécureuil, A.; Scagnelli, A.; Jasinski, S.; Lepiniec, L.; Baud, S. WRINKLED transcription factors orchestrate tissue-specific regulation of fatty acid biosynthesis in Arabidopsis. *Plant Cell* **2012**, *24*, 5007–5023. [CrossRef] [PubMed]

97. Meinke, D.W.; Franzmann, L.H.; Nickle, T.C.; Yeung, E.C. Leafy Cotyledon Mutants of Arabidopsis. *Plant Cell* **1994**, *6*, 1049–1064. [CrossRef]

98. Gubler, F.; Millar, A.A.; Jacobsen, J.V. Dormancy release, ABA and pre-harvest sprouting. *Curr. Opin. Plant Biol.* **2005**, *8*, 183–187. [CrossRef]

99. Mendes, A.; Kelly, A.A.; van Erp, H.; Shaw, E.; Powers, S.J.; Kurup, S.; Eastmond, P.J. bZIP67 regulates the omega-3 fatty acid content of Arabidopsis seed oil by activating fatty acid desaturase3. *Plant Cell* **2013**, *25*, 3104–3116. [CrossRef]

100. Sall, K.; Dekkers, B.J.; Nonogaki, M.; Katsuragawa, Y.; Koyari, R.; Hendrix, D.; Willems, L.A.; Bentsink, L.; Nonogaki, H. DELAY OF GERMINATION 1-LIKE 4 acts as an inducer of seed reserve accumulation. *Plant J.* **2019**, *100*, 7–19. [CrossRef]

101. Armstead, I.; Donnison, I.; Aubry, S.; Harper, J.; Hörtensteiner, S.; James, C.; Mani, J.; Moffet, M.; Ougham, H.; Roberts, L.; et al. Cross-species identification of Mendel's I locus. *Science* **2007**, *315*, 73. [CrossRef] [PubMed]

102. Sato, Y.; Morita, R.; Nishimura, M.; Yamaguchi, H.; Kusaba, M. Mendel's green cotyledon gene encodes a positive regulator of the chlorophyll-degrading pathway. *Proc. Natl. Acad. Sci. USA* **2007**, *104*, 14169. [CrossRef] [PubMed]

103. Delmas, F.; Sankaranarayanan, S.; Deb, S.; Widdup, E.; Bournonville, C.; Bollier, N.; Northey, J.G.; McCourt, P.; Samuel, M.A. ABI3 controls embryo degreening through Mendel's I locus. *Proc. Natl. Acad. Sci. USA* **2013**, *110*, E3888–E3894. [CrossRef] [PubMed]

104. Roscoe, T.T.; Guilleminot, J.; Bessoule, J.-J.; Berger, F.; Devic, M. Complementation of seed maturation phenotypes by ectopic expression of ABSCISIC ACID INSENSITIVE3, FUSCA3 and LEAFY COTYLEDON2 in Arabidopsis. *Plant Cell Physiol.* **2015**, *56*, 1215–1228. [CrossRef] [PubMed]

105. Manfre, A.J.; LaHatte, G.A.; Climer, C.R.; Marcotte Jr, W.R. Seed dehydration and the establishment of desiccation tolerance during seed maturation is altered in the Arabidopsis thaliana mutant atem6-1. *Plant Cell Physiol.* **2009**, *50*, 243–253. [CrossRef] [PubMed]

106. Delahaie, J.; Hundertmark, M.; Bove, J.; Leprince, O.; Rogniaux, H.; Buitink, J. LEA polypeptide profiling of recalcitrant and orthodox legume seeds reveals ABI3-regulated LEA protein abundance linked to desiccation tolerance. *J. Exp. Bot.* **2013**, *64*, 4559–4573. [CrossRef]

107. Wehmeyer, N.; Vierling, E. The expression of small heat shock proteins in seeds responds to discrete developmental signals and suggests a general protective role in desiccation tolerance. *Plant Physiol.* **2000**, *122*, 1099–1108. [CrossRef]

108. Bailly, C. Active oxygen species and antioxidants in seed biology. *Seed Sci. Res.* **2004**, *14*, 93–107. [CrossRef]

109. Verdier, J.; Lalanne, D.; Pelletier, S.; Torres-Jerez, I.; Righetti, K.; Bandyopadhyay, K.; Leprince, O.; Chatelain, E.; Vu, B.L.; Gouzy, J.; et al. A Regulatory Network-Based Approach Dissects Late Maturation Processes Related to the Acquisition of Desiccation Tolerance and Longevity of Medicago truncatula Seeds. *Plant Physiol.* **2013**, *163*, 757–774. [CrossRef]

110. Mène-Saffrané, L.; Jones, A.D.; DellaPenna, D. Plastochromanol-8 and tocopherols are essential lipid-soluble antioxidants during seed desiccation and quiescence in Arabidopsis. *Proc. Natl. Acad. Sci. USA* **2010**, *107*, 17815–17820. [CrossRef]

111. Koornneef, M.; Bentsink, L.; Hilhorst, H. Seed dormancy and germination. *Curr. Opin. Plant Biol.* **2002**, *5*, 33–36. [CrossRef]

112. Battaglia, M.; Olvera-Carrillo, Y.; Garciarrubio, A.; Campos, F.; Covarrubias, A.A. The enigmatic LEA proteins and other hydrophilins. *Plant Physiol.* **2008**, *148*, 6–24. [CrossRef] [PubMed]

113. Hong-Bo, S.; Zong-Suo, L.; Ming-An, S. LEA proteins in higher plants: Structure, function, gene expression and regulation. *Colloids Surf. B Biointerfaces* **2005**, *45*, 131–135. [CrossRef] [PubMed]

114. Olvera-Carrillo, Y.; Campos, F.; Reyes, J.L.; Garciarrubio, A.; Covarrubias, A.A. Functional analysis of the group 4 late embryogenesis abundant proteins reveals their relevance in the adaptive response during water deficit in Arabidopsis. *Plant Physiol.* **2010**, *154*, 373–390. [CrossRef] [PubMed]

115. Kijak, H.; Ratajczak, E. What Do We Know About the Genetic Basis of Seed Desiccation Tolerance and Longevity? *Int. J. Mol. Sci.* **2020**, *21*, 3612. [CrossRef] [PubMed]

116. Bies-Ethève, N.; Gaubier-Comella, P.; Debures, A.; Lasserre, E.; Jobet, E.; Raynal, M.; Cooke, R.; Delseny, M. Inventory, evolution and expression profiling diversity of the LEA (late embryogenesis abundant) protein gene family in Arabidopsis thaliana. *Plant Mol. Biol.* **2008**, *67*, 107–124. [CrossRef]

117. Finkelstein, R.R.; Lynch, T.J. The Arabidopsis abscisic acid response gene ABI5 encodes a basic leucine zipper transcription factor. *Plant Cell* **2000**, *12*, 599–609. [CrossRef]

118. Bensmihen, S.; Rippa, S.; Lambert, G.; Jublot, D.; Pautot, V.; Granier, F.; Giraudat, J.; Parcy, F. The homologous ABI5 and EEL transcription factors function antagonistically to fine-tune gene expression during late embryogenesis. *Plant Cell* **2002**, *14*, 1391–1403. [CrossRef]

119. Carles, C.; Bies-Etheve, N.; Aspart, L.; Léon-Kloosterziel, K.M.; Koornneef, M.; Echeverria, M.; Delseny, M. Regulation of Arabidopsis thaliana Em genes: Role of ABI5. *Plant J.* **2002**, *30*, 373–383. [CrossRef]

120. Kotak, S.; Vierling, E.; Baumlein, H.; von Koskull-Doring, P. A novel transcriptional cascade regulating expression of heat stress proteins during seed development of Arabidopsis. *Plant Cell* **2007**, *19*, 182–195. [CrossRef]

121. Dekkers, B.J.; He, H.; Hanson, J.; Willems, L.A.; Jamar, D.C.; Cueff, G.; Rajjou, L.; Hilhorst, H.W.; Bentsink, L. The Arabidopsis DELAY OF GERMINATION 1 gene affects ABSCISIC ACID INSENSITIVE 5 (ABI 5) expression and genetically interacts with ABI 3 during Arabidopsis seed development. *Plant J.* **2016**, *85*, 451–465. [CrossRef] [PubMed]

122. Bentsink, L.; Jowett, J.; Hanhart, C.J.; Koornneef, M. Cloning of DOG1, a quantitative trait locus controlling seed dormancy in Arabidopsis. *Proc. Natl. Acad. Sci. USA* **2006**, *103*, 17042–17047. [CrossRef] [PubMed]

123. Zinsmeister, J.; Lalanne, D.; Terrasson, E.; Chatelain, E.; Vandecasteele, C.; Vu, B.L.; Dubois-Laurent, C.; Geoffriau, E.; Signor, C.L.; Dalmais, M.; et al. ABI5 Is a Regulator of Seed Maturation and Longevity in Legumes. *Plant Cell* **2016**, *28*, 2735–2754. [CrossRef] [PubMed]

124. Finkelstein, R.; Reeves, W.; Ariizumi, T.; Steber, C. Molecular aspects of seed dormancy. *Annu. Rev. Plant Biol.* **2008**, *59*, 387–415. [CrossRef] [PubMed]

125. Koornneef, M.; Jorna, M.L.; Brinkhorst-van der Swan, D.L.C.; Karssen, C.M. The isolation of abscisic acid (ABA) deficient mutants by selection of induced revertants in non-germinating gibberellin sensitive lines of *Arabidopsis thaliana* (L.) heynh. *Theor. Appl. Genet.* **1982**, *61*, 385–393. [CrossRef] [PubMed]

126. Léon-Kloosterziel, K.M.; Gil, M.A.; Ruijs, G.J.; Jacobsen, S.E.; Olszewski, N.E.; Schwartz, S.H.; Zeevaart, J.A.D.; Koornneef, M. Isolation and characterization of abscisic acid-deficient Arabidopsis mutants at two new loci. *Plant J.* **1996**, *10*, 655–661. [CrossRef] [PubMed]

127. Chono, M.; Matsunaka, H.; Seki, M.; Fujita, M.; Kiribuchi-Otobe, C.; Oda, S.; Kojima, H.; Kobayashi, D.; Kawakami, N. Isolation of a wheat (*Triticum aestivum* L.) mutant in ABA 8'-hydroxylase gene: Effect of reduced ABA catabolism on germination inhibition under field condition. *Breed. Sci.* **2013**, *63*, 104–115. [CrossRef]

128. Lee, H.G.; Lee, K.; Seo, P.J. The Arabidopsis MYB96 transcription factor plays a role in seed dormancy. *Plant Mol. Biol.* **2015**, *87*, 371–381. [CrossRef]

129. Shu, K.; Chen, Q.; Wu, Y.; Liu, R.; Zhang, H.; Wang, P.; Li, Y.; Wang, S.; Tang, S.; Liu, C. ABI 4 mediates antagonistic effects of abscisic acid and gibberellins at transcript and protein levels. *Plant J.* **2016**, *85*, 348–361. [CrossRef]

130. Shu, K.; Zhang, H.; Wang, S.; Chen, M.; Wu, Y.; Tang, S.; Liu, C.; Feng, Y.; Cao, X.; Xie, Q. ABI4 regulates primary seed dormancy by regulating the biogenesis of abscisic acid and gibberellins in Arabidopsis. *PLoS Genet.* **2013**, *9*, e1003577. [CrossRef]

131. Miao, C.; Xiao, L.; Hua, K.; Zou, C.; Zhao, Y.; Bressan, R.A.; Zhu, J.-K. Mutations in a subfamily of abscisic acid receptor genes promote rice growth and productivity. *Proc. Natl. Acad. Sci. USA* **2018**, *115*, 6058–6063. [CrossRef] [PubMed]

132. West, M.A.; Yee, K.M.; Danao, J.; Zimmerman, J.L.; Fischer, R.L.; Goldberg, R.B.; Harada, J.J. LEAFY COTYLEDON1 is an essential regulator of late embryogenesis and cotyledon identity in Arabidopsis. *Plant Cell* **1994**, *6*, 1731–1745. [CrossRef] [PubMed]

133. McCarty, D.R.; Hattori, T.; Carson, C.B.; Vasil, V.; Lazar, M.; Vasil, I.K. The Viviparous-1 developmental gene of maize encodes a novel transcriptional activator. *Cell* **1991**, *66*, 895–905. [CrossRef]

134. Hoecker, U.; Vasil, I.K.; McCarty, D.R. Integrated control of seed maturation and germination programs by activator and repressor functions of Viviparous-1 of maize. *Genes Dev.* **1995**, *9*, 2459–2469. [CrossRef] [PubMed]

135. McKibbin, R.S.; Wilkinson, M.D.; Bailey, P.C.; Flintham, J.E.; Andrew, L.M.; Lazzeri, P.A.; Gale, M.D.; Lenton, J.R.; Holdsworth, M.J. Transcripts of *Vp-1* homeologues are misspliced in modern wheat and ancestral species. *Proc. Natl. Acad. Sci. USA* **2002**, *99*, 10203–10208. [CrossRef] [PubMed]

136. Fan, J.; Niu, X.; Wang, Y.; Ren, G.; Zhuo, T.; Yang, Y.; Lu, B.-R.; Liu, Y. Short, direct repeats (SDRs)-mediated post-transcriptional processing of a transcription factor gene OsVP1 in rice (*Oryza sativa*). *J. Exp. Bot.* **2007**, *58*, 3811–3817. [CrossRef] [PubMed]

137. Carrari, F.; Perez-Flores, L.; Lijavetzky, D.; Enciso, S.; Sanchez, R.; Benech-Arnold, R.; Iusem, N. Cloning and expression of a sorghum gene with homology to maize vp1. Its potential involvement in pre-harvest sprouting resistance. *Plant Mol. Biol.* **2001**, *45*, 631–640. [CrossRef]

138. Suzuki, M.; Latshaw, S.; Sato, Y.; Settles, A.M.; Koch, K.E.; Hannah, L.C.; Kojima, M.; Sakakibara, H.; McCarty, D.R. The maize Viviparous8 locus, encoding a putative ALTERED MERISTEM PROGRAM1-like peptidase, regulates abscisic acid accumulation and coordinates embryo and endosperm development. *Plant Physiol.* **2008**, *146*, 1193–1206. [CrossRef]

139. Griffiths, J.; Barrero, J.M.; Taylor, J.; Helliwell, C.A.; Gubler, F. ALTERED MERISTEM PROGRAM 1 is involved in development of seed dormancy in Arabidopsis. *PLoS ONE* **2011**, *6*, e20408. [CrossRef]

140. Utsugi, S.; Ashikawa, I.; Nakamura, S.; Shibasaka, M. TaABI5, a wheat homolog of Arabidopsis thaliana ABA insensitive 5, controls seed germination. *J. Plant Res.* **2020**, *133*, 245–256. [CrossRef]

141. Yamasaki, Y.; Gao, F.; Jordan, M.C.; Ayele, B.T. Seed maturation associated transcriptional programs and regulatory networks underlying genotypic difference in seed dormancy and size/weight in wheat (*Triticum aestivum* L.). *BMC Plant Biol.* **2017**, *17*, 1–18. [CrossRef] [PubMed]

142. Cantoro, R.; Crocco, C.D.; Benech-Arnold, R.L.; Rodríguez, M.V. In vitro binding of Sorghum bicolor transcription factors ABI4 and ABI5 to a conserved region of a GA 2-OXIDASE promoter: Possible role of this interaction in the expression of seed dormancy. *J. Exp. Bot.* **2013**, *64*, 5721–5735. [CrossRef] [PubMed]

143. Xiang, Y.; Nakabayashi, K.; Ding, J.; He, F.; Bentsink, L.; Soppe, W.J.J. REDUCED DORMANCY5 Encodes a Protein Phosphatase 2C That Is Required for Seed Dormancy in Arabidopsis. *Plant Cell* **2014**, *26*, 4362–4375. [CrossRef] [PubMed]

144. Nakabayashi, K.; Bartsch, M.; Xiang, Y.; Miatton, E.; Pellengahr, S.; Yano, R.; Seo, M.; Soppe, W.J. The time required for dormancy release in Arabidopsis is determined by DELAY OF GERMINATION1 protein levels in freshly harvested seeds. *Plant Cell* **2012**, *24*, 2826–2838. [CrossRef] [PubMed]

145. Née, G.; Kramer, K.; Nakabayashi, K.; Yuan, B.; Xiang, Y.; Miatton, E.; Finkemeier, I.; Soppe, W.J.J. DELAY OF GERMINATION1 requires PP2C phosphatases of the ABA signalling pathway to control seed dormancy. *Nat. Commun.* **2017**, *8*, 72. [CrossRef] [PubMed]

146. Antoni, R.; Gonzalez-Guzman, M.; Rodriguez, L.; Rodrigues, A.; Pizzio, G.A.; Rodriguez, P.L. Selective inhibition of clade A phosphatases type 2C by PYR/PYL/RCAR abscisic acid receptors. *Plant Physiol.* **2012**, *158*, 970–980. [CrossRef]

147. Sugimoto, K.; Takeuchi, Y.; Ebana, K.; Miyao, A.; Hirochika, H.; Hara, N.; Ishiyama, K.; Kobayashi, M.; Ban, Y.; Hattori, T.; et al. Molecular cloning of *Sdr4*, a regulator involved in seed dormancy and domestication of rice. *Proc. Natl. Acad. Sci. USA* **2010**, *107*, 5792–5797. [CrossRef] [PubMed]

148. Cao, H.; Han, Y.; Li, J.; Ding, M.; Li, Y.; Li, X.; Chen, F.; Soppe, W.J.; Liu, Y. Arabidopsis thaliana SEED DORMANCY 4-LIKE regulates dormancy and germination by mediating the gibberellin pathway. *J. Exp. Bot.* **2019**, *71*, 919–933. [CrossRef]

149. Liu, F.; Zhang, H.; Ding, L.; Soppe, W.J.; Xiang, Y. Reversal of rdo5 1, a homolog of rice seed dormancy4, interacts with bhlh57 and controls aba biosynthesis and seed dormancy in Arabidopsis. *Plant Cell* **2020**, *32*, 1933–1948. [CrossRef]

150. Lee, S.-J.; Lee, M.H.; Kim, J.-I.; Kim, S.Y. Arabidopsis putative MAP kinase kinase kinases Raf10 and Raf11 are positive regulators of seed dormancy and ABA response. *Plant Cell Physiol.* **2015**, *56*, 84–97. [CrossRef]

151. Nguyen, Q.T.C.; Lee, S.-J.; Choi, S.-W.; Na, Y.-J.; Song, M.-R.; Hoang, Q.T.N.; Sim, S.Y.; Kim, M.-S.; Kim, J.-I.; Soh, M.-S. Arabidopsis raf-like kinase Raf10 is a regulatory component of core ABA signaling. *Mol. Cells* **2019**, *42*, 646. [PubMed]

152. Wang, Z.; Cao, H.; Sun, Y.; Li, X.; Chen, F.; Carles, A.; Li, Y.; Ding, M.; Zhang, C.; Deng, X. Arabidopsis paired amphipathic helix proteins SNL1 and SNL2 redundantly regulate primary seed dormancy via abscisic acid–ethylene antagonism mediated by histone deacetylation. *Plant Cell* **2013**, *25*, 149–166. [CrossRef] [PubMed]

153. Liu, Y.; Koornneef, M.; Soppe, W.J. The absence of histone H2B monoubiquitination in the Arabidopsis hub1 (rdo4) mutant reveals a role for chromatin remodeling in seed dormancy. *Plant Cell* **2007**, *19*, 433–444. [CrossRef] [PubMed]

154. Liu, Y.; Geyer, R.; Van Zanten, M.; Carles, A.; Li, Y.; Hörold, A.; van Nocker, S.; Soppe, W.J. Identification of the Arabidopsis REDUCED DORMANCY 2 gene uncovers a role for the polymerase associated factor 1 complex in seed dormancy. *PLoS ONE* **2011**, *6*, e22241. [CrossRef] [PubMed]

155. Sano, N.; Marion-Poll, A. ABA Metabolism and Homeostasis in Seed Dormancy and Germination. *Int. J. Mol. Sci.* **2021**, *22*, 5069. [CrossRef] [PubMed]

156. Davière, J.M.; Achard, P. A Pivotal Role of DELLAs in Regulating Multiple Hormone Signals. *Mol. Plant* **2016**, *9*, 10–20. [CrossRef]

157. Ito, T.; Okada, K.; Fukazawa, J.; Takahashi, Y. DELLA-dependent and -independent gibberellin signaling. *Plant Signal. Behav.* **2018**, *13*, e1445933. [CrossRef]

158. Chen, B.; Fiers, M.; Dekkers, B.J.W.; Maas, L.; van Esse, G.W.; Angenent, G.C.; Zhao, Y.; Boutilier, K. ABA signalling promotes cell totipotency in the shoot apex of germinating embryos. *J. Exp. Bot.* **2021**, *72*, 6418–6436. [CrossRef]

159. Liu, X.; Zhang, H.; Zhao, Y.; Feng, Z.; Li, Q.; Yang, H.Q.; Luan, S.; Li, J.; He, Z.H. Auxin controls seed dormancy through stimulation of abscisic acid signaling by inducing ARF-mediated ABI3 activation in Arabidopsis. *Proc. Natl. Acad. Sci. USA* **2013**, *110*, 15485–15490. [CrossRef]

160. Emenecker, R.J.; Strader, L.C. Auxin-Abscisic Acid Interactions in Plant Growth and Development. *Biomolecules* **2020**, *10*, 281. [CrossRef]

MDPI

St. Alban-Anlage 66

4052 Basel

Switzerland

Tel. +41 61 683 77 34

Fax +41 61 302 89 18

www.mdpi.com

International Journal of Molecular Sciences Editorial Office

E-mail: ijms@mdpi.com

www.mdpi.com/journal/ijms